The Higgs Boson

Edited by Paul F. Kisak

Contents

Chapter 1

Higgs boson

The **Higgs boson** or **Higgs particle** is an elementary particle in the Standard Model of particle physics. It is the quantum excitation of the **Higgs field**[6][7]—a fundamental field of crucial importance to particle physics theory,[7] first suspected to exist in the 1960s, that unlike other known fields such as the electromagnetic field, takes a non-zero constant value almost everywhere. The question of the Higgs field's existence has been the last unverified part of the Standard Model of particle physics and, according to some, "the central problem in particle physics".[8][9] The presence of this field, now believed to be confirmed, explains why some fundamental particles have mass when, based on the symmetries controlling their interactions, they should be massless. The existence of the Higgs field would also resolve several other long-standing puzzles, such as the reason for the weak force's extremely short range.

Although it is hypothesized that the Higgs field permeates the entire Universe, evidence for its existence has been very difficult to obtain. In principle, the Higgs field can be detected through its excitations, manifest as Higgs particles, but these are extremely difficult to produce and detect. The importance of this fundamental question led to a 40 year search, and the construction of one of the world's most expensive and complex experimental facilities to date, CERN's Large Hadron Collider,[10] able to create Higgs bosons and other particles for observation and study. On 4 July 2012, the discovery of a new particle with a mass between 125 and 127 GeV/c^2 was announced; physicists suspected that it was the Higgs boson.[11][12][13] Since then, however, the particle had been shown to behave, interact, and decay in many of the ways predicted by the Standard Model. It was also tentatively confirmed to have even parity and zero spin,[1] two fundamental attributes of a Higgs boson. This appears to be the first elementary scalar particle discovered in nature.[14] More data are needed to verify that the discovered particle has properties matching those predicted for the Higgs boson by the Standard Model, or whether, as predicted by some theories, multiple Higgs bosons exist.[3]

The Higgs boson is named after Peter Higgs, one of six physicists who, in 1964, proposed the mechanism that suggested the existence of such a particle. On December 10, 2013, two of them, Peter Higgs and François Englert, were awarded the Nobel Prize in Physics for their work and prediction (Englert's co-researcher Robert Brout had died in 2011 and the Nobel Prize is not ordinarily given posthumously).[15] Although Higgs's name has come to be associated with this theory, several researchers between about 1960 and 1972 each independently developed different parts of it. In mainstream media the Higgs boson has often been called the "God particle", from a 1993 book on the topic; the nickname is strongly disliked by many physicists, including Higgs, who regard it as sensationalistic.[16][17][18]

In the Standard Model, the Higgs particle is a boson with no spin, electric charge, or colour charge. It is also very unstable, decaying into other particles almost immediately. It is a quantum excitation of one of the four components of the Higgs field. The latter constitutes a scalar field, with two neutral and two electrically charged components that form a complex doublet of the weak isospin SU(2) symmetry. The Higgs field is tachyonic (this does not refer to faster-than-light speeds, it means that symmetry-breaking through condensation of a particle must occur under certain conditions), and has a "Mexican hat" shaped potential with nonzero strength everywhere (including otherwise empty space), which in its vacuum state breaks the weak isospin symmetry of the electroweak interaction. When this happens, three components of the Higgs field are "absorbed" by the SU(2) and U(1) gauge bosons (the "Higgs mechanism") to become the longitudinal components of the now-massive W and Z bosons of the weak force. The remaining electrically neutral component separately couples to other particles known as fermions (via Yukawa couplings), causing these to acquire mass as well. Some versions of the theory predict more than one kind of Higgs fields and bosons. Alternative "Higgsless" models would have been considered if the Higgs boson was not discovered.

1.1   A non-technical summary

1.1.1   "Higgs" terminology

1.1.2   Overview

Physicists explain the properties and forces between elementary particles in terms of the Standard Model—a widely accepted and "remarkably" accurate[21] framework based on gauge invariance and symmetries, believed to explain almost everything in the known universe, other than gravity.[22] But by around 1960 all attempts to create a gauge invariant theory for two of the four fundamental forces had consistently failed at one crucial point: although gauge invariance seemed extremely important, it seemed to make any theory of electromagnetism and the weak force go haywire, by demanding that either many particles with mass were massless or that non-existent forces and massless particles had to exist. Scientists had no idea how to get past this point.

In 1962 physicist Philip Anderson wrote a paper that built upon work by Yoichiro Nambu concerning "broken symmetries" in superconductivity and particle physics. He suggested that "broken symmetries" might also be the missing piece needed to solve the problems of gauge invariance. In 1964 a theory was created almost simultaneously by 3 different groups of researchers, that showed Anderson's suggestion was possible - the gauge theory and "mass problems" could indeed be resolved if an unusual kind of field, now generally called the "Higgs field", existed throughout the universe; if the Higgs field did exist, it would apparently cause existing particles to acquire mass instead of new massless particles being formed. Although these ideas did not gain much initial support or attention, by 1972 they had been developed into a comprehensive theory and proved capable of giving "sensible" results that accurately described particles known at the time, and which accurately predicted of several other particles discovered during the following years.[Note 7] During the 1970s these theories rapidly became the "standard model". There was not yet any direct evidence that the Higgs field actually existed, but even without proof of the field, the accuracy of its predictions led scientists to believe the theory might be true. By the 1980s the question whether or not the Higgs field existed had come to be regarded as one of the most important unanswered questions in particle physics.

If Higgs field could be shown to exist, it would be a monumental discovery for science and human knowledge, and would open doorways to new knowledge in many disciplines. If not, then other more complicated theories would need to be considered. The simplest means to test the existence of the Higgs field would be a search for a new elementary particle that the field would have to give off, a particle known as "Higgs bosons" or the "Higgs particle". This particle would be extremely difficult to find. After significant technological advancements, by the 1990s two large experimental installations were being designed and constructed that allowed to search for the Higgs boson.

While several symmetries in nature are spontaneously broken through a form of the Higgs mechanism, in the context of the Standard Model the term "Higgs mechanism" almost always means symmetry breaking of the electroweak field. It is considered confirmed, but revealing the exact cause has been difficult. Various analogies have also been invented to describe the Higgs field and boson, including analogies with well-known symmetry breaking effects such as the rainbow and prism, electric fields, ripples, and resistance of macro objects moving through media, like people moving through crowds or some objects moving through syrup or molasses. However, analogies based on simple resistance to motion are inaccurate as the Higgs field does not work by resisting motion.

1.2   Significance

1.2.1 Scientific impact

Evidence of the Higgs field and its properties has been extremely significant scientifically, for many reasons. The Higgs boson's importance is largely that it is able to be examined using existing knowledge and experimental technology, as a way to confirm and study the entire Higgs field theory.[6][7] Conversely, proof that the Higgs field and boson do not exist would also have been significant. In discussion form, the relevance includes:

1.2.2 Practical and technological impact of discovery

As yet, there are no known immediate technological benefits of finding the Higgs particle. However, a common pattern for fundamental discoveries is for practical applications to follow later, once the discovery has been explored further, at which point they become the basis for new technologies of importance to society.[44][45][46]

The challenges in particle physics have furthered major technological of widespread importance. For example, the World Wide Web began as a project to improve CERN's communication system. CERN's requirement to process massive amounts of data produced by the Large Hadron Collider also led to contributions to the fields of distributed and cloud computing.

1.3 History

See also: 1964 PRL symmetry breaking papers, Higgs mechanism and History of quantum field theory
Particle physicists study matter made from fundamental particles whose interactions are mediated by exchange particles - gauge bosons - acting as force carriers. At the beginning of the 1960s a number of these particles had been discovered or proposed, along with theories suggesting how they relate to each other, some of which had already been reformulated as field theories in which the objects of study are not particles and forces, but quantum fields and their symmetries.[47]:150 However, attempts to unify known fundamental forces such as the electromagnetic force and the weak nuclear force were known to be incomplete. One known omission was that gauge invariant approaches, including non-abelian models such as Yang–Mills theory (1954), which held great promise for unified theories, also seemed to predict known massive particles as massless.[48] Goldstone's theorem, relating to continuous symmetries within some theories, also appeared to rule out many obvious solutions,[49] since it appeared to show that zero-mass particles would have to also exist that were "simply not seen".[50] According to Guralnik, physicists had "no understanding" how these problems could be overcome.[50]

Particle physicist and mathematician Peter Woit summarised the state of research at the time:

> "Yang and Mills work on non-abelian gauge theory had one huge problem: in perturbation theory it has massless particles which don't correspond to anything we see. One way of getting rid of this problem is now fairly well-understood, the phenomenon of confinement realized in QCD, where the strong interactions get rid of the massless "gluon" states at long distances. By the very early sixties, people had begun to understand another source of massless particles: spontaneous symmetry breaking of a continuous symmetry. What Philip Anderson realized and worked out in the summer of 1962 was that, when you have *both* gauge symmetry *and* spontaneous symmetry breaking, the Nambu–Goldstone massless mode can combine with the massless gauge field modes to produce a physical massive vector field. This is what happens in superconductivity, a subject about which Anderson was (and is) one of the leading experts." *[text condensed]* [48]

The Higgs mechanism is a process by which vector bosons can get rest mass *without* explicitly breaking gauge invariance, as a byproduct of spontaneous symmetry breaking.[51][52] The mathematical theory behind spontaneous symmetry breaking was initially conceived and published within particle physics by Yoichiro Nambu in 1960,[53] the concept that such a mechanism could offer a possible solution for the "mass problem" was originally suggested in 1962 by Philip Anderson (who had previously written papers on broken symmetry and its outcomes in superconductivity[54] and concluded in his 1963 paper on Yang-Mills theory that *"considering the superconducting analog... [t]hese two types of bosons seem capable of canceling each other out... leaving finite mass bosons"*),[55]:4–5[56] and Abraham Klein and Benjamin Lee showed in March 1964 that Goldstone's theorem could be avoided this way in at least some non-relativistic cases and speculated it might be possible in truly relativistic cases.[57]

Nobel Prize Laureate Peter Higgs in Stockholm, December 2013

These approaches were quickly developed into a full relativistic model, independently and almost simultaneously, by three groups of physicists: by François Englert and Robert Brout in August 1964;[58] by Peter Higgs in October 1964;[59] and by Gerald Guralnik, Carl Hagen, and Tom Kibble (GHK) in November 1964.[60] Higgs also wrote a short but important[51] response published in September 1964 to an objection by Gilbert,[61] which showed that if calculating within the radiation gauge, Goldstone's theorem and Gilbert's objection would become inapplicable.[Note 11] (Higgs later described Gilbert's objection as prompting his own paper.[62]) Properties of the model were further considered by Guralnik in 1965,[63] by Higgs in 1966,[64] by Kibble in 1967,[65] and further by GHK in 1967.[66] The original three 1964 papers showed that when a gauge theory is combined with an additional field that spontaneously breaks the symmetry, the gauge bosons can consistently acquire a finite mass.[51][52][67] In 1967, Steven Weinberg[68] and Abdus Salam[69] independently showed how a Higgs mechanism could be used to break the electroweak symmetry of Sheldon Glashow's unified model for the weak and electromagnetic interactions[70] (itself an extension of work by Schwinger), forming what became the Standard Model of particle physics. Weinberg was the first to observe that this would also provide mass terms for the fermions.[71] [Note 12]

However, the seminal papers on spontaneous breaking of gauge symmetries were at first largely ignored, because it was widely believed that the (non-Abelian gauge) theories in question were a dead-end, and in particular that they could not be renormalised. In 1971–72, Martinus Veltman and Gerard 't Hooft proved renormalisation of Yang–Mills was possible in two papers covering massless, and then massive, fields.[71] Their contribution, and others' work on the renormalization group - including "substantial" theoretical work by Russian physicists Ludvig Faddeev, Andrei Slavnov, Efim Fradkin and Igor Tyutin[72] - was eventually "enormously profound and influential",[73] but even with all key elements of the eventual theory published there was still almost no wider interest. For example, Coleman found in a study that "essentially no-one paid any attention" to Weinberg's paper prior to 1971[74] and discussed by David Politzer in his 2004 Nobel speech.[73] – now the most cited in particle physics[75] – and even in 1970 according to Politzer, Glashow's teaching of the weak interaction contained no mention of Weinberg's, Salam's, or Glashow's own work.[73] In practice, Politzer states, almost everyone learned of the theory due to physicist Benjamin Lee, who combined the work of Veltman and 't Hooft with insights by others, and popularised the completed theory.[73] In this way, from 1971, interest and acceptance "exploded"[73] and the ideas were quickly absorbed in the mainstream.[71][73]

The resulting electroweak theory and Standard Model have accurately predicted (among other things) weak neutral currents, three bosons, the top and charm quarks, and with great precision, the mass and other properties of some of these.[Note 7] Many of those involved eventually won Nobel Prizes or other renowned awards. A 1974 paper and comprehensive review in *Reviews of Modern Physics* commented that "while no one doubted the [mathematical] correctness of these arguments, no one quite believed that nature was diabolically clever enough to take advantage of them",[76]:9 adding that the theory had so far produced accurate answers that accorded with experiment, but it was unknown whether the theory was fundamentally correct.[76]:9,36(footnote),43–44,47 By 1986 and again in the 1990s it became possible to write that understanding and proving the Higgs sector of the Standard Model was "the central problem today in particle physics".[8][9]

1.3.1 Summary and impact of the PRL papers

The three papers written in 1964 were each recognised as milestone papers during *Physical Review Letters*'s 50th anniversary celebration.[67] Their six authors were also awarded the 2010 J. J. Sakurai Prize for Theoretical Particle Physics for this work.[77] (A controversy also arose the same year, because in the event of a Nobel Prize only up to three scientists could be recognised, with six being credited for the papers.[78]) Two of the three PRL papers (by Higgs and by GHK) contained equations for the hypothetical field that eventually would become known as the Higgs field and its hypothetical quantum, the Higgs boson.[59][60] Higgs' subsequent 1966 paper showed the decay mechanism of the boson; only a massive boson can decay and the decays can prove the mechanism.

In the paper by Higgs the boson is massive, and in a closing sentence Higgs writes that "an essential feature" of the theory "is the prediction of incomplete multiplets of scalar and vector bosons".[59] (Frank Close comments that 1960s gauge theorists were focused on the problem of massless *vector* bosons, and the implied existence of a massive *scalar* boson was not seen as important; only Higgs directly addressed it.[79]:154, 166, 175) In the paper by GHK the boson is massless and decoupled from the massive states.[60] In reviews dated 2009 and 2011, Guralnik states that in the GHK model the boson is massless only in a lowest-order approximation, but it is not subject to any constraint and acquires mass at higher orders, and adds that the GHK paper was the only one to show that there are no massless Goldstone bosons in the model and to give a complete analysis of the general Higgs mechanism.[50][80] All three reached similar conclusions, despite their very different approaches: Higgs' paper essentially used classical techniques, Englert and Brout's involved calculating vacuum polarization in perturbation theory around an assumed symmetry-breaking vacuum state, and GHK used operator formalism and conservation laws to explore in depth the ways in which Goldstone's theorem may be worked around.[51]

1.4 Theoretical properties

Main article: Higgs mechanism

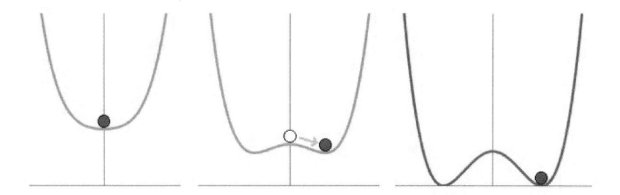

"Symmetry breaking illustrated": – At high energy levels (left) *the ball settles in the center, and the result is symmetrical. At lower energy levels* (right), *the overall "rules" remain symmetrical, but the "Mexican hat" potential comes into effect: "local" symmetry inevitably becomes broken since eventually the ball must at random roll one way or another.*

1.4.1   Theoretical need for the Higgs

Gauge invariance is an important property of modern particle theories such as the Standard Model, partly due to its success in other areas of fundamental physics such as electromagnetism and the strong interaction (quantum chromodynamics). However, there were great difficulties in developing gauge theories for the weak nuclear force or a possible unified electroweak interaction. Fermions with a mass term would violate gauge symmetry and therefore cannot be gauge invariant. (This can be seen by examining the Dirac Lagrangian for a fermion in terms of left and right handed components; we find none of the spin-half particles could ever flip helicity as required for mass, so they must be massless.[Note 13]) W and Z bosons are observed to have mass, but a boson mass term contains terms, which clearly depend on the choice of gauge and therefore these masses too cannot be gauge invariant. Therefore, it seems that *none* of the standard model fermions *or* bosons could "begin" with mass as an inbuilt property except by abandoning gauge invariance. If gauge invariance were to be retained, then these particles had to be acquiring their mass by some other mechanism or interaction. Additionally, whatever was giving these particles their mass, had to not "break" gauge invariance as the basis for other parts of the theories where it worked well, *and* had to not require or predict unexpected massless particles and long-range forces (seemingly an inevitable consequence of Goldstone's theorem) which did not actually seem to exist in nature.

A solution to all of these overlapping problems came from the discovery of a previously unnoticed borderline case hidden in the mathematics of Goldstone's theorem,[Note 11] that under certain conditions it *might* theoretically be possible for a symmetry to be broken *without* disrupting gauge invariance and *without* any new massless particles or forces, and having "sensible" (renormalisable) results mathematically: this became known as the Higgs mechanism.

The Standard Model hypothesizes a field which is responsible for this effect, called the Higgs field (symbol: ϕ), which has the unusual property of a non-zero amplitude in its ground state; i.e., a non-zero vacuum expectation value. It can have this effect because of its unusual "Mexican hat" shaped potential whose lowest "point" is not at its "centre". Below a certain extremely high energy level the existence of this non-zero vacuum expectation spontaneously breaks electroweak gauge symmetry which in turn gives rise to the Higgs mechanism and triggers the acquisition of mass by those particles interacting with the field. This effect occurs because scalar field components of the Higgs field are "absorbed" by the massive bosons as degrees of freedom, and couple to the fermions via Yukawa coupling, thereby producing the expected mass terms. In effect when symmetry breaks under these conditions, the Goldstone bosons that arise *interact* with the Higgs field (and with other particles capable of interacting with the Higgs field) instead of becoming new massless particles, the intractable problems of both underlying theories "neutralise" each other, and the residual outcome is that elementary particles acquire a consistent mass based on how strongly they interact with the Higgs field. It is the simplest known process capable of giving mass to the gauge bosons while remaining compatible with gauge theories.[81] Its quantum would be a scalar boson, known as the Higgs boson.[82]

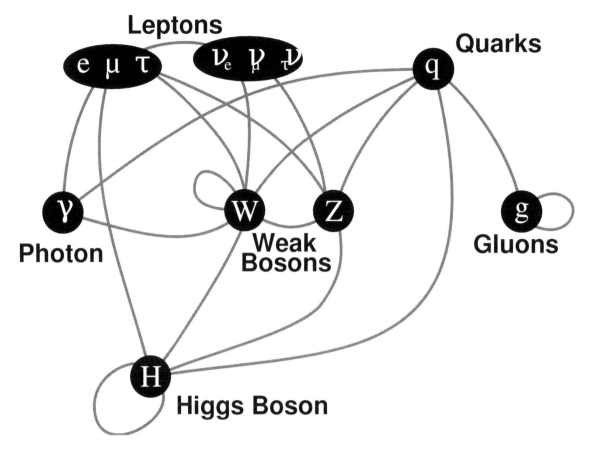

Summary of interactions between certain particles described by the Standard Model.

1.4.2 Properties of the Higgs field

In the Standard Model, the Higgs field is a scalar tachyonic field – 'scalar' meaning it does not transform under Lorentz transformations, and 'tachyonic' meaning the field (but not the particle) has imaginary mass and in certain configurations must undergo symmetry breaking. It consists of four components, two neutral ones and two charged component fields. Both of the charged components and one of the neutral fields are Goldstone bosons, which act as the longitudinal third-polarization components of the massive W^+, W^-, and Z bosons. The quantum of the remaining neutral component corresponds to (and is theoretically realised as) the massive Higgs boson,[83] this component can interact with fermions via Yukawa coupling to give them mass, as well.

Mathematically, the Higgs field has imaginary mass and is therefore a tachyonic field.[84] While tachyons (particles that move faster than light) are a purely hypothetical concept, fields with imaginary mass have come to play an important role in modern physics.[85][86] Under no circumstances do any excitations ever propagate faster than light in such theories — the presence or absence of a tachyonic mass has no effect whatsoever on the maximum velocity of signals (there is no violation of causality).[87] Instead of faster-than-light particles, the imaginary mass creates an instability:- any configuration in which one or more field excitations are tachyonic must spontaneously decay, and the resulting configuration contains no physical tachyons. This process is known as tachyon condensation, and is now believed to be the explanation for how the Higgs mechanism itself arises in nature, and therefore the reason behind electroweak symmetry breaking.

Although the notion of imaginary mass might seem troubling, it is only the field, and not the mass itself, that is quantized. Therefore, the field operators at spacelike separated points still commute (or anticommute), and information and particles still do not propagate faster than light.[88] Tachyon condensation drives a physical system that has reached a local limit and might naively be expected to produce physical tachyons, to an alternate stable state where no physical tachyons exist. Once a tachyonic field such as the Higgs field reaches the minimum of the potential, its quanta are not tachyons any more but rather are ordinary particles such as the Higgs boson.[89]

1.4.3 Properties of the Higgs boson

Since the Higgs field is scalar, the Higgs boson has no spin. The Higgs boson is also its own antiparticle and is CP-even, and has zero electric and colour charge.[90]

The Minimal Standard Model does not predict the mass of the Higgs boson.[91] If that mass is between 115 and 180 GeV/c^2, then the Standard Model can be valid at energy scales all the way up to the Planck scale (10^{19} GeV).[92] Many theorists expect new physics beyond the Standard Model to emerge at the TeV-scale, based on unsatisfactory properties of the Standard Model.[93] The highest possible mass scale allowed for the Higgs boson (or some other electroweak symmetry breaking mechanism) is 1.4 TeV; beyond this point, the Standard Model becomes inconsistent without such a mechanism, because unitarity is violated in certain scattering processes.[94]

It is also possible, although experimentally difficult, to estimate the mass of the Higgs boson indirectly. In the Standard Model, the Higgs boson has a number of indirect effects; most notably, Higgs loops result in tiny corrections to masses of W and Z bosons. Precision measurements of electroweak parameters, such as the Fermi constant and masses of W/Z bosons, can be used to calculate constraints on the mass of the Higgs. As of July 2011, the precision electroweak measurements tell us that the mass of the Higgs boson is likely to be less than about 161 GeV/c^2 at 95% confidence level (this upper limit would increase to 185 GeV/c^2 if the lower bound of 114.4 GeV/c^2 from the LEP-2 direct search is allowed for[95]). These indirect constraints rely on the assumption that the Standard Model is correct. It may still be possible to discover a Higgs boson above these masses if it is accompanied by other particles beyond those predicted by the Standard Model.[96]

1.4.4 Production

If Higgs particle theories are valid, then a Higgs particle can be produced much like other particles that are studied, in a particle collider. This involves accelerating a large number of particles to extremely high energies and extremely close to the speed of light, then allowing them to smash together. Protons and lead ions (the bare nuclei of lead atoms) are used at the LHC. In the extreme energies of these collisions, the desired esoteric particles will occasionally be produced and this can be detected and studied; any absence or difference from theoretical expectations can also be used to improve the theory. The relevant particle theory (in this case the Standard Model) will determine the necessary kinds of collisions and detectors. The Standard Model predicts that Higgs bosons could be formed in a number of ways,[97][98][99] although the probability of producing a Higgs boson in any collision is always expected to be very small—for example, only 1 Higgs boson per 10 billion collisions in the Large Hadron Collider.[Note 14] The most common expected processes for Higgs boson production are:

- *Gluon fusion.* If the collided particles are hadrons such as the proton or antiproton—as is the case in the LHC and Tevatron—then it is most likely that two of the gluons binding the hadron together collide. The easiest way to produce a Higgs particle is if the two gluons combine to form a loop of virtual quarks. Since the coupling of particles to the Higgs boson is proportional to their mass, this process is more likely for heavy particles. In practice it is enough to consider the contributions of virtual top and bottom quarks (the heaviest quarks). This process is the dominant contribution at the LHC and Tevatron being about ten times more likely than any of the other processes.[97][98]

- *Higgs Strahlung.* If an elementary fermion collides with an anti-fermion—e.g., a quark with an anti-quark or an electron with a positron—the two can merge to form a virtual W or Z boson which, if it carries sufficient energy, can then emit a Higgs boson. This process was the dominant production mode at the LEP, where an electron and a positron collided to form a virtual Z boson, and it was the second largest contribution for Higgs production at the Tevatron. At the LHC this process is only the third largest, because the LHC collides protons with protons, making a quark-antiquark collision less likely than at the Tevatron. Higgs Strahlung is also known as *associated production*.[97][98][99]

- *Weak boson fusion.* Another possibility when two (anti-)fermions collide is that the two exchange a virtual W or Z boson, which emits a Higgs boson. The colliding fermions do not need to be the same type. So, for example, an up quark may exchange a Z boson with an anti-down quark. This process is the second most important for the production of Higgs particle at the LHC and LEP.[97][99]

- *Top fusion.* The final process that is commonly considered is by far the least likely (by two orders of magnitude). This process involves two colliding gluons, which each decay into a heavy quark–antiquark pair. A quark and antiquark from each pair can then combine to form a Higgs particle.[97][98]

1.4.5 Decay

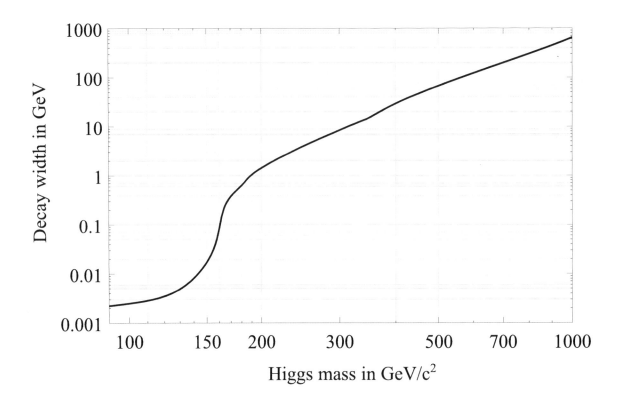

The Standard Model prediction for the decay width of the Higgs particle depends on the value of its mass.

Quantum mechanics predicts that if it is possible for a particle to decay into a set of lighter particles, then it will eventually do so.[101] This is also true for the Higgs boson. The likelihood with which this happens depends on a variety of factors including: the difference in mass, the strength of the interactions, etc. Most of these factors are fixed by the Standard Model, except for the mass of the Higgs boson itself. For a Higgs boson with a mass of 126 GeV/c^2 the SM predicts a mean life time of about 1.6×10^{-22} s.[Note 2]

Since it interacts with all the massive elementary particles of the SM, the Higgs boson has many different processes through which it can decay. Each of these possible processes has its own probability, expressed as the *branching ratio*; the fraction of the total number decays that follows that process. The SM predicts these branching ratios as a function of the Higgs mass (see plot).

One way that the Higgs can decay is by splitting into a fermion–antifermion pair. As general rule, the Higgs is more likely to decay into heavy fermions than light fermions, because the mass of a fermion is proportional to the strength of its interaction with the Higgs.[102] By this logic the most common decay should be into a top–antitop quark pair. However, such a decay is only possible if the Higgs is heavier than ~346 GeV/c^2, twice the mass of the top quark. For a Higgs mass of 126 GeV/c^2 the SM predicts that the most common decay is into a bottom–antibottom quark pair, which happens 56.1% of the time.[5] The second most common fermion decay at that mass is a tau–antitau pair, which happens only about 6% of the time.[5]

Another possibility is for the Higgs to split into a pair of massive gauge bosons. The most likely possibility is for the Higgs to decay into a pair of W bosons (the light blue line in the plot), which happens about 23.1% of the time for a

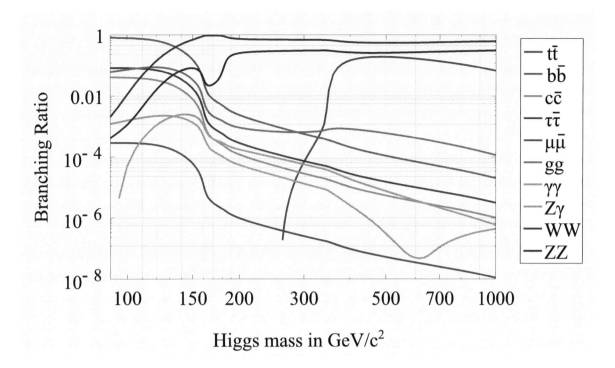

The Standard Model prediction for the branching ratios of the different decay modes of the Higgs particle depends on the value of its mass.

Higgs boson with a mass of 126 GeV/c^2.[5] The W bosons can subsequently decay either into a quark and an antiquark or into a charged lepton and a neutrino. However, the decays of W bosons into quarks are difficult to distinguish from the background, and the decays into leptons cannot be fully reconstructed (because neutrinos are impossible to detect in particle collision experiments). A cleaner signal is given by decay into a pair of Z-bosons (which happens about 2.9% of the time for a Higgs with a mass of 126 GeV/c^2),[5] if each of the bosons subsequently decays into a pair of easy-to-detect charged leptons (electrons or muons).

Decay into massless gauge bosons (i.e., gluons or photons) is also possible, but requires intermediate loop of virtual heavy quarks (top or bottom) or massive gauge bosons.[102] The most common such process is the decay into a pair of gluons through a loop of virtual heavy quarks. This process, which is the reverse of the gluon fusion process mentioned above, happens approximately 8.5% of the time for a Higgs boson with a mass of 126 GeV/c^2.[5] Much rarer is the decay into a pair of photons mediated by a loop of W bosons or heavy quarks, which happens only twice for every thousand decays.[5] However, this process is very relevant for experimental searches for the Higgs boson, because the energy and momentum of the photons can be measured very precisely, giving an accurate reconstruction of the mass of the decaying particle.[102]

1.4.6 Alternative models

Main article: Alternatives to the Standard Model Higgs

The Minimal Standard Model as described above is the simplest known model for the Higgs mechanism with just one Higgs field. However, an extended Higgs sector with additional Higgs particle doublets or triplets is also possible, and many extensions of the Standard Model have this feature. The non-minimal Higgs sector favoured by theory are the two-Higgs-doublet models (2HDM), which predict the existence of a quintet of scalar particles: two CP-even neutral Higgs bosons h^0 and H^0, a CP-odd neutral Higgs boson A^0, and two charged Higgs particles $H^{\pm}$. Supersymmetry ("SUSY") also predicts relations between the Higgs-boson masses and the masses of the gauge bosons, and could accommodate a 125 GeV/c^2 neutral Higgs boson.

The key method to distinguish between these different models involves study of the particles' interactions ("coupling") and exact decay processes ("branching ratios"), which can be measured and tested experimentally in particle collisions. In the Type-I 2HDM model one Higgs doublet couples to up and down quarks, while the second doublet does not couple to quarks. This model has two interesting limits, in which the lightest Higgs couples to just fermions ("gauge-phobic") or just gauge bosons ("fermiophobic"), but not both. In the Type-II 2HDM model, one Higgs doublet only couples to up-type quarks, the other only couples to down-type quarks.[103] The heavily researched Minimal Supersymmetric Standard Model (MSSM) includes a Type-II 2HDM Higgs sector, so it could be disproven by evidence of a Type-I 2HDM Higgs.

In other models the Higgs scalar is a composite particle. For example, in technicolor the role of the Higgs field is played by strongly bound pairs of fermions called techniquarks. Other models, feature pairs of top quarks (see top quark condensate). In yet other models, there is no Higgs field at all and the electroweak symmetry is broken using extra dimensions.[104][105]

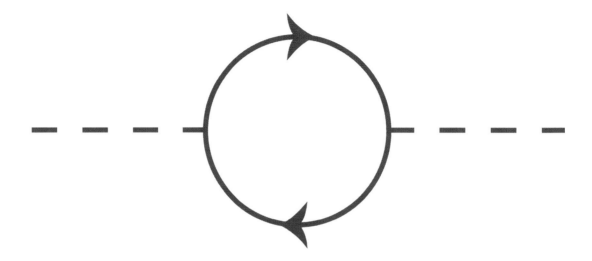

A one-loop Feynman diagram of the first-order correction to the Higgs mass. In the Standard Model the effects of these corrections are potentially enormous, giving rise to the so-called hierarchy problem.

1.4.7   Further theoretical issues and hierarchy problem

Main articles: Hierarchy problem and Hierarchy problem § The Higgs mass

The Standard Model leaves the mass of the Higgs boson as a parameter to be measured, rather than a value to be calculated. This is seen as theoretically unsatisfactory, particularly as quantum corrections (related to interactions with virtual particles) should apparently cause the Higgs particle to have a mass immensely higher than that observed, but at the same time the Standard Model requires a mass of the order of 100 to 1000 GeV to ensure unitarity (in this case, to unitarise longitudinal vector boson scattering).[106] Reconciling these points appears to require explaining why there is an almost-perfect cancellation resulting in the visible mass of ~ 125 GeV, and it is not clear how to do this. Because the weak force is about 10^{32} times stronger than gravity, and (linked to this) the Higgs boson's mass is so much less than the Planck mass or the grand unification energy, it appears that either there is some underlying connection or reason for these observations which is unknown and not described by the Standard Model, or some unexplained and extremely precise fine-tuning of parameters – however at present neither of these explanations is proven. This is known as a hierarchy problem.[107] More broadly, the hierarchy problem amounts to the worry that a future theory of fundamental particles and interactions should not have excessive fine-tunings or unduly delicate cancellations, and should allow masses of particles such as the Higgs boson to be calculable. The problem is in some ways unique to spin-0 particles (such as the Higgs boson), which can give rise to issues related to quantum corrections that do not affect particles with spin.[106] A number of solutions have been proposed, including supersymmetry, conformal solutions and solutions via extra dimensions such as braneworld models.

There are also issues of quantum triviality, which suggests that it may not be possible to create a consistent quantum field theory involving elementary scalar particles.

1.5 Experimental search

Main article: Search for the Higgs boson

To produce Higgs bosons, two beams of particles are accelerated to very high energies and allowed to collide within a particle detector. Occasionally, although rarely, a Higgs boson will be created fleetingly as part of the collision byproducts. Because the Higgs boson decays very quickly, particle detectors cannot detect it directly. Instead the detectors register all the decay products (the *decay signature*) and from the data the decay process is reconstructed. If the observed decay products match a possible decay process (known as a *decay channel*) of a Higgs boson, this indicates that a Higgs boson may have been created. In practice, many processes may produce similar decay signatures. Fortunately, the Standard Model precisely predicts the likelihood of each of these, and each known process, occurring. So, if the detector detects more decay signatures consistently matching a Higgs boson than would otherwise be expected if Higgs bosons did not exist, then this would be strong evidence that the Higgs boson exists.

Because Higgs boson production in a particle collision is likely to be very rare (1 in 10 billion at the LHC),[Note 14] and many other possible collision events can have similar decay signatures, the data of hundreds of trillions of collisions needs to be analysed and must "show the same picture" before a conclusion about the existence of the Higgs boson can be reached. To conclude that a new particle has been found, particle physicists require that the statistical analysis of two independent particle detectors each indicate that there is lesser than a one-in-a-million chance that the observed decay signatures are due to just background random Standard Model events—i.e., that the observed number of events is more than 5 standard deviations (sigma) different from that expected if there was no new particle. More collision data allows better confirmation of the physical properties of any new particle observed, and allows physicists to decide whether it is indeed a Higgs boson as described by the Standard Model or some other hypothetical new particle.

To find the Higgs boson, a powerful particle accelerator was needed, because Higgs bosons might not be seen in lower-energy experiments. The collider needed to have a high luminosity in order to ensure enough collisions were seen for conclusions to be drawn. Finally, advanced computing facilities were needed to process the vast amount of data (25 petabytes per year as at 2012) produced by the collisions.[109] For the announcement of 4 July 2012, a new collider known as the Large Hadron Collider was constructed at CERN with a planned eventual collision energy of 14 TeV—over seven times any previous collider—and over 300 trillion (3×10^{14}) LHC proton–proton collisions were analysed by the LHC Computing Grid, the world's largest computing grid (as of 2012), comprising over 170 computing facilities in a worldwide network across 36 countries.[109][110][111]

1.5.1 Search prior to 4 July 2012

The first extensive search for the Higgs boson was conducted at the Large Electron–Positron Collider (LEP) at CERN in the 1990s. At the end of its service in 2000, LEP had found no conclusive evidence for the Higgs.[Note 15] This implied that if the Higgs boson were to exist it would have to be heavier than 114.4 GeV/c^2.[112]

The search continued at Fermilab in the United States, where the Tevatron—the collider that discovered the top quark in 1995—had been upgraded for this purpose. There was no guarantee that the Tevatron would be able to find the Higgs, but it was the only supercollider that was operational since the Large Hadron Collider (LHC) was still under construction and the planned Superconducting Super Collider had been cancelled in 1993 and never completed. The Tevatron was only able to exclude further ranges for the Higgs mass, and was shut down on 30 September 2011 because it no longer could keep up with the LHC. The final analysis of the data excluded the possibility of a Higgs boson with a mass between 147 GeV/c^2 and 180 GeV/c^2. In addition, there was a small (but not significant) excess of events possibly indicating a Higgs boson with a mass between 115 GeV/c^2 and 140 GeV/c^2.[113]

The Large Hadron Collider at CERN in Switzerland, was designed specifically to be able to either confirm or exclude the existence of the Higgs boson. Built in a 27 km tunnel under the ground near Geneva originally inhabited by LEP, it was designed to collide two beams of protons, initially at energies of 3.5 TeV per beam (7 TeV total), or almost 3.6 times that of the Tevatron, and upgradeable to 2×7 TeV (14 TeV total) in future. Theory suggested if the Higgs boson existed, collisions at these energy levels should be able to reveal it. As one of the most complicated scientific instruments ever built, its operational readiness was delayed for 14 months by a magnet quench event nine days after its inaugural tests, caused by a faulty electrical connection that damaged over 50 superconducting magnets and contaminated the vacuum system.[114][115][116]

Data collection at the LHC finally commenced in March 2010.[117] By December 2011 the two main particle detectors at the LHC, ATLAS and CMS, had narrowed down the mass range where the Higgs could exist to around 116-130 GeV (ATLAS) and 115-127 GeV (CMS).[118][119] There had also already been a number of promising event excesses that had "evaporated" and proven to be nothing but random fluctuations. However, from around May 2011,[120] both experiments had seen among their results, the slow emergence of a small yet consistent excess of gamma and 4-lepton decay signatures and several other particle decays, all hinting at a new particle at a mass around 125 GeV.[120] By around November 2011, the anomalous data at 125 GeV was becoming "too large to ignore" (although still far from conclusive), and the team leaders at both ATLAS and CMS each privately suspected they might have found the Higgs.[120] On November 28, 2011, at an internal meeting of the two team leaders and the director general of CERN, the latest analyses were discussed outside their teams for the first time, suggesting both ATLAS and CMS might be converging on a possible shared result at 125 GeV, and initial preparations commenced in case of a successful finding.[120] While this information was not known publicly at the time, the narrowing of the possible Higgs range to around 115–130 GeV and the repeated observation of small but consistent event excesses across multiple channels at both ATLAS and CMS in the 124-126 GeV region (described as "tantalising hints" of around 2-3 sigma) were public knowledge with "a lot of interest".[121] It was therefore widely anticipated around the end of 2011, that the LHC would provide sufficient data to either exclude or confirm the finding of a Higgs boson by the end of 2012, when their 2012 collision data (with slightly higher 8 TeV collision energy) had been examined.[121][122]

1.5.2 Discovery of candidate boson at CERN

On 22 June 2012 CERN announced an upcoming seminar covering tentative findings for 2012,[126][127] and shortly afterwards (from around 1 July 2012 according to an analysis of the spreading rumour in social media[128]) rumours began to spread in the media that this would include a major announcement, but it was unclear whether this would be a stronger signal or a formal discovery.[129][130] Speculation escalated to a "fevered" pitch when reports emerged that Peter Higgs, who proposed the particle, was to be attending the seminar,[131][132] and that "five leading physicists" had been invited – generally believed to signify the five living 1964 authors – with Higgs, Englert, Guralnik, Hagen attending and Kibble confirming his invitation (Brout having died in 2011).[133][134]

On 4 July 2012 both of the CERN experiments announced they had independently made the same discovery:[135] CMS of a previously unknown boson with mass 125.3 ± 0.6 GeV/c^2[136][137] and ATLAS of a boson with mass 126.0 ± 0.6 GeV/c^2.[138][139] Using the combined analysis of two interaction types (known as 'channels'), both experiments independently reached a local significance of 5 sigma — implying that the probability of getting at least as strong a result by chance alone is less than 1 in 3 million. When additional channels were taken into account, the CMS significance was reduced to 4.9 sigma.[137]

The two teams had been working 'blinded' from each other from around late 2011 or early 2012,[120] meaning they did not discuss their results with each other, providing additional certainty that any common finding was genuine validation of a particle.[109] This level of evidence, confirmed independently by two separate teams and experiments, meets the formal level of proof required to announce a confirmed discovery.

On 31 July 2012, the ATLAS collaboration presented additional data analysis on the "observation of a new particle", including data from a third channel, which improved the significance to 5.9 sigma (1 in 588 million chance of obtaining at least as strong evidence by random background effects alone) and mass 126.0 ± 0.4 (stat) ± 0.4 (sys) GeV/c^2, [139] and CMS improved the significance to 5-sigma and mass 125.3 ± 0.4 (stat) ± 0.5 (sys) GeV/c^2.[136]

1.5.3 The new particle tested as a possible Higgs boson

Following the 2012 discovery, it was still unconfirmed whether or not the 125 GeV/c^2 particle was a Higgs boson. On one hand, observations remained consistent with the observed particle being the Standard Model Higgs boson, and the particle decayed into at least some of the predicted channels. Moreover, the production rates and branching ratios for the observed channels broadly matched the predictions by the Standard Model within the experimental uncertainties. However, the experimental uncertainties currently still left room for alternative explanations, meaning an announcement of the discovery of a Higgs boson would have been premature.[102] To allow more opportunity for data collection, the LHC's proposed 2012 shutdown and 2013–14 upgrade were postponed by 7 weeks into 2013.[140]

In November 2012, in a conference in Kyoto researchers said evidence gathered since July was falling into line with the basic Standard Model more than its alternatives, with a range of results for several interactions matching that theory's predictions.[141] Physicist Matt Strassler highlighted "considerable" evidence that the new particle is not a pseudoscalar negative parity particle (consistent with this required finding for a Higgs boson), "evaporation" or lack of increased significance for previous hints of non-Standard Model findings, expected Standard Model interactions with W and Z bosons, absence of "significant new implications" for or against supersymmetry, and in general no significant deviations to date from the results expected of a Standard Model Higgs boson.[142] However some kinds of extensions to the Standard Model would also show very similar results;[143] so commentators noted that based on other particles that are still being understood long after their discovery, it may take years to be sure, and decades to fully understand the particle that has been found.[141][142]

These findings meant that as of January 2013, scientists were very sure they had found an unknown particle of mass ~ 125 GeV/c^2, and had not been misled by experimental error or a chance result. They were also sure, from initial observations, that the new particle was some kind of boson. The behaviours and properties of the particle, so far as examined since July 2012, also seemed quite close to the behaviours expected of a Higgs boson. Even so, it could still have been a Higgs boson or some other unknown boson, since future tests could show behaviours that do not match a Higgs boson, so as of December 2012 CERN still only stated that the new particle was "consistent with" the Higgs boson,[11][13] and scientists did not yet positively say it was the Higgs boson.[144] Despite this, in late 2012, widespread media reports announced (incorrectly) that a Higgs boson had been confirmed during the year.[Note 16]

In January 2013, CERN director-general Rolf-Dieter Heuer stated that based on data analysis to date, an answer could be possible 'towards' mid-2013,[150] and the deputy chair of physics at Brookhaven National Laboratory stated in February 2013 that a "definitive" answer might require "another few years" after the collider's 2015 restart.[151] In early March 2013, CERN Research Director Sergio Bertolucci stated that confirming spin-0 was the major remaining requirement to determine whether the particle is at least some kind of Higgs boson.[152]

1.5.4 Preliminary confirmation of existence and current status

On 14 March 2013 CERN confirmed that:

> "CMS and ATLAS have compared a number of options for the spin-parity of this particle, and these all prefer no spin and even parity [two fundamental criteria of a Higgs boson consistent with the Standard Model]. This, coupled with the measured interactions of the new particle with other particles, strongly indicates that it is a Higgs boson." [1]

This also makes the particle the first elementary scalar particle to be discovered in nature.[14]

Examples of tests used to validate whether the 125 GeV particle is a Higgs boson:[142][153]

1.6 Public discussion

1.6.1 Naming

Names used by physicists

The name most strongly associated with the particle and field is the Higgs boson[79]:168 and Higgs field. For some time the particle was known by a combination of its PRL author names (including at times Anderson), for example the Brout–Englert–Higgs particle, the Anderson-Higgs particle, or the Englert–Brout–Higgs–Guralnik–Hagen–Kibble mechanism,[Note 17] and these are still used at times.[51][160] Fueled in part by the issue of recognition and a potential shared Nobel Prize,[160][161] the most appropriate name is still occasionally a topic of debate as at 2012.[160] (Higgs himself prefers to call the particle either by an acronym of all those involved, or "the scalar boson", or "the so-called Higgs particle".[161])

A considerable amount has been written on how Higgs' name came to be exclusively used. Two main explanations are offered.

Nickname

The Higgs boson is often referred to as the "God particle" in popular media outside the scientific community.[170][171][172][173][174] The nickname comes from the title of the 1993 book on the Higgs boson and particle physics -The God Particle: If theUniverse Is the Answer, What Is the Question? byNobel Physics prizewinner and Fermi lab director Leon Lederman.

[21]

Lederman wrote it in the context of failing US government support for the Superconducting Super Collider,[175] a part-constructed titanic[176][177] competitor to the Large Hadron Collider with planned collision energies of 2×20 TeV that was championed by Lederman since its 1983 inception[175][178][179] and shut down in 1993. The book sought in part to promote awareness of the significance and need for such a project in the face of its possible loss of funding.[180] Lederman, a leading researcher in the field, wanted to title his book "The Goddamn Particle: If the Universe is the Answer, What is the Question?" But his editor decided that the title was too controversial and convinced Lederman to change the title to "The God Particle: If the Universe is the Answer, What is the Question?"[181]

While media use of this term may have contributed to wider awareness and interest,[182] many scientists feel the name is inappropriate[16][17][183] since it is sensational hyperbole and misleads readers;[184] the particle also has nothing to do with God, leaves open numerous questions in fundamental physics, and does not explain the ultimate origin of the universe. Higgs, an atheist, was reported to be displeased and stated in a 2008 interview that he found it "embarrassing" because it was "the kind of misuse... which I think might offend some people".[184][185][186] Science writer Ian Sample stated in his 2010 book on the search that the nickname is "universally hate[d]" by physicists and perhaps the "worst derided" in the history of physics, but that (according to Lederman) the publisher rejected all titles mentioning "Higgs" as unimaginative and too unknown.[187]

Lederman begins with a review of the long human search for knowledge, and explains that his tongue-in-cheek title draws an analogy between the impact of the Higgs field on the fundamental symmetries at the Big Bang, and the apparent chaos of structures, particles, forces and interactions that resulted and shaped our present universe, with the biblical story of Babel in which the primordial single language of early Genesis was fragmented into many disparate languages and cultures.[188]

> Today ... we have the standard model, which reduces all of reality to a dozen or so particles and four forces. ... It's a hard-won simplicity [...and...] remarkably accurate. But it is also incomplete and, in fact, internally inconsistent... This boson is so central to the state of physics today, so crucial to our final understanding of the structure of matter, yet so elusive, that I have given it a nickname: the God Particle. Why God Particle? Two reasons. One, the publisher wouldn't let us call it the Goddamn Particle, though that might be a more appropriate title, given its villainous nature and the expense it is causing. And two, there is a connection, of sorts, to another book, a *much* older one...
> — Leon M. Lederman and Dick Teresi, *The God Particle: If the Universe is the Answer, What is the Question*[21] p. 22

Lederman asks whether the Higgs boson was added just to perplex and confound those seeking knowledge of the universe, and whether physicists will be confounded by it as recounted in that story, or ultimately surmount the challenge and understand "how beautiful is the universe [God has] made".[189]

Other proposals

A renaming competition by British newspaper *The Guardian* in 2009 resulted in their science correspondent choosing the name "the champagne bottle boson" as the best submission: "The bottom of a champagne bottle is in the shape of the Higgs potential and is often used as an illustration in physics lectures. So it's not an embarrassingly grandiose name, it is memorable, and [it] has some physics connection too."[190] The name *Higgson* was suggested as well, in an opinion piece in the Institute of Physics' online publication *physicsworld.com*.[191]

1.6.2   Media explanations and analogies

There has been considerable public discussion of analogies and explanations for the Higgs particle and how the field creates mass,[192][193] including coverage of explanatory attempts in their own right and a competition in 1993 for the best popular explanation by then-UK Minister for Science Sir William Waldegrave[194] and articles in newspapers worldwide.

An educational collaboration involving an LHC physicist and a High School Teachers at CERN educator suggests that dispersion of light – responsible for the rainbow and dispersive prism – is a useful analogy for the Higgs field's symmetry breaking and mass-causing effect.[195]

Matt Strassler uses electric fields as an analogy:[196]

> Some particles interact with the Higgs field while others don't. Those particles that feel the Higgs field act as if they have mass. Something similar happens in an electric field – charged objects are pulled around and neutral objects can sail through unaffected. So you can think of the Higgs search as an attempt to make waves in the Higgs field *[create Higgs bosons]* to prove it's really there.

A similar explanation was offered by *The Guardian*:[197]

> The Higgs boson is essentially a ripple in a field said to have emerged at the birth of the universe and to span the cosmos to this day ... The particle is crucial however: it is the smoking gun, the evidence required to show the theory is right.

The Higgs field's effect on particles was famously described by physicist David Miller as akin to a room full of political party workers spread evenly throughout a room: the crowd gravitates to and slows down famous people but does not slow down others.[Note 18] He also drew attention to well-known effects in solid state physics where an electron's effective mass can be much greater than usual in the presence of a crystal lattice.[198]

Analogies based on drag effects, including analogies of "syrup" or "molasses" are also well known, but can be somewhat misleading since they may be understood (incorrectly) as saying that the Higgs field simply resists some particles' motion but not others' – a simple resistive effect could also conflict with Newton's third law.[200]

1.6.3   Recognition and awards

There has been considerable discussion of how to allocate the credit if the Higgs boson is proven, made more pointed as a Nobel prize had been expected, and the very wide basis of people entitled to consideration. These include a range of theoreticians who made the Higgs mechanism theory possible, the theoreticians of the 1964 PRL papers (including Higgs himself), the theoreticians who derived from these, a working electroweak theory and the Standard Model itself, and also the experimentalists at CERN and other institutions who made possible the proof of the Higgs field and boson in reality. The Nobel prize has a limit of 3 persons to share an award, and some possible winners are already prize holders for other work, or are deceased (the prize is only awarded to persons in their lifetime). Existing prizes for works relating to the Higgs field, boson, or mechanism include:

Photograph of light passing through a dispersive prism: the rainbow effect arises because photons are not all affected to the same degree by the dispersive material of the prism.

- Nobel Prize in Physics (1979) – Glashow, Salam, and Weinberg, *for contributions to the theory of the unified weak and electromagnetic interaction between elementary particles* [201]

- Nobel Prize in Physics (1999) – 't Hooft and Veltman, *for elucidating the quantum structure of electroweak inter-*

actions in physics [202]

- Nobel Prize in Physics (2008) – Nambu (shared), *for the discovery of the mechanism of spontaneous broken symmetry in subatomic physics* [53]

- J. J. Sakurai Prize for Theoretical Particle Physics (2010) – Hagen, Englert, Guralnik, Higgs, Brout, and Kibble, *for elucidation of the properties of spontaneous symmetry breaking in four-dimensional relativistic gauge theory and of the mechanism for the consistent generation of vector boson masses* [77] (for the 1964 papers described above)

- Wolf Prize (2004) – Englert, Brout, and Higgs

- Nobel Prize in Physics (2013) - Peter Higgs and François Englert, *for the theoretical discovery of a mechanism that contributes to our understanding of the origin of mass of subatomic particles, and which recently was confirmed through the discovery of the predicted fundamental particle, by the ATLAS and CMS experiments at CERN's Large Hadron Collider* [203]

Additionally Physical Review Letters' 50-year review (2008) recognized the 1964 PRL symmetry breaking papers and Weinberg's 1967 paper *A model of Leptons* (the most cited paper in particle physics, as of 2012) "milestone Letters".[75]

Following reported observation of the Higgs-like particle in July 2012, several Indian media outlets reported on the supposed neglect of credit to Indian physicist Satyendra Nath Bose after whose work in the 1920s the class of particles "bosons" is named[204][205] (although physicists have described Bose's connection to the discovery as tenuous).[206]

1.7   Technical aspects and mathematical formulation

See also: Standard Model (mathematical formulation)

In the Standard Model, the Higgs field is a four-component scalar field that forms a complex doublet of the weak isospin SU(2) symmetry:

while the field has charge +1/2 under the weak hypercharge U(1) symmetry (in the convention where the electric charge, Q, the weak isospin, I_3, and the weak hypercharge, Y, are related by $Q = I_3 + Y$).[207]

The Higgs part of the Lagrangian is[207]

where W_μ^a and B_μ are the gauge bosons of the SU(2) and U(1) symmetries, g and g' their respective coupling constants, $\tau^a = \sigma^a/2$ (where σ^a are the Pauli matrices) a complete set generators of the SU(2) symmetry, and $\lambda > 0$ and $\mu^2 > 0$, so that the ground state breaks the SU(2) symmetry (see figure). The ground state of the Higgs field (the bottom of the potential) is degenerate with different ground states related to each other by a SU(2) gauge transformation. It is always possible to pick a gauge such that in the ground state $\phi^1 = \phi^2 = \phi^3 = 0$. The expectation value of ϕ^0 in the ground state (the vacuum expectation value or vev) is then $\langle \phi^0 \rangle = v$, where $v = \frac{|\mu|}{\sqrt{\lambda}}$. The measured value of this parameter is ~246 GeV/c^2.[102] It has units of mass, and is the only free parameter of the Standard Model that is not a dimensionless number. Quadratic terms in W_μ and B_μ arise, which give masses to the W and Z bosons:[207]

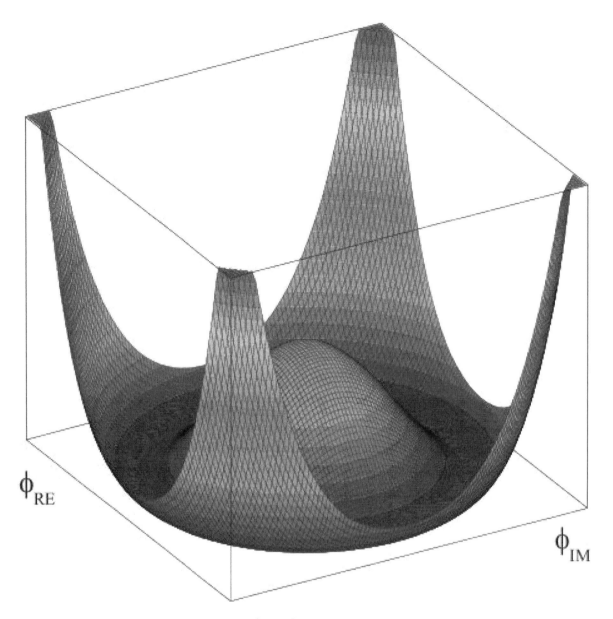

The potential for the Higgs field, plotted as function of ϕ^0 and ϕ^3. It has a Mexican-hat *or* champagne-bottle profile *at the ground.*

with their ratio determining the Weinberg angle, $\cos\theta_W = \frac{M_W}{M_Z} = \frac{|g|}{\sqrt{g^2+g'^2}}$, and leave a massless U(1) photon, γ .

The quarks and the leptons interact with the Higgs field through Yukawa interaction terms:

where $(d, u, e, \nu)^i_{L,R}$ are left-handed and right-handed quarks and leptons of the ith generation, $\lambda^{ij}_{u,d,e}$ are matrices of Yukawa couplings where h.c. denotes the hermitian conjugate terms. In the symmetry breaking ground state, only the terms containing ϕ^0 remain, giving rise to mass terms for the fermions. Rotating the quark and lepton fields to the basis where the matrices of Yukawa couplings are diagonal, one gets

where the masses of the fermions are $m^i_{u,d,e} = \lambda^i_{u,d,e} v/\sqrt{2}$, and $\lambda^i_{u,d,e}$ denote the eigenvalues of the Yukawa matrices.[207]

1.8   See also

Standard Model

- Quantum gauge theory

- History of quantum field theory

- Introduction to quantum mechanics

- Noncommutative standard model and noncommutative geometry generally

- Standard Model (mathematical formulation) (and especially Standard Model fields overview and mass terms and the Higgs mechanism)

Other

- Bose–Einstein statistics

- Dalitz plot

- Higgs boson in fiction

- Quantum triviality

- ZZ diboson

- Scalar boson

- Stueckelberg action

- Tachyonic field

1.9   Notes

[1] Note that such events also occur due to other processes. Detection involves a statistically significant excess of such events at specific energies.

[2] In the Standard Model, the total decay width of a Higgs boson with a mass of 126 GeV/c^2 is predicted to be 4.21×10^{-3} GeV.[5] The mean lifetime is given by $\tau = \hbar/\Gamma$.

[3] The range of a force is inversely proportional to the mass of the particles transmitting it.[19] In the Standard Model, forces are carried by virtual particles. These particles' movement and interactions with each other are limited by the energy–time uncertainty principle. As a result, the more massive a single virtual particle is, the greater its energy, and therefore the shorter the distance it can travel. A particle's mass therefore determines the maximum distance at which it can interact with other particles and on any force it mediates. By the same token, the reverse is also true: massless and near-massless particles can carry long distance forces. *(See also: Compton wavelength and Static forces and virtual-particle exchange)* Since experiments have shown that the weak force acts over only a very short range, this implies that there must exist massive gauge bosons. And indeed, their masses have since been confirmed by measurement.

[4] It is quite common for a law of physics to hold true only if certain assumptions held true or only under certain conditions. For example, Newton's laws of motion apply only at speeds where relativistic effects are negligible; and laws related to conductivity, gases, and classical physics (as opposed to quantum mechanics) may apply only within certain ranges of size, temperature, pressure, or other conditions.

[5] Electroweak symmetry is broken by the Higgs field in its lowest energy state, called its "ground state". At high energy levels this does not happen, and the gauge bosons of the weak force would therefore be expected to be massless.

[6] By the 1960s, many had already started to see gauge theories as failing to explain particle physics because theorists had been unable to solve the mass problem or even explain how gauge theory could provide a solution. So the idea that the Standard Model – which relied on a Higgs field, not yet proved to exist – could be fundamentally incorrect. Against this, once the model was developed around 1972, no better theory existed, and its predictions and solutions were so accurate, that it became the preferred theory anyway. It then became crucial to science, to know whether it was *correct*.

[7] The success of the Higgs-based electroweak theory and Standard Model is illustrated by their predictions of the mass of two particles later detected: the W boson (predicted mass: 80.390 ± 0.018 GeV, experimental measurement: 80.387 ± 0.019 GeV), and the Z boson (predicted mass: 91.1874 ± 0.0021, experimental measurement: 91.1876 ± 0.0021 GeV). The existence of the Z boson was itself another prediction. Other accurate predictions included the weak neutral current, the gluon, and the top and charm quarks, all later proven to exist as the theory said.

[8] For example, Huffington Post/Reuters[35] and others[36][37]

[9] The bubble's effects would be expected to propagate across the universe at the speed of light from wherever it occurred. However space is vast – with even the nearest galaxy being over 2 million lightyears from us, and others being many billions of lightyears distant, so the effect of such an event would be unlikely to arise here for billions of years after first occurring.[39][40]

[10] If the Standard Model is valid, then the particles and forces we observe in our universe exist as they do, because of underlying quantum fields. Quantum fields can have states of differing stability, including 'stable', 'unstable' and 'metastable' states (the latter remain stable unless sufficiently perturbed). If a more stable vacuum state were able to arise, then existing particles and forces would no longer arise as they presently do. Different particles or forces would arise from (and be shaped by) whatever new quantum states arose. The world we know depends upon these particles and forces, so if this happened, everything around us, from subatomic particles to galaxies, and all fundamental forces, would be reconstituted into new fundamental particles and forces and structures. The universe would potentially lose all of its present structures and become inhabited by new ones (depending upon the exact states involved) based upon the same quantum fields.

[11] Goldstone's theorem only applies to gauges having manifest Lorentz covariance, a condition that took time to become questioned. But the process of quantisation requires a gauge to be fixed and at this point it becomes possible to choose a gauge such as the 'radiation' gauge which is not invariant over time, so that these problems can be avoided. According to Bernstein (1974, p.8):

> "the "radiation gauge" condition $\nabla \cdot A(x) = 0$ is clearly noncovariant, which means that if we wish to maintain transversality of the photon in all Lorentz frames, the photon field $A\mu(x)$ cannot transform like a four-vector. This is no catastrophe, since the photon *field* is not an observable, and one can readily show that the S-matrix elements, which *are* observable have covariant structures .... in gauge theories one might arrange things so that one had a symmetry breakdown because of the noninvariance of the vacuum; but, because the Goldstone *et al.* proof breaks down, the zero mass Goldstone mesons need not appear." [Emphasis in original]

Bernstein (1974) contains an accessible and comprehensive background and review of this area, see external links

[12] A field with the "Mexican hat" potential $V(\phi) = \mu^2 \phi^2 + \lambda \phi^4$ and $\mu^2 < 0$ has a minimum not at zero but at some non-zero value ϕ_0. By expressing the action in terms of the field $\tilde{\phi} = \phi - \phi_0$ (where ϕ_0 is a constant independent of position), we find the Yukawa term has a component $g\phi_0\bar{\psi}\psi$. Since both g and ϕ_0 are constants, this looks exactly like the mass term for a fermion of mass $g\phi_0$. The field $\tilde{\phi}$ is then the Higgs field.

[13] In the Standard Model, the mass term arising from the Dirac Lagrangian for any fermion ψ is $-m\bar{\psi}\psi$. This is *not* invariant under the electroweak symmetry, as can be seen by writing ψ in terms of left and right handed components:

$$-m\bar{\psi}\psi = -m(\bar{\psi}_L\psi_R + \bar{\psi}_R\psi_L)$$

i.e., contributions from $\bar{\psi}_L\psi_L$ and $\bar{\psi}_R\psi_R$ terms do not appear. We see that the mass-generating interaction is achieved by constant flipping of particle chirality. Since the spin-half particles have no right/left helicity pair with the same SU(2) and SU(3) representation and the same weak hypercharge, then assuming these gauge charges are conserved in the vacuum, none of the spin-half particles could ever swap helicity. Therefore, in the absence of some other cause, all fermions must be massless.

[14] The example is based on the production rate at the LHC operating at 7 TeV. The total cross-section for producing a Higgs boson at the LHC is about 10 picobarn,[97] while the total cross-section for a proton–proton collision is 110 millibarn.[100]

[15] Just before LEP's shut down, some events that hinted at a Higgs were observed, but it was not judged significant enough to extend its run and delay construction of the LHC.

[16] Announced in articles in *Time*,[145] Forbes,[146] *Slate*,[147] *NPR*,[148] and others.[149]

[17] Other names have included: the "Anderson–Higgs" mechanism,[159] "Higgs–Kibble" mechanism (by Abdus Salam)[79] and "ABEGHHK'tH" mechanism [for Anderson, Brout, Englert, Guralnik, Hagen, Higgs, Kibble and 't Hooft] (by Peter Higgs).[79]

[18] In Miller's analogy, the Higgs field is compared to political party workers spread evenly throughout a room. There will be some people (in Miller's example an anonymous person) who pass through the crowd with ease, paralleling the interaction between the field and particles that do not interact with it, such as massless photons. There will be other people (in Miller's example the British prime minister) who would find their progress being continually slowed by the swarm of admirers crowding around, paralleling the interaction for particles that do interact with the field and by doing so, acquire a finite mass.[198][199]

1.10 References

[1] O'Luanaigh, C. (14 March 2013). "New results indicate that new particle is a Higgs boson". CERN. Retrieved 2013-10-09.

[2] Bryner, J. (14 March 2013). "Particle confirmed as Higgs boson". *NBC News*. Retrieved 2013-03-14.

[3] Heilprin, J. (14 March 2013). "Higgs Boson Discovery Confirmed After Physicists Review Large Hadron Collider Data at CERN". *The Huffington Post*. Retrieved 2013-03-14.

[4] ATLAS; CMS (26 March 2015). "Combined Measurement of the Higgs Boson Mass in pp Collisions at $\sqrt{s}$=7 and 8 TeV with the ATLAS and CMS Experiments". arXiv:1503.07589.

[5] LHC Higgs Cross Section Working Group; Dittmaier; Mariotti; Passarino; Tanaka; Alekhin; Alwall; Bagnaschi; Banfi (2012). "Handbook of LHC Higgs Cross Sections: 2. Differential Distributions". *CERN Report 2 (Tables A.1 – A.20)* **1201**: 3084. arXiv:1201.3084. Bibcode:2012arXiv1201.3084L.

[6] Onyisi, P. (23 October 2012). "Higgs boson FAQ". University of Texas ATLAS group. Retrieved 2013-01-08.

[7] Strassler, M. (12 October 2012). "The Higgs FAQ 2.0". *ProfMattStrassler.com*. Retrieved 2013-01-08. [Q] Why do particle physicists care so much about the Higgs particle?
[A] Well, actually, they don't. What they really care about is the Higgs *field*, because it is *so* important. [emphasis in original]

[8] José Luis Lucio and Arnulfo Zepeda (1987). *Proceedings of the II Mexican School of Particles and Fields, Cuernavaca-Morelos, 1986*. World Scientific. p. 29. ISBN 9971504340.

[9] Gunion, Dawson, Kane, and Haber (199). *The Higgs Hunter's Guide (1st ed.)*. pp. 11 (?). ISBN 9780786743186. – quoted as being in the first (1990) edition of the book by Peter Higgs in his talk "My Life as a Boson", 2001, ref#25.

[10] Strassler, M. (8 October 2011). "The Known Particles – If The Higgs Field Were Zero". *ProfMattStrassler.com*. Retrieved 13 November 2012. The Higgs field: so important it merited an entire experimental facility, the Large Hadron Collider, dedicated to understanding it.

[11] Biever, C. (6 July 2012). "It's a boson! But we need to know if it's the Higgs". *New Scientist*. Retrieved 2013-01-09. 'As a layman, I would say, I think we have it,' said Rolf-Dieter Heuer, director general of CERN at Wednesday's seminar announcing the results of the search for the Higgs boson. But when pressed by journalists afterwards on what exactly 'it' was, things got more complicated. 'We have discovered a boson – now we have to find out what boson it is'
Q: 'If we don't know the new particle is a Higgs, what do we know about it?' We know it is some kind of boson, says Vivek Sharma of CMS [...]
Q: 'are the CERN scientists just being too cautious? What would be enough evidence to call it a Higgs boson?' As there could be many different kinds of Higgs bosons, there's no straight answer.
[emphasis in original]

[12] Siegfried, T. (20 July 2012). "Higgs Hysteria". *Science News*. Retrieved 2012-12-09. In terms usually reserved for athletic achievements, news reports described the finding as a monumental milestone in the history of science.

[13] Del Rosso, A. (19 November 2012). "Higgs: The beginning of the exploration". *CERN Bulletin* (47–48). Retrieved 2013-01-09. Even in the most specialized circles, the new particle discovered in July is not yet being called the "Higgs boson". Physicists still hesitate to call it that before they have determined that its properties fit with those the Higgs theory predicts the Higgs boson has.

[14] Naik, G. (14 March 2013). "New Data Boosts Case for Higgs Boson Find". *The Wall Street Journal*. Retrieved 2013-03-15. 'We've never seen an elementary particle with spin zero,' said Tony Weidberg, a particle physicist at the University of Oxford who is also involved in the CERN experiments.

[15] Overbye, D. (8 October 2013). "For Nobel, They Can Thank the 'God Particle'". *The New York Times*. Retrieved 2013-11-03.

[16] Sample, I. (29 May 2009). "Anything but the God particle". *The Guardian*. Retrieved 2009-06-24.

[17] Evans, R. (14 December 2011). "The Higgs boson: Why scientists hate that you call it the 'God particle'". *National Post*. Retrieved 2013-11-03.

[18] The nickname occasionally has been satirised in mainstream media as well. Borowitz, Andy (July 13, 2012). "5 questions for the Higgs boson". *The New Yorker*.

[19] Shu, F. H. (1982). *The Physical Universe: An Introduction to Astronomy*. University Science Books. pp. 107–108. ISBN 978-0-935702-05-7.

[20] Shu, F. H. (1982). *The Physical Universe: An Introduction to Astronomy*. University Science Books. pp. 107–108. ISBN 978-0-935702-05-7.

[21] Leon M. Lederman and Dick Teresi (1993). *The God Particle: If the Universe is the Answer, What is the Question*. Houghton Mifflin Company.

[22] Heath, Nick, *The Cern tech that helped track down the God particle*, TechRepublic, 4 July 2012

[23] Rao, Achintya (2 July 2012). "Why would I care about the Higgs boson?". *CMS Public Website*. CERN. Retrieved 18 July 2012.

[24] Max Jammer, *Concepts of Mass in Contemporary Physics and Philosophy* (Princeton, NJ: Princeton University Press, 2000) pp.162–163, who provides many references in support of this statement.

[25] The Large Hadron Collider: Shedding Light on the Early Universe – lecture by R.-D. Heuer, CERN, Chios, Greece, 28 September 2011

[26] Alekhin, Djouadi and Moch, S.; Djouadi, A.; Moch, S. (2012-08-13). "The top quark and Higgs boson masses and the stability of the electroweak vacuum". *Physics Letters B* **716**: 214. arXiv:1207.0980. Bibcode:2012PhLB..716..214A. doi:10.1016/j.physletb.2012.0 Retrieved 20 February 2013.

[27] M.S. Turner, F. Wilczek (1982). "Is our vacuum metastable?". *Nature* **298** (5875): 633–634. Bibcode:1982Natur.298..633T. doi:10.1038/298633a0.

[28] S. Coleman and F. De Luccia (1980). "Gravitational effects on and of vacuum decay". *Physical Review* **D21** (12): 3305. Bibcode:1980PhRvD..21.3305C. doi:10.1103/PhysRevD.21.3305.

[29] M. Stone (1976). "Lifetime and decay of excited vacuum states". *Phys. Rev.* **D14**(12): 3568–3573.Bibcode:1976PhRvD..14.. doi:10.1103/PhysRevD.14.3568.

[30] P.H. Frampton (1976). "Vacuum Instability and Higgs Scalar Mass".*Phys. Rev. Lett.***37**(21): 1378–1380.Bibcode:19F. doi:10.1103/PhysRevLett.37.1378.

[31] P.H. Frampton (1977). "Consequences of Vacuum Instability in Quantum Field Theory". *Phys. Rev.* **D15** (10): 2922–28. Bibcode:1977PhRvD..15.2922F. doi:10.1103/PhysRevD.15.2922.

[32] Ellis, Espinosa, Giudice, Hoecker, & Riotto, J.; Espinosa, J.R.; Giudice, G.F.; Hoecker, A.; Riotto, A. (2009). "The Probable Fate of the Standard Model". *Phys. Lett. B* **679** (4): 369–375. arXiv:0906.0954. Bibcode:2009PhLB..679..369E. doi:10.1016/j.physletb.2009.07.054.

[33] Masina, Isabella (2013-02-12). "Higgs boson and top quark masses as tests of electroweak vacuum stability". *Phys. Rev. D* **87** (5): 53001. arXiv:1209.0393. Bibcode:2013PhRvD..87e3001M. doi:10.1103/PhysRevD.87.053001.

[34] Buttazzo, Degrassi, Giardino, Giudice, Sala, Salvio, Strumia (2013-07-12). "Investigating the near-criticality of the Higgs boson". *JHEP 1312 (2013) 089*. arXiv:1307.3536. Bibcode:2013JHEP...12..089B. doi:10.1007/JHEP12(2013)089.

[35] Irene Klotz (editing by David Adams and Todd Eastham) (2013-02-18). "Universe Has Finite Lifespan, Higgs Boson Calculations Suggest". *Huffington Post*. Reuters. Retrieved 21 February 2013. Earth will likely be long gone before any Higgs boson particles set off an apocalyptic assault on the universe

[36] Hoffman, Mark (2013-02-19). "Higgs Boson Will Destroy The Universe Eventually". *ScienceWorldReport*. Retrieved 21 February 2013.

[37] "Higgs boson will aid in creation of the universe – and how it will end". *Catholic Online/NEWS CONSORTIUM*. 2013-02-20. Retrieved 21 February 2013. [T]he Earth will likely be long gone before any Higgs boson particles set off an apocalyptic assault on the universe

[38] Salvio, Alberto (2015-04-09). "A Simple Motivated Completion of the Standard Model below the Planck Scale: Axions and Right-Handed Neutrinos". *Physics Letters B* **743**: 428. arXiv:1501.03781. Bibcode:2015PhLB..743..428S. doi:10.1016/j.

[39] Boyle, Alan (2013-02-19). "Will our universe end in a 'big slurp'? Higgs-like particle suggests it might". *NBC News' Cosmic log*. Retrieved 21 February 2013. [T]he bad news is that its mass suggests the universe will end in a fast-spreading bubble of doom. The good news? It'll probably be tens of billions of years. The article quotes Fermilab's Joseph Lykken: "[T]he parameters for our universe, including the Higgs [and top quark's masses] suggest that we're just at the edge of stability, in a "metastable" state. Physicists have been contemplating such a possibility for more than 30 years. Back in 1982, physicists Michael Turner and Frank Wilczek wrote in Nature that "without warning, a bubble of true vacuum could nucleate somewhere in the universe and move outwards...".

[40] Peralta, Eyder (2013-02-19). "If Higgs Boson Calculations Are Right, A Catastrophic 'Bubble' Could End Universe". *npr – two way*. Retrieved 21 February 2013. Article cites Fermilab's Joseph Lykken: "The bubble forms through an unlikely quantum fluctuation, at a random time and place," Lykken tells us. "So in principle it could happen tomorrow, but then most likely in a very distant galaxy, so we are still safe for billions of years before it gets to us."

[41] Bezrukov; Shaposhnikov (2007-10-19). "The Standard Model Higgs boson as the inflaton". *Phys.Lett. B659 (2008) 703-706*. arXiv:0710.3755. Bibcode:2008PhLB..659..703B. doi:10.1016/j.physletb.2007.11.072.

[42] Salvio, Alberto (2013-08-09). "Higgs Inflation at NNLO after the Boson Discovery". *Phys.Lett. B727 (2013) 234-239*. arXiv:1308.2244. Bibcode:2013PhLB..727..234S. doi:10.1016/j.physletb.2013.10.042.

[43] Cole, K. (2000-12-14). "One Thing Is Perfectly Clear: Nothingness Is Perfect". *Los Angeles Times*. p. 'Science File'. Retrieved 17 January 2013. [T]he Higgs' influence (or the influence of something like it) could reach much further. For example, something like the Higgs—if not exactly the Higgs itself—may be behind many other unexplained "broken symmetries" in the universe as well ... In fact, something very much like the Higgs may have been behind the collapse of the symmetry that led to the Big Bang, which created the universe. When the forces first began to separate from their primordial sameness—taking on the distinct characters they have today—they released energy in the same way as water releases energy when it turns to ice. Except in this case, the freezing packed enough energy to blow up the universe. ... However it happened, the moral is clear: Only when the perfection shatters can everything else be born.

[44] Higgs Matters – Kathy Sykes, 30 Nove 2012

[45] Why the public should care about the Higgs Boson – Jodi Lieberman, American Physical Society (APS)

[46] Matt Strassler's blog – Why the Higgs particle matters 2 July 2012

[47] Sean Carroll (13 November 2012). *The Particle at the End of the Universe: How the Hunt for the Higgs Boson Leads Us to the Edge of a New World*. Penguin Group US. ISBN 978-1-101-60970-5.

[48] Woit, Peter (13 November 2010). "The Anderson–Higgs Mechanism". Dr. Peter Woit (Senior Lecturer in Mathematics Columbia University and Ph.D. particle physics). Retrieved 12 November 2012.

[49] Goldstone, J; Salam, Abdus; Weinberg, Steven (1962). "Broken Symmetries".*Physical Review***127**(3): 965–970.Bibcode:1962P. doi:10.1103/PhysRev.127.965.

[50] Guralnik, G. S. (2011). "The Beginnings of Spontaneous Symmetry Breaking in Particle Physics". arXiv:1110.2253 [physics.hist-ph].

[51] Kibble, T. W. B. (2009). "Englert–Brout–Higgs–Guralnik–Hagen–Kibble Mechanism". *Scholarpedia* **4** (1): 6441. Bibcode:. doi:10.4249/scholarpedia.6441. Retrieved 2012-11-23.

[52] Kibble, T. W. B. "History of Englert–Brout–Higgs–Guralnik–Hagen–Kibble Mechanism (history)". *Scholarpedia* **4** (1): 8741. Bibcode:2009SchpJ...4.8741K. doi:10.4249/scholarpedia.8741. Retrieved 2012-11-23.

[53] The Nobel Prize in Physics 2008 – official Nobel Prize website.

[54] List of Anderson 1958–1959 papers referencing 'symmetry', at APS Journals

[55] Higgs, Peter (2010-11-24). "My Life as a Boson" (PDF). Talk given by Peter Higgs at Kings College, London, Nov 24 2010, expanding on a paper originally presented in 2001. Retrieved 17 January 2013. – the original 2001 paper can be found at: Duff and Liu, ed. (2003) [year of publication]. *2001 A Spacetime Odyssey: Proceedings of the Inaugural Conference of the Michigan Center for Theoretical Physics, Michigan, USA, 21–25 May 2001.* World Scientific. pp. 86–88. ISBN 9812382313. Retrieved 17 January 2013.

[56] Anderson, P. (1963). "Plasmons, gauge invariance and mass". *Physical Review* **130**: 439. Bibcode:1963PhRv..130..439A. doi:10.1103/PhysRev.130.439.

[57] Klein, A.; Lee, B. (1964). "Does Spontaneous Breakdown of Symmetry Imply Zero-Mass Particles?". *Physical Review Letters* **12** (10): 266. Bibcode:1964PhRvL..12..266K. doi:10.1103/PhysRevLett.12.266.

[58] Englert, François; Brout, Robert (1964). "Broken Symmetry and the Mass of Gauge Vector Mesons". *Physical Review Letters* **13** (9): 321–23. Bibcode:1964PhRvL..13..321E. doi:10.1103/PhysRevLett.13.321.

[59] Higgs, Peter (1964). "Broken Symmetries and the Masses of Gauge Bosons". *Physical Review Letters* **13** (16): 508–509. Bibcode:1964PhRvL..13..508H. doi:10.1103/PhysRevLett.13.508.

[60] Guralnik, Gerald; Hagen, C. R.; Kibble, T. W. B. (1964). "Global Conservation Laws and Massless Particles". *Physical Review Letters* **13** (20): 585–587. Bibcode:1964PhRvL..13..585G. doi:10.1103/PhysRevLett.13.585.

[61] Higgs, Peter (1964). "Broken symmetries, massless particles and gauge fields". *Physics Letters* **12** (2): 132–133. Bibcode:1964Ph. doi:10.1016/0031-9163(64)91136-9.

[62] Higgs, Peter (2010-11-24). "My Life as a Boson" (PDF). Talk given by Peter Higgs at Kings College, London, Nov 24 2010. Retrieved 17 January 2013. Gilbert ... wrote a response to [Klein and Lee's paper] saying 'No, you cannot do that in a relativistic theory. You cannot have a preferred unit time-like vector like that.' This is where I came in, because the next month was when I responded to Gilbert's paper by saying 'Yes, you can have such a thing' but only in a gauge theory with a gauge field coupled to the current.

[63] G.S. Guralnik (2011). "Gauge invariance and the Goldstone theorem – 1965 Feldafing talk". *Modern Physics Letters A* **26** (19): 1381–1392. arXiv:1107.4592. Bibcode:2011MPLA...26.1381G. doi:10.1142/S0217732311036188.

[64] Higgs, Peter (1966). "Spontaneous Symmetry Breakdown without Massless Bosons". *Physical Review* **145** (4): 1156–1163. Bibcode:1966PhRv..145.1156H. doi:10.1103/PhysRev.145.1156.

[65] Kibble, Tom (1967). "Symmetry Breaking in Non-Abelian Gauge Theories". *Physical Review* **155** (5): 1554–1561. BibcodK. doi:10.1103/PhysRev.155.1554.

[66] "Guralnik, G S; Hagen, C R and Kibble, T W B (1967). Broken Symmetries and the Goldstone Theorem. Advances in Physics, vol. 2" (PDF).

[67] "Physical Review Letters – 50th Anniversary Milestone Papers". Physical Review Letters.

[68] S. Weinberg (1967). "A Model of Leptons". *Physical Review Letters* **19** (21): 1264–1266. Bibcode:1967PhRvL...19.1264W. doi:10.1103/PhysRevLett.19.1264.

[69] A. Salam (1968). N. Svartholm, ed. *Elementary Particle Physics: Relativistic Groups and Analyticity.* Eighth Nobel Symposium. Stockholm: Almquvist and Wiksell. p. 367.

[70] S.L. Glashow (1961). "Partial-symmetries of weak interactions". *Nuclear Physics* **22** (4): 579–588. Bibcode:1961NucPh... doi:10.1016/0029-5582(61)90469-2.

[71] Ellis, John; Gaillard, Mary K.; Nanopoulos, Dimitri V. (2012). "A Historical Profile of the Higgs Boson". arXiv:1201.6045 [hep-ph].

[72] "Martin Veltman Nobel Lecture, December 12, 1999, p.391" (PDF). Retrieved 2013-10-09.

[73] Politzer, David. "The Dilemma of Attribution". *Nobel Prize lecture, 2004.* Nobel Prize. Retrieved 22 January 2013. Sidney Coleman published in Science magazine in 1979 a citation search he did documenting that essentially no one paid any attention to Weinberg's Nobel Prize winning paper until the work of 't Hooft (as explicated by Ben Lee). In 1971 interest in Weinberg's paper exploded. I had a parallel personal experience: I took a one-year course on weak interactions from Shelly Glashow in 1970, and he never even mentioned the Weinberg–Salam model or his own contributions.

[74]Coleman, Sidney(1979-12-14)."The 1979 Nobel Prize in Physics".*Science***206**(4424): 1290–1292.Bibcode:1979Sci....
     doi:10.1126/science.206.4424.1290.

[75] Letters from the Past – A PRL Retrospective (50 year celebration, 2008)

[76] Jeremy Bernstein (January 1974). "Spontaneous symmetry breaking, gauge theories, the Higgs mechanism and all that" (PDF).
     Reviews of Modern Physics **46** (1): 7. Bibcode:1974RvMP...46....7B. doi:10.1103/RevModPhys.46.7. Retrieved 2012-12-10.

[77] American Physical Society – "J. J. Sakurai Prize for Theoretical Particle Physics".

[78] Merali, Zeeya (4 August 2010). "Physicists get political over Higgs". *Nature Magazine*. Retrieved 28 December 2011.

[79] Close, Frank (2011). *The Infinity Puzzle: Quantum Field Theory and the Hunt for an Orderly Universe*. Oxford: Oxford
     University Press. ISBN 978-0-19-959350-7.

[80] G.S. Guralnik (2009). "The History of the Guralnik, Hagen and Kibble development of the Theory of Spontaneous Sym-
     metry Breaking and Gauge Particles". *International Journal of Modern Physics A* **24** (14): 2601–2627. arXiv:0907.3466.
     Bibcode:2009IJMPA..24.2601G. doi:10.1142/S0217751X09045431.

[81] Peskin, Michael E.; Schroeder, Daniel V. (1995). *Introduction to Quantum Field Theory*. Reading, MA: Addison-Wesley
     Publishing Company. pp. 717–719 and 787–791. ISBN 0-201-50397-2.

[82] Peskin & Schroeder 1995, pp. 715–716

[83] Gunion, John (2000). *The Higgs Hunter's Guide* (illustrated, reprint ed.). Westview Press. pp. 1–3. ISBN 9780738203058.

[84] Lisa Randall, *Warped Passages: Unraveling the Mysteries of the Universe's Hidden Dimensions*, p.286: "People initially thought
     of tachyons as particles travelling faster than the speed of light...But we now know that a tachyon indicates an instability in a
     theory that contains it. Regrettably for science fiction fans, tachyons are not real physical particles that appear in nature."

[85] Sen, Ashoke (April 2002). "Rolling Tachyon". *J. High Energy Phys.* **2002** (0204): 048. arXiv:hep-th/0203211.Bibc.
     doi:10.1088/1126-6708/2002/04/048.

[86] Kutasov, David; Marino, Marcos & Moore, Gregory W. (2000). "Some exact results on tachyon condensation in string field
     theory". *JHEP* **0010**: 045.

[87] Aharonov, Y.; Komar, A.; Susskind, L. (1969). "Superluminal Behavior, Causality, and Instability". *Phys. Rev.* (American
     Physical Society) **182** (5): 1400–1403. Bibcode:1969PhRv..182.1400A. doi:10.1103/PhysRev.182.1400.

[88]Feinberg, Gerald(1967). "Possibility of Faster-Than-Light Particles".*Physical Review***159**(5): 1089–1105.Bibcode:1967Ph.
     doi:10.1103/PhysRev.159.1089.

[89] Michael E. Peskin and Daniel V. Schroeder (1995). *An Introduction to Quantum Field Theory*, Perseus books publishing.

[90] Flatow, Ira (6 July 2012). "At Long Last, The Higgs Particle... Maybe". *NPR*. Retrieved 10 July 2012.

[91] "Explanatory Figures for the Higgs Boson Exclusion Plots". *ATLAS News*. CERN. Retrieved 6 July 2012.

[92] Bernardi, G.; Carena, M.; Junk, T. (2012). "Higgs Bosons: Theory and Searches" (PDF). p. 7.

[93] Lykken, Joseph D. (2009). "Beyond the Standard Model". *Proceedings of the 2009 European School of High-Energy Physics,
     Bautzen, Germany, 14 – 27 June 2009*. arXiv:1005.1676.

[94] Plehn, Tilman (2012). *Lectures on LHC Physics*. Lecture Notes is Physics **844**. Springer. Sec. 1.2.2. arXiv:0910.4122. ISBN
     3642240399.

[95] "LEP Electroweak Working Group".

[96] Peskin, Michael E.; Wells, James D. (2001). "How Can a Heavy Higgs Boson be Consistent with the Precision Electroweak Mea-
     surements?". *Physical Review D* **64** (9): 093003. arXiv:hep-ph/0101342. Bibcode:2001PhRvD..64i3003P. doi:10.110.

[97] Baglio, Julien; Djouadi, Abdelhak (2011). "Higgs production at the lHC". *Journal of High Energy Physics* **1103** (3): 055.
     arXiv:1012.0530. Bibcode:2011JHEP...03..055B. doi:10.1007/JHEP03(2011)055.

[98] Baglio, Julien; Djouadi, Abdelhak (2010). "Predictions for Higgs production at the Tevatron and the associated uncertainties".
     Journal of High Energy Physics **1010** (10): 063. arXiv:1003.4266. Bibcode:2010JHEP...10..064B. doi:10.1007/JHEP10.

[99] Teixeira-Dias (LEP Higgs working group), P. (2008). "Higgs boson searches at LEP". *Journal of.Physics: Conference Series* **110** (4): 042030. arXiv:0804.4146. Bibcode:2008JPhCS.110d2030T. doi:10.1088/1742-6596/110/4/042030.

[100] "Collisions". *LHC Machine Outreach*. CERN. Retrieved 26 July 2012.

[101] Asquith, Lily (22 June 2012). "Why does the Higgs decay?". *Life and Physics* (London: The Guardian). Retrieved 14 August 2012.

[102] "Higgs bosons: theory and searches" (PDF). *PDGLive*. Particle Data Group. 12 July 2012. Retrieved 15 August 2012.

[103] Branco, G. C.; Ferreira, P.M.; Lavoura, L.; Rebelo, M.N.; Sher, Marc; Silva, João P. (July 2012). "Theory and phenomenology of two-Higgs-doublet models". *Physics Reports* (Elsevier) **516** (1): 1–102. arXiv:1106.0034. Bibcode:2012PhR...516....1B. doi:10.1016/j.physrep.2012.02.002.

[104] Csaki, C.; Grojean, C.; Pilo, L.; Terning, J. (2004). "Towards a realistic model of Higgsless electroweak symmetry breaking". *Physical Review Letters* **92** (10): 101802. arXiv:hep-ph/0308038. Bibcode:2004PhRvL..92j1802C. doi:10.1103/PhysRev. PMID 15089195.

[105] Csaki, C.; Grojean, C.; Pilo, L.; Terning, J.; Terning, John (2004). "Gauge theories on an interval: Unitarity without a Higgs". *Physical Review D* **69** (5): 055006. arXiv:hep-ph/0305237. Bibcode:2004PhRvD..69e5006C. doi:10.1103/PhysRevD.69.055006.

[106] "The Hierarchy Problem: why the Higgs has a snowball's chance in hell". Quantum Diaries. 2012-07-01. Retrieved 19 March 2013.

[107] "The Hierarchy Problem | Of Particular Significance". Profmattstrassler.com. Retrieved 2013-10-09.

[108] "Collisions". *LHC Machine Outreach*. CERN. Retrieved 26 July 2012.

[109] "Hunt for Higgs boson hits key decision point". MSNBC. 2012-12-06. Retrieved 2013-01-19.

[110] Worldwide LHC Computing Grid main page 14 November 2012: *"[A] global collaboration of more than 170 computing centres in 36 countries ... to store, distribute and analyse the ~25 Petabytes (25 million Gigabytes) of data annually generated by the Large Hadron Collider"*

[111] What is the Worldwide LHC Computing Grid? (Public 'About' page) 14 November 2012: *"Currently WLCG is made up of more than 170 computing centers in 36 countries...The WLCG is now the world's largest computing grid"*

[112] W.-M. Yao; et al. (2006). "Review of Particle Physics" (PDF). *Journal of Physics G* **33**: 1. arXiv:astro-ph/0601168.Bibcod. doi:10.1088/0954-3899/33/1/001.

[113] The CDF Collaboration, the D0 Collaboration, the Tevatron New Physics, Higgs Working Group (2012). "Updated Combination of CDF and D0 Searches for Standard Model Higgs Boson Production with up to 10.0 fb^{-1} of Data". arXiv:1207.0449 [hep-ex].

[114] "Interim Summary Report on the Analysis of the 19 September 2008 Incident at the LHC" (PDF). CERN. 15 October 2008. EDMS 973073. Retrieved 28 September 2009.

[115] "CERN releases analysis of LHC incident" (Press release). CERN Press Office. 16 October 2008. Retrieved 28 September 2009.

[116] "LHC to restart in 2009" (Press release). CERN Press Office. 5 December 2008. Retrieved 8 December 2008.

[117] "LHC progress report". *The Bulletin*. CERN. 3 May 2010. Retrieved 7 December 2011.

[118] "ATLAS experiment presents latest Higgs search status". *ATLAS homepage*. CERN. 13 December 2011. Retrieved 13 December 2011.

[119] Taylor, Lucas (13 December 2011). "CMS search for the Standard Model Higgs Boson in LHC data from 2010 and 2011". *CMS public website*. CERN. Retrieved 13 December 2011.

[120] Overbye, D. (5 March 2013). "Chasing The Higgs Boson". *The New York Times*. Retrieved 2013-03-05.

[121] "ATLAS and CMS experiments present Higgs search status" (Press release). CERN Press Office. 13 December 2011. Retrieved 14 September 2012. the statistical significance is not large enough to say anything conclusive. As of today what we see is consistent either with a background fluctuation or with the presence of the boson. Refined analyses and additional data delivered in 2012 by this magnificent machine will definitely give an answer

[122] "WLCG Public Website". CERN. Retrieved 29 October 2012.

[123] CMS collaboration (2014). "Precise determination of the mass of the Higgs boson and tests of compatibility of its couplings with the standard model predictions using proton collisions at 7 and 8 TeV". arXiv:1412.8662.

[124] ATLAS collaboration (2014). "Measurements of Higgs boson production and couplings in the four-lepton channel in pp collisions at center-of-mass energies of 7 and 8 TeV with the ATLAS detector". arXiv:1408.5191.

[125] ATLAS collaboration (2014). "Measurement of Higgs boson production in the diphoton decay channel in pp collisions at center-of-mass energies of 7 and 8 TeV with the ATLAS detector". arXiv:1408.7084.

[126] "Press Conference: Update on the search for the Higgs boson at CERN on 4 July 2012". Indico.cern.ch. 22 June 2012. Retrieved 4 July 2012.

[127] "CERN to give update on Higgs search". CERN. 22 June 2012. Retrieved 2 July 2011.

[128] "Scientists analyse global Twitter gossip around Higgs boson discovery". *phys.org (from arXiv)*. 2013-01-23. Retrieved 6 February 2013. – stated to be *" the first time scientists have been able to analyse the dynamics of social media on a global scale before, during and after the announcement of a major scientific discovery."* For the paper itself see: De Domenico, M.; Lima, A.; Mougel, P.; Musolesi, M. (2013). "The Anatomy of a Scientific Gossip". arXiv:1301.2952. Bibcode:2013NatSR...3E2980D. doi:10.1038/srep02980.

[129] "Higgs boson particle results could be a quantum leap". Times LIVE. 28 June 2012. Retrieved 4 July 2012.

[130] CERN prepares to deliver Higgs particle findings, Australian Broadcasting Corporation. Retrieved 4 July 2012.

[131] "God Particle Finally Discovered? Higgs Boson News At Cern Will Even Feature Scientist It's Named After". Huffingtonpost.co.uk. Retrieved 2013-01-19.

[132] Our Bureau (2012-07-04). "Higgs on way, theories thicken". Calcutta, India: Telegraphindia.com. Retrieved 2013-01-19.

[133] Thornhill, Ted (2013-07-03). "God Particle Finally Discovered? Higgs Boson News At Cern Will Even Feature Scientist It's Named After". *Huffington Post*. Retrieved 23 July 2013.

[134] Cooper, Rob (2013-07-01) [updated subsequently]. "God particle is 'found': Scientists at Cern expected to announce on Wednesday Higgs boson particle has been discovered". *Daily Mail* (London). Retrieved 23 July 2013. - States that "*"Five leading theoretical physicists have been invited to the event on Wednesday - sparking speculation that the particle has been discovered."*, including Higgs and Englert, and that Kibble - who was invited but unable to attend - "told the Sunday Times: 'My guess is that is must be a pretty positive result for them to be asking us out there'."

[135] Adrian Cho (13 July 2012). "Higgs Boson Makes Its Debut After Decades-Long Search". *Science* **337** (6091): 141–143. doi:10.1126/science.337.6091.141. PMID 22798574.

[136] CMS collaboration (2012). "Observation of a new boson at a mass of 125 GeV with the CMS experiment at the LHC". *Physics Letters B* **716** (1): 30–61. arXiv:1207.7235. Bibcode:2012PhLB..716...30C. doi:10.1016/j.physletb.2012.08.021.

[137] Taylor, Lucas (4 July 2012). "Observation of a New Particle with a Mass of 125 GeV". *CMS Public Website*. CERN. Retrieved 4 July 2012.

[138] "Latest Results from ATLAS Higgs Search". *ATLAS News*. CERN. 4 July 2012. Retrieved 4 July 2012.

[139] ATLAS collaboration (2012). "Observation of a New Particle in the Search for the Standard Model Higgs Boson with the ATLAS Detector at the LHC". *Physics Letters B* **716** (1): 1–29. arXiv:1207.7214. Bibcode:2012PhLB..716....1A. doi:10.1016.

[140] Gillies, James (23 July 2012). "LHC 2012 proton run extended by seven weeks". *CERN bulletin*. Retrieved 29 August 2012.

[141] "Higgs boson behaving as expected". *3 News NZ*. 15 November 2012.

[142] Strassler, Matt (2012-11-14). "Higgs Results at Kyoto". *Of Particular Significance: Conversations About Science with Theoretical Physicist Matt Strassler*. Prof. Matt Strassler's personal particle physics website. Retrieved 10 January 2013. ATLAS and CMS only just co-discovered this particle in July ... We will not know after today whether it is a Higgs at all, whether it is a Standard Model Higgs or not, or whether any particular speculative idea...is now excluded. [...] Knowledge about nature does not come easy. We discovered the top quark in 1995, and we are still learning about its properties today... we will still be learning important things about the Higgs during the coming few decades. We've no choice but to be patient.

[143] Sample, Ian (14 November 2012). "Higgs particle looks like a bog Standard Model boson, say scientists". *The Guardian* (London). Retrieved 15 November 2012.

[144] "CERN experiments observe particle consistent with long-sought Higgs boson". CERN press release. 4 July 2012. Retrieved 4 July 2012.

[145] "Person Of The Year 2012". *Time*. 19 December 2012.

[146] "Higgs Boson Discovery Has Been Confirmed". Forbes. Retrieved 2013-10-09.

[147] Slate Video Staff (2012-09-11). "Higgs Boson Confirmed; CERN Discovery Passes Test". Slate.com. Retrieved 2013-10-09.

[148] "The Year Of The Higgs, And Other Tiny Advances In Science". NPR. 2013-01-01. Retrieved 2013-10-09.

[149] "Confirmed: the Higgs boson does exist". *The Sydney Morning Herald*. 4 July 2012.

[150] "AP CERN chief: Higgs boson quest could wrap up by midyear". *MSNBC*. Associated Press. 2013-01-27. Retrieved 20 February 2013. Rolf Heuer, director of [CERN], said he is confident that "towards the middle of the year, we will be there." – Interview by AP, at the World Economic Forum, 26 Jan 2013.

[151] Boyle, Alan (2013-02-16). "Will our universe end in a 'big slurp'? Higgs-like particle suggests it might". *NBCNews.com – cosmic log*. Retrieved 20 February 2013. 'it's going to take another few years' after the collider is restarted to confirm definitively that the newfound particle is the Higgs boson.

[152] Gillies, James (2013-03-06). "A question of spin for the new boson". CERN. Retrieved 7 March 2013.

[153] Adam Falkowski (writing as 'Jester') (2013-02-27). "When shall we call it Higgs?". Résonaances particle physics blog. Retrieved 7 March 2013.

[154] CMS Collaboration (February 2013). "Study of the Mass and Spin-Parity of the Higgs Boson Candidate via Its Decays to Z Boson Pairs". *Phys. Rev. Lett.* (American Physical Society) **110** (8): 081803. arXiv:1212.6639. Bibcode:2013PhRvL.110h1803C. doi:10.1103/PhysRevLett.110.081803. Retrieved 15 September 2014.

[155] ATLAS Collaboration (7 October 2013). "Evidence for the spin-0 nature of the Higgs boson using ATLAS data". *Phys. Lett. B* (American Physical Society) **726** (1-3): 120–144. Bibcode:2013PhLB..726..120A. doi:10.1016/j.physletb.2013.08.026. Retrieved 15 September 2014.

[156] "Higgs-like Particle in a Mirror". American Physical Society. Retrieved 26 February 2013.

[157] The CMS Collaboration (2014-06-22). "Evidence for the direct decay of the 125 GeV Higgs boson to fermions". Nature Publishing Group doi= 10.1038/nphys3005.

[158] Adam Falkowski (writing as 'Jester') (2012-12-13). "Twin Peaks in ATLAS". Résonaances particle physics blog. Retrieved 24 February 2013.

[159] Liu, G. Z.; Cheng, G. (2002). "Extension of the Anderson-Higgs mechanism". *Physical Review B* **65** (13): 132513. arXiv:cond-mat/0106070. Bibcode:2002PhRvB..65m2513L. doi:10.1103/PhysRevB.65.132513.

[160] Editorial (2012-03-21). "Mass appeal: As physicists close in on the Higgs boson, they should resist calls to change its name". *Nature*. 483, 374 (7390): 374. Bibcode:2012Natur.483..374.. doi:10.1038/483374a. Retrieved 21 January 2013.

[161] Becker, Kate (2012-03-29). "A Higgs by Any Other Name". "NOVA" (PBS) physics. Retrieved 21 January 2013.

[162] "Frequently Asked Questions: The Higgs!". *The Bulletin*. CERN. Retrieved 18 July 2012.

[163] Woit's physics blog *"Not Even Wrong"*: Anderson on Anderson-Higgs 2013-04-13

[164] Sample, Ian (2012-07-04). "Higgs boson's many great minds cause a Nobel prize headache". *The Guardian* (London). Retrieved 23 July 2013.

[165] "Rochester's Hagen Sakurai Prize Announcement" (Press release). University of Rochester. 2010.

[166] *C.R. Hagen Sakurai Prize Talk* (YouTube). 2010.

[167]Cho, A (2012-09-14)."Particle physics. Why the 'Higgs'?"(PDF).*Science***337**(6100): 1287.doi:10.1126/science.337..PMID229
84044. Lee ... apparently used the term 'Higgs Boson' as early as 1966... but what may have made the term stick isa seminalpaperS
teven Weinberg...published in 1967...Weinberg acknowledged the mix-up in an essay in the*New York Reviewof Books*in May 2012.
(Seealsothe original article in*New York Review of Books*   [168] and Frank Close's 2011 book *The Infinity
Puzzle*[79]:372)

[168] Weinberg, Steven (2012-05-10). "The Crisis of Big Science". *The New York Review of Books* (footnote 1). Retrieved 12
February 2013.

[169] Examples of early papers using the term "Higgs boson" include 'A phenomenological profile of the Higgs boson' (Ellis, Gaillard and Nanopoulos, 1976), 'Weak interaction theory and neutral currents' (Bjorken, 1977), and 'Mass of the Higgs boson'
(Wienberg, received 1975)

[170] Leon Lederman; Dick Teresi (2006). *The God Particle: If the Universe Is the Answer, What Is the Question?*. Houghton Mifflin
Harcourt. ISBN 0-547-52462-5.

[171] Kelly Dickerson (September 8, 2014). "Stephen Hawking Says 'God Particle' Could Wipe Out the Universe". livescience.com.

[172] Jim Baggott (2012). *Higgs: The invention and discovery of the 'God Particle'*. Oxford University Press. ISBN 978-0-19-165003-
1.

[173] Scientific American Editors (2012). *The Higgs Boson: Searching for the God Particle*. Macmillan. ISBN 978-1-4668-2413-3.

[174] Ted Jaeckel (2007). *The God Particle: The Discovery and Modeling of the Ultimate Prime Particle*. Universal-Publishers. ISBN
978-1-58112-959-5.

[175] Aschenbach, Joy (1993-12-05). "No Resurrection in Sight for Moribund Super Collider : Science: Global financial partnerships
could be the only way to salvage such a project. Some feel that Congress delivered a fatal blow". *Los Angeles Times*. Retrieved
16 January 2013. 'We have to keep the momentum and optimism and start thinking about international collaboration,' said
Leon M. Lederman, the Nobel Prize-winning physicist who was the architect of the super collider plan

[176] "A Supercompetition For Illinois". *Chicago Tribune*. 1986-10-31. Retrieved 16 January 2013. The SSC, proposed by the U.S.
Department of Energy in 1983, is a mind-bending project ... this gigantic laboratory ... this titanic project

[177] Diaz, Jesus (2012-12-15). "This Is [The] World's Largest Super Collider That Never Was". *Gizmodo*. Retrieved 16 January
2013. ...this titanic complex...

[178] Abbott, Charles (June 1987). "Illinois Issues journal, June 1987". p. 18. Lederman, who considers himself an unofficial
propagandist for the super collider, said the SSC could reverse the physics brain drain in which bright young physicists have left
America to work in Europe and elsewhere.

[179] Kevles, Dan. "Good-bye to the SSC: On the Life and Death of the Superconducting Super Collider" (PDF). *California Institute
of Technology: "Engineering & Science"*. 58 no. 2 (Winter 1995): 16–25. Retrieved 16 January 2013. Lederman, one of
the principal spokesmen for the SSC, was an accomplished high-energy experimentalist who had made Nobel Prize-winning
contributions to the development of the Standard Model during the 1960s (although the prize itself did not come until 1988).
He was a fixture at congressional hearings on the collider, an unbridled advocate of its merits.

[180] Calder, Nigel (2005). *Magic Universe:A Grand Tour of Modern Science*. pp. 369–370. ISBN 9780191622359. The possibility
that the next big machine would create the Higgs became a carrot to dangle in front of funding agencies and politicians. A
prominent American physicist, Leon lederman *[sic]*, advertised the Higgs as The God Particle in the title of a book published in
1993 ...Lederman was involved in a campaign to persuade the US government to continue funding the Superconducting Super
Collider... the ink was not dry on Lederman's book before the US Congress decided to write off the billions of dollars already
spent

[181] Lederman, Leon (1993). *The God Particle If the Universe Is the Answer, What Is the Question?* (PDF). Dell Publishing. p.
Chapter 2, Page 2. ISBN 0-385-31211-3. Retrieved 30 July 2015.

[182] Alister McGrath, Higgs boson: the particle of faith, *The Daily Telegraph*, Published 15 December 2011. Retrieved 15 December
2011.

[183] Sample, Ian (3 March 2009). "Father of the God particle: Portrait of Peter Higgs unveiled". London: The Guardian. Retrieved
24 June 2009.

[184] Chivers, Tom (2011-12-13). "How the 'God particle' got its name". *The Telegraph* (London). Retrieved 2012-12-03.

[185] Key scientist sure "God particle" will be found soon Reuters news story. 7 April 2008.

[186] "Interview: the man behind the 'God particle'", New Scientist 13 September 2008, pp. 44–5 (original interview in the Guardian: Father of the 'God Particle', June 30, 2008)

[187] Sample, Ian (2010). *Massive: The Hunt for the God Particle*. pp. 148–149 and 278–279. ISBN 9781905264957.

[188] Cole, K. (2000-12-14). "One Thing Is Perfectly Clear: Nothingness Is Perfect". *Los Angeles Times*. p. 'Science File'. Retrieved 17 January 2013. Consider the early universe–a state of pure, perfect nothingness; a formless fog of undifferentiated stuff ... 'perfect symmetry' ... What shattered this primordial perfection? One likely culprit is the so-called Higgs field ... Physicist Leon Lederman compares the way the Higgs operates to the biblical story of Babel [whose citizens] all spoke the same language ... Like God, says Lederman, the Higgs differentiated the perfect sameness, confusing everyone (physicists included) ... [Nobel Prizewinner Richard] Feynman wondered why the universe we live in was so obviously askew ... Perhaps, he speculated, total perfection would have been unacceptable to God. And so, just as God shattered the perfection of Babel, 'God made the laws only nearly symmetrical'

[189] Lederman, p. 22 *et seq*:

> "Something we cannot yet detect and which, one might say, has been put there to test and confuse us ... The issue is whether physicists will be confounded by this puzzle or whether, in contrast to the unhappy Babylonians, we will continue to build the tower and, as Einstein put it, 'know the mind of God'."
>
> "And the Lord said, Behold the people are un-confounding my confounding. And the Lord sighed and said, Go to, let us go down, and there give them the God Particle so that they may see how beautiful is the universe I have made".

[190] Sample, Ian (12 June 2009). "Higgs competition: Crack open the bubbly, the God particle is dead". *The Guardian* (London). Retrieved 4 May 2010.

[191] Gordon, Fraser (5 July 2012). "Introducing the higgson". *physicsworld.com*. Retrieved 25 August 2012.

[192] Wolchover, Natalie (2012-07-03). "Higgs Boson Explained: How 'God Particle' Gives Things Mass". *Huffington Post*. Retrieved 21 January 2013.

[193] Oliver, Laura (2012-07-04). "Higgs boson: how would you explain it to a seven-year-old?". *The Guardian* (London). Retrieved 21 January 2013.

[194] Zimmer, Ben (2012-07-15). "Higgs boson metaphors as clear as molasses". *The Boston Globe*. Retrieved 21 January 2013.

[195] "The Higgs particle: an analogy for Physics classroom (section)". www.lhc-closer.es (a collaboration website of LHCb physicist Xabier Vidal and High School Teachers at CERN educator Ramon Manzano). Retrieved 2013-01-09.

[196] Flam, Faye (2012-07-12). "Finally – A Higgs Boson Story Anyone Can Understand". *The Philadelphia Inquirer (philly.com)*. Retrieved 21 January 2013.

[197] Sample, Ian (2011-04-28). "How will we know when the Higgs particle has been detected?". *The Guardian* (London). Retrieved 21 January 2013.

[198] Miller, David. "A quasi-political Explanation of the Higgs Boson; for Mr Waldegrave, UK Science Minister 1993". Retrieved 10 July 2012.

[199] Kathryn Grim. "Ten things you may not know about the Higgs boson". Symmetry Magazine. Retrieved 10 July 2012.

[200] David Goldberg, Associate Professor of Physics, Drexel University (2010-10-17). "What's the Matter with the Higgs Boson?". io9.com "Ask a physicist". Retrieved 21 January 2013.

[201] The Nobel Prize in Physics 1979 – official Nobel Prize website.

[202] The Nobel Prize in Physics 1999 – official Nobel Prize website.

[203] – official Nobel Prize website.

[204] Daigle, Katy (10 July 2012). "India: Enough about Higgs, let's discuss the boson". *AP News*. Retrieved 10 July 2012.

[205] Bal, Hartosh Singh (19 September 2012). "The Bose in the Boson". New York Times. Retrieved 21 September 2012.

[206] Alikhan, Anvar (16 July 2012). "The Spark In A Crowded Field". *Outlook India*. Retrieved 10 July 2012.

[207] Peskin & Schroeder 1995, Chapter 20

1.11 Further reading

- Nambu, Yoichiro; Jona-Lasinio, Giovanni (1961). "Dynamical Model of Elementary Particles Based on an Analogy with Superconductivity". *Physical Review* **122**: 345–358. Bibcode:1961PhRv..122..345N. doi:10.1103/.

- Klein, Abraham; Lee, Benjamin W. (1964). "Does Spontaneous Breakdown of Symmetry Imply Zero-Mass Particles?". *Physical Review Letters* **12** (10): 266. Bibcode:1964PhRvL..12..266K. doi:10.1103/PhysRevLett.12.266.

- Anderson, Philip W.(1963). "Plasmons, Gauge Invariance, and Mass".*Physical Review***130**: 439.Bibcode:1963. doi:10.1103/PhysRev.130.439.

- Gilbert, Walter (1964). "Broken Symmetries and Massless Particles". *Physical Review Letters* **12** (25): 713. Bibcode:1964PhRvL..12..713G. doi:10.1103/PhysRevLett.12.713.

- Higgs, Peter (1964). "Broken Symmetries, Massless Particles and Gauge Fields". *Physics Letters* **12** (2): 132–133. Bibcode:1964PhL....12..132H. doi:10.1016/0031-9163(64)91136-9.

- Guralnik, Gerald S.; Hagen, C.R.; Kibble, Tom W.B. (1968). "Broken Symmetries and the Goldstone Theorem". In R.L. Cool and R.E. Marshak. *Advances in Physics, Vol. 2*. Interscience Publishers. pp. 567–708. ISBN 978-0470170571.

- The Higgs Boson discovery Karl Jakobs, Chris Seez Scholarpedia 10(9):32413. doi:10.4249/scholarpedia.32413

1.12 External links

1.12.1 Popular science, mass media, and general coverage

- Hunting the Higgs Boson at C.M.S. Experiment, at CERN

- The Higgs Boson" by the CERN exploratorium.

- "Particle Fever", documentary film about the search for the Higgs Boson.

- "The Atom Smashers", documentary film about the search for the Higgs Boson at Fermilab.

- Collected Articles at the *Guardian*

- Video (04:38) – CERN Announcement on 4 July 2012, of the discovery of a particle which is suspected will be a Higgs Boson.

- Video1 (07:44) + Video2 (07:44) – Higgs Boson Explained by CERN Physicist, Dr. Daniel Whiteson (16 June 2011).

- HowStuffWorks: What exactly is the Higgs Boson?

- Carroll, Sean. "Higgs Boson with Sean Carroll". *Sixty Symbols*. University of Nottingham.

- Overbye, Dennis (2013-03-05). "Chasing the Higgs Boson: How 2 teams of rivals at CERN searched for physics' most elusive particle". *New York Times Science pages*. Retrieved 22 July 2013. - New York Times "behind the scenes" style article on the Higgs' search at ATLAS and CMS

- The story of the Higgs theory by the authors of the PRL papers and others closely associated:

  - Higgs, Peter (2010). "My Life as a Boson" (PDF). Talk given at Kings College, London, Nov 24 2010. Retrieved 17 January 2013. (also:)

  - Kibble, Tom (2009). "Englert–Brout–Higgs–Guralnik–Hagen–Kibble mechanism (history)". Scholarpedia. Retrieved 17 January 2013. (also:)

- Guralnik, Gerald (2009). "The History of the Guralnik, Hagen and Kibble development of the Theory of Spontaneous Symmetry Breaking and Gauge Particles". *International Journal of Modern Physics A* **24** (14): 2601–2627. arXiv:0907.3466. Bibcode:2009IJMPA..24.2601G. doi:10.1142/S0217751X09045431., Guralnik, Gerald (2011). "The Beginnings of Spontaneous Symmetry Breaking in Particle Physics. Proceedings of the DPF-2011 Conference, Providence, RI, 8–13 August 2011". arXiv:1110.2253v1 [physics.hist-ph]., and Guralnik, Gerald (2013). "Heretical Ideas that Provided the Cornerstone for the Standard Model of Particle Physics". SPG MITTEILUNGEN March 2013, No. 39, (p. 14), and Talk at Brown University about the 1964 PRL papers

  - Philip Anderson (not one of the PRL authors) on symmetry breaking in superconductivity and its migration into particle physics and the PRL papers

- Cartoon about the search

- Cham, Jorge (2014-02-19). "True Tales from the Road: The Higgs Boson Re-Explained". *Piled Higher and Deeper*. Retrieved 2014-02-25.

1.12.2 Significant papers and other

- Observation of a new particle in the search for the Standard Model Higgs Boson with the ATLAS detector at the LHC

- Observation of a new Boson at a mass of 125 GeV with the CMS experiment at the LHC

- Particle Data Group: Review of searches for Higgs Bosons.

- 2001, a spacetime odyssey: proceedings of the Inaugural Conference of the Michigan Center for Theoretical Physics : Michigan, USA, 21–25 May 2001, (p.86 – 88), ed. Michael J. Duff, James T. Liu, ISBN 978-981-238-231-3, containing Higgs' story of the Higgs Boson.

- A.A. Migdal & A.M. Polyakov, *Spontaneous Breakdown of Strong Interaction Symmetry and the Absence of Massless Particles*, Sov.J.-JETP 24,91 (1966) - example of a 1966 Russian paper on the subject.

1.12.3 Introductions to the field

- Spontaneous symmetry breaking, gauge theories, the Higgs mechanism and all that (Bernstein, *Reviews of Modern Physics* Jan 1974) - an introduction of 47 pages covering the development, history and mathematics of Higgs theories from around 1950 to 1974.

Chapter 2

Elementary particle

This article is about the physics concept. For the novel, see The Elementary Particles.

In particle physics, an **elementary particle** or **fundamental particle** is a particle whose substructure is unknown, thus

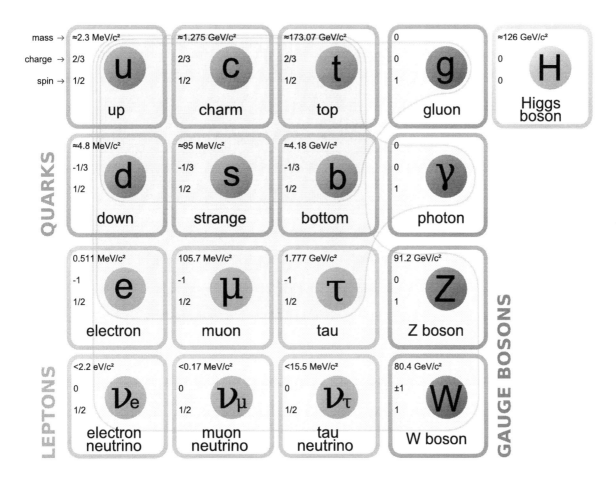

Elementary particles included in the Standard Model

it is unknown whether it is composed of other particles.[1] Known elementary particles include the fundamental fermions (quarks, leptons, antiquarks, and antileptons), which generally are "matter particles" and "antimatter particles", as well as the fundamental bosons (gauge bosons and Higgs boson), which generally are "force particles" that mediate interactions among fermions.[1] A particle containing two or more elementary particles is a *composite particle*.

Everyday matter is composed of atoms, once presumed to be matter's elementary particles—*atom* meaning "indivisible" in Greek—although the atom's existence remained controversial until about 1910, as some leading physicists regarded molecules as mathematical illusions, and matter as ultimately composed of energy.[1][2] Soon, subatomic constituents of the atom were identified. As the 1930s opened, the electron and the proton had been observed, along with the photon, the particle of electromagnetic radiation.[1] At that time, the recent advent of quantum mechanics was radically altering the conception of particles, as a single particle could seemingly span a field as would a wave, a paradox still eluding satisfactory explanation.[3][4][5]

Via quantum theory, protons and neutrons were found to contain quarks—up quarks and down quarks—now considered elementary particles.[1] And within a molecule, the electron's three degrees of freedom (charge, spin, orbital) can separate via wavefunction into three quasiparticles (holon, spinon, orbiton).[6] Yet a free electron—which, not orbiting an atomic nucleus, lacks orbital motion—appears unsplittable and remains regarded as an elementary particle.[6]

Around 1980, an elementary particle's status as indeed elementary—an *ultimate constituent* of substance—was mostly discarded for a more practical outlook,[1] embodied in particle physics' Standard Model, science's most experimentally successful theory.[5][7] Many elaborations upon and theories beyond the Standard Model, including the extremely popular supersymmetry, double the number of elementary particles by hypothesizing that each known particle associates with a "shadow" partner far more massive,[8][9] although all such superpartners remain undiscovered.[7][10] Meanwhile, an elementary boson mediating gravitation—the graviton—remains hypothetical.[1]

2.1 Overview

Main article: Standard Model
See also: Physics beyond the Standard Model

All elementary particles are—depending on their *spin*—either bosons or fermions. These are differentiated via the spin–

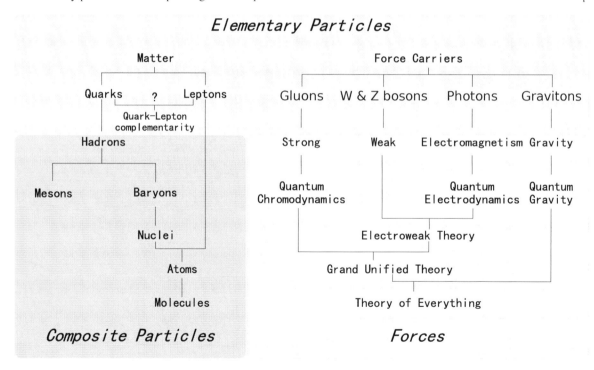

An overview of the various families of elementary and composite particles, and the theories describing their interactions

statistics theorem of quantum statistics. Particles of *half-integer* spin exhibit Fermi–Dirac statistics and are fermions.[1] Particles of *integer* spin, in other words full-integer, exhibit Bose–Einstein statistics and are bosons.[1]

Elementary fermions:

- Matter particles
  - Quarks:
    - up, down
    - charm, strange
    - top, bottom
  - Leptons:
    - electron, electron neutrino (a.k.a., "neutrino")
    - muon, muon neutrino
    - tau, tau neutrino
- Antimatter particles
  - Antiquarks
  - Antileptons

Elementary bosons:

- Force particles (gauge bosons):
  - photon
  - gluon (numbering eight)[1]
  - W^+, W^-, and Z^0 bosons
  - graviton (hypothetical)[1]
- Scalar boson
  - Higgs boson

A particle's mass is quantified in units of energy versus the electron's (electronvolts). Through conversion of energy into mass, any particle can be produced through collision of other particles at high energy,[1][11] although the output particle might not contain the input particles, for instance matter creation from colliding photons. Likewise, the composite fermions protons were collided at nearly light speed to produce a Higgs boson, which elementary boson is far more massive.[11] The most massive elementary particle, the top quark, rapidly decays, but apparently does not contain, lighter particles.

When probed at energies available in experiments, particles exhibit spherical sizes. In operating particle physics' Standard Model, elementary particles are usually represented for predictive utility as point particles, which, as zero-dimensional, lack spatial extension. Though extremely successful, the Standard Model is limited to the microcosm by its omission of gravitation, and has some parameters arbitrarily added but unexplained.[12] Seeking to resolve those shortcomings, string theory posits that elementary particles are ultimately composed of one-dimensional energy strings whose absolute minimum size is the Planck length.

2.2   Common elementary particles

Main article: cosmic abundance of elements

According to the current models of big bang nucleosynthesis, the primordial composition of visible matter of the universe should be about 75% hydrogen and 25% helium-4 (in mass). Neutrons are made up of one up and two down quark, while protons are made of two up and one down quark. Since the other common elementary particles (such as electrons, neutrinos, or weak bosons) are so light or so rare when compared to atomic nuclei, we can neglect their mass contribution to the observable universe's total mass. Therefore, one can conclude that most of the visible mass of the universe consists of protons and neutrons, which, like all baryons, in turn consist of up quarks and down quarks.

Some estimates imply that there are roughly 10^{80} baryons (almost entirely protons and neutrons) in the observable universe.[13][14][15]

The number of protons in the observable universe is called the Eddington number.

In terms of number of particles, some estimates imply that nearly all the matter, excluding dark matter, occurs in neutrinos, and that roughly 10^{86} elementary particles of matter exist in the visible universe, mostly neutrinos.[15] Other estimates imply that roughly 10^{97} elementary particles exist in the visible universe (not including dark matter), mostly photons, gravitons, and other massless force carriers.[15]

2.3 Standard Model

Main article: Standard Model
 The Standard Model of particle physics contains 12 flavors of elementary fermions, plus their corresponding antiparticles, as well as elementary bosons that mediate the forces and the Higgs boson, which was reported on July 4, 2012, as having been likely detected by the two main experiments at the LHC (ATLAS and CMS). However, the Standard Model is widely considered to be a provisional theory rather than a truly fundamental one, since it is not known if it is compatible with Einstein's general relativity. There may be hypothetical elementary particles not described by the Standard Model, such as the graviton, the particle that would carry the gravitational force, and sparticles, supersymmetric partners of the ordinary particles.

2.3.1 Fundamental fermions

Main article: Fermion

The 12 fundamental fermionic flavours are divided into three generations of four particles each. Six of the particles are quarks. The remaining six are leptons, three of which are neutrinos, and the remaining three of which have an electric charge of −1: the electron and its two cousins, the muon and the tau.

Antiparticles

Main article: Antimatter

There are also 12 fundamental fermionic antiparticles that correspond to these 12 particles. For example, the antielectron (positron) $e+$ is the electron's antiparticle and has an electric charge of +1.

Quarks

Main article: Quark

Isolated quarks and antiquarks have never been detected, a fact explained by confinement. Every quark carries one of three color charges of the strong interaction; antiquarks similarly carry anticolor. Color-charged particles interact via gluon exchange in the same way that charged particles interact via photon exchange. However, gluons are themselves color-charged, resulting in an amplification of the strong force as color-charged particles are separated. Unlike the electromagnetic force, which diminishes as charged particles separate, color-charged particles feel increasing force.

However, color-charged particles may combine to form color neutral composite particles called hadrons. A quark may pair up with an antiquark: the quark has a color and the antiquark has the corresponding anticolor. The color and anticolor cancel out, forming a color neutral meson. Alternatively, three quarks can exist together, one quark being "red",

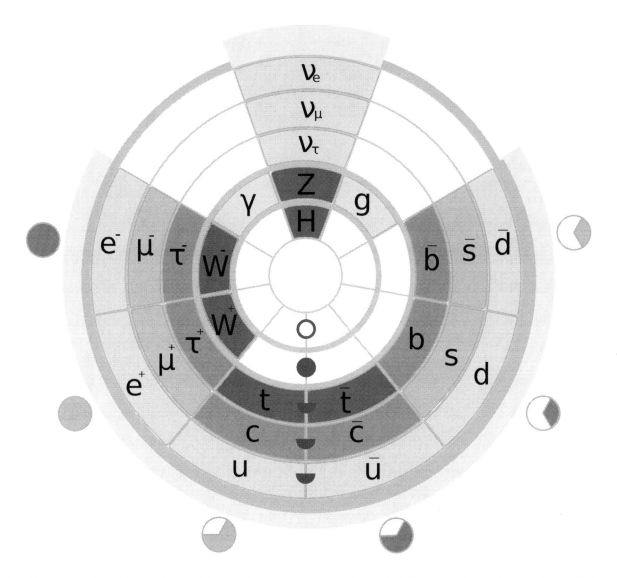

Graphic representation of the standard model. Spin, charge, mass and participation in different force interactions are shown. Click on the image to see the full description

another "blue", another "green". These three colored quarks together form a color-neutral baryon. Symmetrically, three antiquarks with the colors "antired", "antiblue" and "antigreen" can form a color-neutral antibaryon.

Quarks also carry fractional electric charges, but, since they are confined within hadrons whose charges are all integral, fractional charges have never been isolated. Note that quarks have electric charges of either +2/3 or −1/3, whereas antiquarks have corresponding electric charges of either −2/3 or +1/3.

Evidence for the existence of quarks comes from deep inelastic scattering: firing electrons at nuclei to determine the distribution of charge within nucleons (which are baryons). If the charge is uniform, the electric field around the proton should be uniform and the electron should scatter elastically. Low-energy electrons do scatter in this way, but, above a particular energy, the protons deflect some electrons through large angles. The recoiling electron has much less energy and a jet of particles is emitted. This inelastic scattering suggests that the charge in the proton is not uniform but split among smaller charged particles: quarks.

2.3.2 Fundamental bosons

Main article: Boson

In the Standard Model, vector (spin−1) bosons (gluons, photons, and the W and Z bosons) mediate forces, whereas the Higgs boson (spin-0) is responsible for the intrinsic mass of particles. Bosons differ from fermions in the fact that multiple bosons can occupy the same quantum state (Pauli exclusion principle). Also, bosons can be either elementary, like photons, or a combination, like mesons. The spin of bosons are integers instead of half integers.

Gluons

Main article: Gluon

Gluons mediate the strong interaction, which join quarks and thereby form hadrons, which are either baryons (three quarks) or mesons (one quark and one antiquark). Protons and neutrons are baryons, joined by gluons to form the atomic nucleus. Like quarks, gluons exhibit colour and anticolour—unrelated to the concept of visual color—sometimes in combinations, altogether eight variations of gluons.

Electroweak bosons

Main articles: W and Z bosons and Photon

There are three weak gauge bosons: W^+, W^-, and Z^0; these mediate the weak interaction. The W bosons are known for their mediation in nuclear decay. The W^- converts a neutron into a proton then decay into an electron and electron antineutrino pair. The Z^0 does not convert charge but rather changes momentum and is the only mechanism for elastically scattering neutrinos. The weak gauge bosons were discovered due to momentum change in electrons from neutrino-Z exchange. The massless photon mediates the electromagnetic interaction. These four gauge bosons form the electroweak interaction among elementary particles.

Higgs boson

Main article: Higgs boson

Although the weak and electromagnetic forces appear quite different to us at everyday energies, the two forces are theorized to unify as a single electroweak force at high energies. This prediction was clearly confirmed by measurements of cross-sections for high-energy electron-proton scattering at the HERA collider at DESY. The differences at low energies is a consequence of the high masses of the W and Z bosons, which in turn are a consequence of the Higgs mechanism. Through the process of spontaneous symmetry breaking, the Higgs selects a special direction in electroweak space that causes three electroweak particles to become very heavy (the weak bosons) and one to remain massless (the photon). On 4 July 2012, after many years of experimentally searching for evidence of its existence, the Higgs boson was announced to have been observed at CERN's Large Hadron Collider. Peter Higgs who first posited the existence of the Higgs boson was present at the announcement.[16] The Higgs boson is believed to have a mass of approximately 125 GeV.[17] The statistical significance of this discovery was reported as 5-sigma, which implies a certainty of roughly 99.99994%. In particle physics, this is the level of significance required to officially label experimental observations as a discovery. Research into the properties of the newly discovered particle continues.

Graviton

Main article: Graviton

The graviton is hypothesized to mediate gravitation, but remains undiscovered and yet is sometimes included in tables of elementary particles.[1] Its spin would be two—thus a boson—and it would lack charge or mass. Besides mediating an extremely feeble force, the graviton would have its own antiparticle and rapidly annihilate, rendering its detection extremely difficult even if it exists.

2.4   Beyond the Standard Model

Although experimental evidence overwhelmingly confirms the predictions derived from the Standard Model, some of its parameters were added arbitrarily, not determined by a particular explanation, which remain mysteries, for instance the hierarchy problem. Theories beyond the Standard Model attempt to resolve these shortcomings.

2.4.1   Grand unification

Main article: Grand Unified Theory

One extension of the Standard Model attempts to combine the electroweak interaction with the strong interaction into a single 'grand unified theory' (GUT). Such a force would be spontaneously broken into the three forces by a Higgs-like mechanism. The most dramatic prediction of grand unification is the existence of X and Y bosons, which cause proton decay. However, the non-observation of proton decay at the Super-Kamiokande neutrino observatory rules out the simplest GUTs, including SU(5) and SO(10).

2.4.2   Supersymmetry

Main article: Supersymmetry

Supersymmetry extends the Standard Model by adding another class of symmetries to the Lagrangian. These symmetries exchange fermionic particles with bosonic ones. Such a symmetry predicts the existence of supersymmetric particles, abbreviated as *sparticles*, which include the sleptons, squarks, neutralinos, and charginos. Each particle in the Standard Model would have a superpartner whose spin differs by 1/2 from the ordinary particle. Due to the breaking of supersymmetry, the sparticles are much heavier than their ordinary counterparts; they are so heavy that existing particle colliders would not be powerful enough to produce them. However, some physicists believe that sparticles will be detected by the Large Hadron Collider at CERN.

2.4.3   String theory

Main article: String theory

String theory is a model of physics where all "particles" that make up matter are composed of strings (measuring at the Planck length) that exist in an 11-dimensional (according to M-theory, the leading version) universe. These strings vibrate at different frequencies that determine mass, electric charge, color charge, and spin. A string can be open (a line) or closed in a loop (a one-dimensional sphere, like a circle). As a string moves through space it sweeps out something called a *world sheet*. String theory predicts 1- to 10-branes (a 1-brane being a string and a 10-brane being a 10-dimensional object) that prevent tears in the "fabric" of space using the uncertainty principle (E.g., the electron orbiting a hydrogen atom has the probability, albeit small, that it could be anywhere else in the universe at any given moment).

String theory proposes that our universe is merely a 4-brane, inside which exist the 3 space dimensions and the 1 time dimension that we observe. The remaining 6 theoretical dimensions either are very tiny and curled up (and too small to be macroscopically accessible) or simply do not/cannot exist in our universe (because they exist in a grander scheme called the "multiverse" outside our known universe).

Some predictions of the string theory include existence of extremely massive counterparts of ordinary particles due to vibrational excitations of the fundamental string and existence of a massless spin-2 particle behaving like the graviton.

2.4.4 Technicolor

Main article: Technicolor (physics)

Technicolor theories try to modify the Standard Model in a minimal way by introducing a new QCD-like interaction. This means one adds a new theory of so-called Techniquarks, interacting via so called Technigluons. The main idea is that the Higgs-Boson is not an elementary particle but a bound state of these objects.

2.4.5 Preon theory

Main article: Preon

According to preon theory there are one or more orders of particles more fundamental than those (or most of those) found in the Standard Model. The most fundamental of these are normally called preons, which is derived from "pre-quarks". In essence, preon theory tries to do for the Standard Model what the Standard Model did for the particle zoo that came before it. Most models assume that almost everything in the Standard Model can be explained in terms of three to half a dozen more fundamental particles and the rules that govern their interactions. Interest in preons has waned since the simplest models were experimentally ruled out in the 1980s.

2.4.6 Acceleron theory

Accelerons are the hypothetical subatomic particles that integrally link the newfound mass of the neutrino and to the dark energy conjectured to be accelerating the expansion of the universe.[18]

In theory, neutrinos are influenced by a new force resulting from their interactions with accelerons. Dark energy results as the universe tries to pull neutrinos apart.[18]

2.5 See also

- Asymptotic freedom
- List of particles
- Physical ontology
- Quantum field theory
- Quantum gravity
- Quantum triviality
- UV fixed point

2.6 Notes

[1] Sylvie Braibant; Giorgio Giacomelli; Maurizio Spurio (2012). *Particles and Fundamental Interactions: An Introduction to Particle Physics* (2nd ed.). Springer. pp. 1–3. ISBN 978-94-007-2463-1.

[2] Ronald Newburgh; Joseph Peidle; Wolfgang Rueckner (2006). "Einstein, Perrin, and the reality of atoms: 1905 revisited" (PDF). *American Journal of Physics*. **74** (6): 478–481. Bibcode:2006AmJPh..74..478N. doi:10.1119/1.2188962.

[3] Friedel Weinert (2004). *The Scientist as Philosopher: Philosophical Consequences of Great Scientific Discoveries*. Springer. p. 43. ISBN 978-3-540-20580-7.

[4] Friedel Weinert (2004). *The Scientist as Philosopher: Philosophical Consequences of Great Scientific Discoveries*. Springer. pp. 57–59. ISBN 978-3-540-20580-7.

[5] Meinard Kuhlmann (24 Jul 2013). "Physicists debate whether the world is made of particles or fields—or something else entirely". *Scientific American*.

[6] Zeeya Merali (18 Apr 2012). "Not-quite-so elementary, my dear electron: Fundamental particle 'splits' into quasiparticles, including the new 'orbiton'". *Nature*. doi:10.1038/nature.2012.10471.

[7] Ian O'Neill (24 Jul 2013). "LHC discovery maims supersymmetry, again". *Discovery News*. Retrieved 2013-08-28.

[8] Particle Data Group. "Unsolved mysteries—supersymmetry". *The Particle Adventure*. Berkeley Lab. Retrieved 2013-08-28.

[9] National Research Council (2006). *Revealing the Hidden Nature of Space and Time: Charting the Course for Elementary Particle Physics*. National Academies Press. p. 68. ISBN 978-0-309-66039-6.

[10] "CERN latest data shows no sign of supersymmetry—yet". *Phys.Org*. 25 Jul 2013. Retrieved 2013-08-28.

[11] Ryan Avent (19 Jul 2012). "The Q&A: Brian Greene—Life after the Higgs". *The Economist*. Retrieved 2013-08-28.

[12] Sylvie Braibant; Giorgio Giacomelli; Maurizio Spurio (2012). *Particles and Fundamental Interactions: An Introduction to Particle Physics* (2nd ed.). Springer. p. 384. ISBN 978-94-007-2463-1.

[13] Frank Heile. "Is the Total Number of Particles in the Universe Stable Over Long Periods of Time?". 2014.

[14] Jared Brooks. "Galaxies and Cosmology". 2014. p. 4, equation 16.

[15] Robert Munafo (24 Jul 2013). "Notable Properties of Specific Numbers". Retrieved 2013-08-28.

[16] Lizzy Davies (4 July 2014). "Higgs boson announcement live: CERN scientists discover subatomic particle". *The Guardian*. Retrieved 2012-07-06.

[17] Lucas Taylor (4 Jul 2014). "Observation of a new particle with a mass of 125 GeV". CMS. Retrieved 2012-07-06.

[18] "New theory links neutrino's slight mass to accelerating Universe expansion". *ScienceDaily*. 28 Jul 2004. Retrieved 2008-06-05.

2.7   Further reading

2.7.1   General readers

- Feynman, R.P. & Weinberg, S. (1987) *Elementary Particles and the Laws of Physics: The 1986 Dirac Memorial Lectures*. Cambridge Univ. Press.

- Ford, Kenneth W. (2005) *The Quantum World*. Harvard Univ. Press.

- Brian Greene (1999). *The Elegant Universe*. W.W.Norton & Company. ISBN 0-393-05858-1.

- John Gribbin (2000) *Q is for Quantum – An Encyclopedia of Particle Physics*. Simon & Schuster. ISBN 0-684-85578-X.

- Oerter, Robert (2006) *The Theory of Almost Everything: The Standard Model, the Unsung Triumph of Modern Physics*. Plume.

- Schumm, Bruce A. (2004) *Deep Down Things: The Breathtaking Beauty of Particle Physics*. Johns Hopkins University Press. ISBN 0-8018-7971-X.

- Martinus Veltman (2003). *Facts and Mysteries in Elementary Particle Physics.* World Scientific. ISBN 981-238-149-X.

- Frank Close (2004). *Particle Physics: A Very Short Introduction.* Oxford: Oxford University Press. ISBN 0-19-280434-0.

- Seiden, Abraham (2005). *Particle Physics – A Comprehensive Introduction.* Addison Wesley. ISBN 0-8053-8736-6.

2.7.2 Textbooks

- Bettini, Alessandro (2008) *Introduction to Elementary Particle Physics.* Cambridge Univ. Press. ISBN 978-0-521-88021-3

- Coughlan, G. D., J. E. Dodd, and B. M. Gripaios (2006) *The Ideas of Particle Physics: An Introduction for Scientists,* 3rd ed. Cambridge Univ. Press. An undergraduate text for those not majoring in physics.

- Griffiths, David J. (1987) *Introduction to Elementary Particles.* John Wiley & Sons. ISBN 0-471-60386-4.

- Kane, Gordon L. (1987). *Modern Elementary Particle Physics.* Perseus Books. ISBN 0-201-11749-5.

- Perkins, Donald H. (2000) *Introduction to High Energy Physics,* 4th ed. Cambridge Univ. Press.

2.8 External links

The most important address about the current experimental and theoretical knowledge about elementary particle physics is the Particle Data Group, where different international institutions collect all experimental data and give short reviews over the contemporary theoretical understanding.

- Particle Data Group

other pages are:

- Greene, Brian, "*Elementary particles*", The Elegant Universe, NOVA (PBS)

- particleadventure.org, a well-made introduction also for non physicists

- CERNCourier: Season of Higgs and melodrama

- Pentaquark information page

- Interactions.org, particle physics news

- Symmetry Magazine, a joint Fermilab/SLAC publication

- "Sized Matter: perception of the extreme unseen", Michigan University project for artistic visualisation of sub-atomic particles

- Elementary Particles made thinkable, an interactive visualisation allowing physical properties to be compared

Chapter 3

Standard Model

This article is about the Standard Model of particle physics. For other uses, see Standard model (disambiguation).
This article is a non-mathematical general overview of the Standard Model. For a mathematical description, see the article Standard Model (mathematical formulation).
For the Standard Model of Big Bang cosmology, Lambda-CDM model.

The **Standard Model** of particle physics is a theory concerning the electromagnetic, weak, and strong nuclear inter-

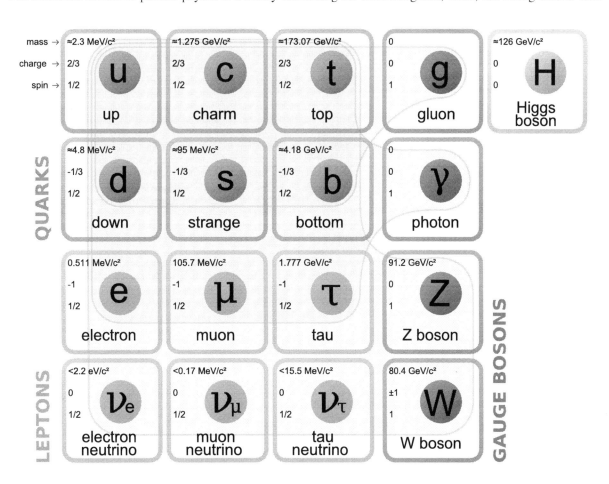

The Standard Model of elementary particles (more schematic depiction), with the three generations of matter, gauge bosons in the fourth column, and the Higgs boson in the fifth.

actions, as well as classifying all the subatomic particles known. It was developed throughout the latter half of the 20th century, as a collaborative effort of scientists around the world.[1] The current formulation was finalized in the mid-1970s upon experimental confirmation of the existence of quarks. Since then, discoveries of the top quark (1995), the tau neutrino (2000), and more recently the Higgs boson (2013), have given further credence to the Standard Model. Because of its success in explaining a wide variety of experimental results, the Standard Model is sometimes regarded as a "theory of almost everything".

Although the Standard Model is believed to be theoretically self-consistent[2] and has demonstrated huge and continued successes in providing experimental predictions, it does leave some phenomena unexplained and it falls short of being a complete theory of fundamental interactions. It does not incorporate the full theory of gravitation[3] as described by general relativity, or account for the accelerating expansion of the universe (as possibly described by dark energy). The model does not contain any viable dark matter particle that possesses all of the required properties deduced from observational cosmology. It also does not incorporate neutrino oscillations (and their non-zero masses).

The development of the Standard Model was driven by theoretical and experimental particle physicists alike. For theorists, the Standard Model is a paradigm of a quantum field theory, which exhibits a wide range of physics including spontaneous symmetry breaking, anomalies, non-perturbative behavior, etc. It is used as a basis for building more exotic models that incorporate hypothetical particles, extra dimensions, and elaborate symmetries (such as supersymmetry) in an attempt to explain experimental results at variance with the Standard Model, such as the existence of dark matter and neutrino oscillations.

3.1 Historical background

The first step towards the Standard Model was Sheldon Glashow's discovery in 1961 of a way to combine the electromagnetic and weak interactions. [4] In 1967 Steven Weinberg[5] and Abdus Salam[6] incorporated the

Higgs mechanism into Glashow's electroweak theory, giving it its modern form.

The Higgs mechanism is believed to give rise to the masses of all the elementary particles in the Standard Model. This includes the masses of the W and Z bosons, and the masses of the fermions, i.e. the quarks and leptons.

After the neutral weak currents caused by Z boson exchange were discovered at CERN in 1973,[10][11][12][13] the electroweak theory became widely accepted and Glashow, Salam, and Weinberg shared the 1979 Nobel Prize in Physics for discovering it. The W and Z bosons were discovered experimentally in 1981, and their masses were found to be as the Standard Model predicted.

The theory of the strong interaction, to which many contributed, acquired its modern form around 1973–74, when experiments confirmed that the hadrons were composed of fractionally charged quarks.

3.2 Overview

At present, matter and energy are best understood in terms of the kinematics and interactions of elementary particles. To date, physics has reduced the laws governing the behavior and interaction of all known forms of matter and energy to a small set of fundamental laws and theories. A major goal of physics is to find the "common ground" that would unite all of these theories into one integrated theory of everything, of which all the other known laws would be special cases, and from which the behavior of all matter and energy could be derived (at least in principle).[14]

3.3 Particle content

The Standard Model includes members of several classes of elementary particles (fermions, gauge bosons, and the Higgs boson), which in turn can be distinguished by other characteristics, such as color charge.

3.3.1   Fermions

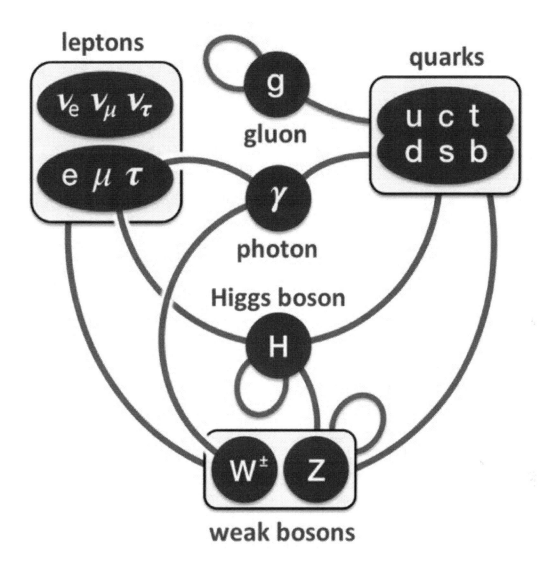

Summary of interactions between particles described by the Standard Model.

The Standard Model includes 12 elementary particles of spin-½ known as fermions. According to the spin-statistics theorem, fermions respect the Pauli exclusion principle. Each fermion has a corresponding antiparticle.

The fermions of the Standard Model are classified according to how they interact (or equivalently, by what charges they carry). There are six quarks (up, down, charm, strange, top, bottom), and six leptons (electron, electron neutrino, muon, muon neutrino, tau, tau neutrino). Pairs from each classification are grouped together to form a generation, with corresponding particles exhibiting similar physical behavior (see table).

The defining property of the quarks is that they carry color charge, and hence, interact via the strong interaction. A phenomenon called color confinement results in quarks being very strongly bound to one another, forming color-neutral composite particles (hadrons) containing either a quark and an antiquark (mesons) or three quarks (baryons). The familiar proton and the neutron are the two baryons having the smallest mass. Quarks also carry electric charge and weak isospin. Hence they interact with other fermions both electromagnetically and via the weak interaction.

The remaining six fermions do not carry colour charge and are called leptons. The three neutrinos do not carry electric charge either, so their motion is directly influenced only by the weak nuclear force, which makes them notoriously difficult to detect. However, by virtue of carrying an electric charge, the electron, muon, and tau all interact electromagnetically.

Each member of a generation has greater mass than the corresponding particles of lower generations. The first generation charged particles do not decay; hence all ordinary (baryonic) matter is made of such particles. Specifically, all atoms consist of electrons orbiting around atomic nuclei, ultimately constituted of up and down quarks. Second and third generation charged particles, on the other hand, decay with very short half lives, and are observed only in very high-energy environments. Neutrinos of all generations also do not decay, and pervade the universe, but rarely interact with baryonic matter.

3.3.2 Gauge bosons

In the Standard Model, gauge bosons are defined as force carriers that mediate the strong, weak, and electromagnetic fundamental interactions.

Interactions in physics are the ways that particles influence other particles. At a macroscopic level, electromagnetism allows particles to interact with one another via electric and magnetic fields, and gravitation allows particles with mass to attract one another in accordance with Einstein's theory of general relativity. The Standard Model explains such forces as resulting from matter particles exchanging other particles, generally referred to as *force mediating particles*. When a force-mediating particle is exchanged, at a macroscopic level the effect is equivalent to a force influencing both of them, and the particle is therefore said to have *mediated* (i.e., been the agent of) that force. The Feynman diagram calculations, which are a graphical representation of the perturbation theory approximation, invoke "force mediating particles", and when applied to analyze high-energy scattering experiments are in reasonable agreement with the data. However, perturbation theory (and with it the concept of a "force-mediating particle") fails in other situations. These include low-energy quantum chromodynamics, bound states, and solitons.

The gauge bosons of the Standard Model all have spin (as do matter particles). The value of the spin is 1, making them bosons. As a result, they do not follow the Pauli exclusion principle that constrains fermions: thus bosons (e.g. photons) do not have a theoretical limit on their spatial density (number per volume). The different types of gauge bosons are described below.

- Photons mediate the electromagnetic force between electrically charged particles. The photon is massless and is well-described by the theory of quantum electrodynamics.

- The W+, W−, and Z gauge bosons mediate the weak interactions between particles of different flavors (all quarks and leptons). They are massive, with the Z being more massive than the W±. The weak interactions involving the W± exclusively act on *left-handed* particles and *right-handed* antiparticles. Furthermore, the W± carries an electric charge of +1 and −1 and couples to the electromagnetic interaction. The electrically neutral Z boson interacts with both left-handed particles and antiparticles. These three gauge bosons along with the photons are grouped together, as collectively mediating the electroweak interaction.

- The eight gluons mediate the strong interactions between color charged particles (the quarks). Gluons are massless. The eightfold multiplicity of gluons is labeled by a combination of color and anticolor charge (e.g. red–antigreen).[nb 11] Because the gluons have an effective color charge, they can also interact among themselves. The gluons and their interactions are described by the theory of quantum chromodynamics.

The interactions between all the particles described by the Standard Model are summarized by the diagrams on the right of this section.

3.3.3 Higgs boson

Main article: Higgs boson

Standard Model Interactions
(Forces Mediated by Gauge Bosons)

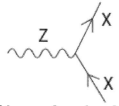

X is any fermion in
the Standard Model.

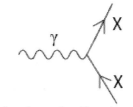

X is electrically charged.

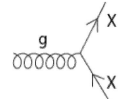

X is any quark.

U is a up-type quark;
D is a down-type quark.

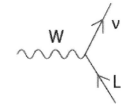

L is a lepton and ν is the
corresponding neutrino.

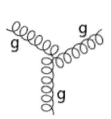

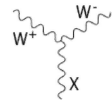

X is a photon or Z-boson.

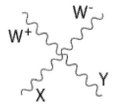

X and Y are any two
electroweak bosons such
that charge is conserved.

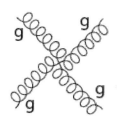

The above interactions form the basis of the standard model. Feynman diagrams in the standard model are built from these vertices. Modifications involving Higgs boson interactions and neutrino oscillations are omitted. The charge of the W bosons is dictated by the fermions they interact with; the conjugate of each listed vertex (i.e. reversing the direction of arrows) is also allowed.

The Higgs particle is a massive scalar elementary particle theorized by Robert Brout, François Englert, Peter Higgs, Gerald Guralnik, C. R. Hagen, and Tom Kibble in 1964 (see 1964 PRL symmetry breaking papers) and is a key building block in the Standard Model.[7][8][9][15] It has no intrinsic spin, and for that reason is classified as a boson (like the gauge bosons, which have integer spin).

The Higgs boson plays a unique role in the Standard Model, by explaining why the other elementary particles, except the photon and gluon, are massive. In particular, the Higgs boson explains why the photon has no mass, while the W and Z bosons are very heavy. Elementary particle masses, and the differences between electromagnetism (mediated by the photon) and the weak force (mediated by the W and Z bosons), are critical to many aspects of the structure of microscopic (and hence macroscopic) matter. In electroweak theory, the Higgs boson generates the masses of the leptons (electron, muon, and tau) and quarks. As the Higgs boson is massive, it must interact with itself.

Because the Higgs boson is a very massive particle and also decays almost immediately when created, only a very high-energy particle accelerator can observe and record it. Experiments to confirm and determine the nature of the Higgs boson using the Large Hadron Collider (LHC) at CERN began in early 2010, and were performed at Fermilab's Tevatron until its closure in late 2011. Mathematical consistency of the Standard Model requires that any mechanism capable of generating the masses of elementary particles become visible at energies above 1.4 TeV;[16] therefore, the LHC (designed to collide two 7 to 8 TeV proton beams) was built to answer the question of whether the Higgs boson actually exists.[17]

On 4 July 2012, the two main experiments at the LHC (ATLAS and CMS) both reported independently that they found a new particle with a mass of about 125 GeV/c^2 (about 133 proton masses, on the order of 10^{-25} kg), which is "consistent with the Higgs boson." Although it has several properties similar to the predicted "simplest" Higgs,[18] they acknowledged that further work would be needed to conclude that it is indeed the Higgs boson, and exactly which version of the Standard Model Higgs is best supported if confirmed.[19][20][21][22][23]

On 14 March 2013 the Higgs Boson was tentatively confirmed to exist.[24]

3.3.4   Total particle count

Counting particles by a rule that distinguishes between particles and their corresponding antiparticles, and among the many color states of quarks and gluons, gives a total of 61 elementary particles.[25]

3.4   Theoretical aspects

Main article: Standard Model (mathematical formulation)

3.4.1   Construction of the Standard Model Lagrangian

Technically, quantum field theory provides the mathematical framework for the Standard Model, in which a Lagrangian controls the dynamics and kinematics of the theory. Each kind of particle is described in terms of a dynamical field that pervades space-time. The construction of the Standard Model proceeds following the modern method of constructing most field theories: by first postulating a set of symmetries of the system, and then by writing down the most general renormalizable Lagrangian from its particle (field) content that observes these symmetries.

The global Poincaré symmetry is postulated for all relativistic quantum field theories. It consists of the familiar translational symmetry, rotational symmetry and the inertial reference frame invariance central to the theory of special relativity. The local SU(3)×SU(2)×U(1) gauge symmetry is an internal symmetry that essentially defines the Standard Model. Roughly, the three factors of the gauge symmetry give rise to the three fundamental interactions. The fields fall into different representations of the various symmetry groups of the Standard Model (see table). Upon writing the most general Lagrangian, one finds that the dynamics depend on 19 parameters, whose numerical values are established by experiment. The parameters are summarized in the table above (note: with the Higgs mass is at 125 GeV, the Higgs self-coupling strength $\lambda \sim 1/8$).

Quantum chromodynamics sector

Main article: Quantum chromodynamics

The quantum chromodynamics (QCD) sector defines the interactions between quarks and gluons, with SU(3) symmetry, generated by T^a. Since leptons do not interact with gluons, they are not affected by this sector. The Dirac Lagrangian of the quarks coupled to the gluon fields is given by

$$\mathcal{L}_{QCD} = i\overline{U}(\partial_\mu - ig_s G_\mu^a T^a)\gamma^\mu U + i\overline{D}(\partial_\mu - ig_s G_\mu^a T^a)\gamma^\mu D.$$

G_μ^a is the SU(3) gauge field containing the gluons, γ^μ are the Dirac matrices, D and U are the Dirac spinors associated with up- and down-type quarks, and g_s is the strong coupling constant.

Electroweak sector

Main article: Electroweak interaction

The electroweak sector is a Yang–Mills gauge theory with the simple symmetry group U(1)×SU(2)L,

$$\mathcal{L}_{EW} = \sum_\psi \bar{\psi}\gamma^\mu \left(i\partial_\mu - g'\frac{1}{2}Y_W B_\mu - g\frac{1}{2}\vec{\tau}_L \vec{W}_\mu \right)\psi$$

where $B\mu$ is the U(1) gauge field; YW is the weak hypercharge—the generator of the U(1) group; $\vec{W}_\mu$ is the three-component SU(2) gauge field; $\vec{\tau}_L$ are the Pauli matrices—infinitesimal generators of the SU(2) group. The subscript L indicates that they only act on left fermions; g' and g are coupling constants.

Higgs sector

Main article: Higgs mechanism

In the Standard Model, the Higgs field is a complex scalar of the group SU(2)L:

$$\varphi = \frac{1}{\sqrt{2}}\begin{pmatrix} \varphi^+ \\ \varphi^0 \end{pmatrix},$$

where the indices + and 0 indicate the electric charge (Q) of the components. The weak isospin (YW) of both components is 1.

Before symmetry breaking, the Higgs Lagrangian is:

$$\mathcal{L}_H = \varphi^\dagger \left(\partial^\mu - \frac{i}{2}\left(g'Y_W B^\mu + g\vec{\tau}\vec{W}^\mu \right) \right)\left(\partial_\mu + \frac{i}{2}\left(g'Y_W B_\mu + g\vec{\tau}\vec{W}_\mu \right) \right)\varphi - \frac{\lambda^2}{4}\left(\varphi^\dagger\varphi - v^2 \right)^2,$$

which can also be written as:

$$\mathcal{L}_H = \left| \left(\partial_\mu + \frac{i}{2}\left(g'Y_W B_\mu + g\vec{\tau}\vec{W}_\mu \right) \right)\varphi \right|^2 - \frac{\lambda^2}{4}\left(\varphi^\dagger\varphi - v^2 \right)^2.$$

3.5 Fundamental forces

Main article: Fundamental interaction

The Standard Model classified all four fundamental forces in nature. In the Standard Model, a force is described as an exchange of bosons between the objects affected, such as a photon for the electromagnetic force and a gluon for the strong interaction. Those particles are called force carriers.[26]

3.6 Tests and predictions

The Standard Model (SM) predicted the existence of the W and Z bosons, gluon, and the top and charm quarks before these particles were observed. Their predicted properties were experimentally confirmed with good precision. To give an idea of the success of the SM, the following table compares the measured masses of the W and Z bosons with the masses predicted by the SM:

The SM also makes several predictions about the decay of Z bosons, which have been experimentally confirmed by the Large Electron-Positron Collider at CERN.

In May 2012 BaBar Collaboration reported that their recently analyzed data may suggest possible flaws in the Standard Model of particle physics.[28][29] These data show that a particular type of particle decay called "B to D-star-tau-nu" happens more often than the Standard Model says it should. In this type of decay, a particle called the B-bar meson decays into a D meson, an antineutrino and a tau-lepton. While the level of certainty of the excess (3.4 sigma) is not enough to claim a break from the Standard Model, the results are a potential sign of something amiss and are likely to impact existing theories, including those attempting to deduce the properties of Higgs bosons.[30]

On December 13, 2012, physicists reported the constancy, over space and time, of a basic physical constant of nature that supports the *standard model of physics*. The scientists, studying methanol molecules in a distant galaxy, found the change ($\Delta\mu/\mu$) in the proton-to-electron mass ratio μ to be equal to "$(0.0 \pm 1.0) \times 10^{-7}$ at redshift $z = 0.89$" and consistent with "a null result".[31][32]

3.7 Challenges

See also: Physics beyond the Standard Model

Self-consistency of the Standard Model (currently formulated as a non-abelian gauge theory quantized through path-integrals) has not been mathematically proven. While regularized versions useful for approximate computations (for example lattice gauge theory) exist, it is not known whether they converge (in the sense of S-matrix elements) in the limit that the regulator is removed. A key question related to the consistency is the Yang–Mills existence and mass gap problem.

Experiments indicate that neutrinos have mass, which the classic Standard Model did not allow.[33] To accommodate this finding, the classic Standard Model can be modified to include neutrino mass.

If one insists on using only Standard Model particles, this can be achieved by adding a non-renormalizable interaction of leptons with the Higgs boson.[34] On a fundamental level, such an interaction emerges in the seesaw mechanism where heavy right-handed neutrinos are added to the theory. This is natural in the left-right symmetric extension of the Standard Model[35][36] and in certain grand unified theories.[37] As long as new physics appears below or around 10^{14} GeV, the neutrino masses can be of the right order of magnitude.

Theoretical and experimental research has attempted to extend the Standard Model into a Unified field theory or a Theory of everything, a complete theory explaining all physical phenomena including constants. Inadequacies of the Standard Model that motivate such research include:

- It does not attempt to explain gravitation, although a theoretical particle known as a graviton would help explain it, and unlike for the strong and electroweak interactions of the Standard Model, there is no known way of describing general relativity, the canonical theory of gravitation, consistently in terms of quantum field theory. The reason for this is, among other things, that quantum field theories of gravity generally break down before reaching the Planck scale. As a consequence, we have no reliable theory for the very early universe;

- Some consider it to be *ad hoc* and inelegant, requiring 19 numerical constants whose values are unrelated and arbitrary. Although the Standard Model, as it now stands, can explain why neutrinos have masses, the specifics of neutrino mass are still unclear. It is believed that explaining neutrino mass will require an additional 7 or 8 constants, which are also arbitrary parameters;

- The Higgs mechanism gives rise to the hierarchy problem if some new physics (coupled to the Higgs) is present at high energy scales. In these cases in order for the weak scale to be much smaller than the Planck scale, severe fine tuning of the parameters is required; there are, however, other scenarios that include quantum gravity in which such fine tuning can be avoided.[38]There are also issues of Quantum triviality, which suggests that it may not be possible to create a consistent quantum field theory involving elementary scalar particles.

- It should be modified so as to be consistent with the emerging "Standard Model of cosmology." In particular, the Standard Model cannot explain the observed amount of cold dark matter (CDM) and gives contributions to dark energy which are many orders of magnitude too large. It is also difficult to accommodate the observed predominance of matter over antimatter (matter/antimatter asymmetry). The isotropy and homogeneity of the visible universe over large distances seems to require a mechanism like cosmic inflation, which would also constitute an extension of the Standard Model.

- The existence of ultra-high-energy cosmic rays are difficult to explain under the Standard Model.

Currently, no proposed Theory of Everything has been widely accepted or verified.

3.8   See also

- Fundamental interaction:

  - Quantum electrodynamics

  - Strong interaction: Color charge, Quantum chromodynamics, Quark model

  - Weak interaction: Electroweak theory, Fermi theory of beta decay, Weak hypercharge, Weak isospin

- Gauge theory: Nontechnical introduction to gauge theory

- Generation

- Higgs mechanism: Higgs boson, Higgsless model

- J. C. Ward

- J. J. Sakurai Prize for Theoretical Particle Physics

- Lagrangian

- Open questions: BTeV experiment, CP violation, Neutrino masses, Quark matter, Quantum triviality

- Penguin diagram

- Quantum field theory

- Standard Model: Mathematical formulation of, Physics beyond the Standard Model

3.9   Notes and references

[1] Technically, there are nine such color–anticolor combinations. However, there is one color-symmetric combination that can be constructed out of a linear superposition of the nine combinations, reducing the count to eight.

3.10 References

[1] R. Oerter (2006). *The Theory of Almost Everything: The Standard Model, the Unsung Triumph of Modern Physics* (Kindle ed.). Penguin Group. p. 2. ISBN 0-13-236678-9.

[2] In fact, there are mathematical issues regarding quantum field theories still under debate (see e.g. Landau pole), but the predictions extracted from the Standard Model by current methods applicable to current experiments are all self-consistent. For a further discussion see e.g. Chapter 25 of R. Mann (2010). *An Introduction to Particle Physics and the Standard Model.* CRC Press. ISBN 978-1-4200-8298-2.

[3] Sean Carroll, Ph.D., Cal Tech, 2007, The Teaching Company, *Dark Matter, Dark Energy: The Dark Side of the Universe*, Guidebook Part 2 page 59, Accessed Oct. 7, 2013, "...Standard Model of Particle Physics: The modern theory of elementary particles and their interactions ... It does not, strictly speaking, include gravity, although it's often convenient to include gravitons among the known particles of nature..."

[4] S.L. Glashow (1961). "Partial-symmetries of weak interactions". *Nuclear Physics* **22** (4): 579–588. Bibcode:1961NucPh..22..G. doi:10.1016/0029-5582(61)90469-2.

[5] S. Weinberg (1967). "A Model of Leptons". *Physical Review Letters* **19** (21): 1264–1266. Bibcode:1967PhRvL..19.1264W. doi:10.1103/PhysRevLett.19.1264.

[6] A. Salam (1968). N. Svartholm, ed. *Elementary Particle Physics: Relativistic Groups and Analyticity.* Eighth Nobel Symposium. Stockholm: Almquvist and Wiksell. p. 367.

[7] F. Englert, R. Brout (1964). "Broken Symmetry and the Mass of Gauge Vector Mesons". *Physical Review Letters* **13** (9): 321–323. Bibcode:1964PhRvL..13..321E. doi:10.1103/PhysRevLett.13.321.

[8] P.W. Higgs (1964). "Broken Symmetries and the Masses of Gauge Bosons". *Physical Review Letters* **13** (16): 508–509. Bibcode:1964PhRvL..13..508H. doi:10.1103/PhysRevLett.13.508.

[9] G.S. Guralnik, C.R. Hagen, T.W.B. Kibble (1964). "Global Conservation Laws and Massless Particles". *Physical Review Letters* **13** (20): 585–587. Bibcode:1964PhRvL..13..585G. doi:10.1103/PhysRevLett.13.585.

[10] F.J. Hasert; et al. (1973). "Search for elastic muon-neutrino electron scattering". *Physics Letters B* **46** (1): 121. Bibcode:197H. doi:10.1016/0370-2693(73)90494-2.

[11] F.J. Hasert; et al. (1973). "Observation of neutrino-like interactions without muon or electron in the Gargamelle neutrino experiment". *Physics Letters B* **46** (1): 138. Bibcode:1973PhLB...46..138H. doi:10.1016/0370-2693(73)90499-1.

[12] F.J. Hasert; et al. (1974). "Observation of neutrino-like interactions without muon or electron in the Gargamelle neutrino experiment". *Nuclear Physics B* **73** (1): 1. Bibcode:1974NuPhB..73....1H. doi:10.1016/0550-3213(74)90038-8.

[13] D. Haidt (4 October 2004). "The discovery of the weak neutral currents". *CERN Courier.* Retrieved 8 May 2008.

[14] "Details can be worked out if the situation is simple enough for us to make an approximation, which is almost never, but often we can understand more or less what is happening." from *The Feynman Lectures on Physics*, Vol 1. pp. 2–7

[15] G.S. Guralnik (2009). "The History of the Guralnik, Hagen and Kibble development of the Theory of Spontaneous Symmetry Breaking and Gauge Particles". *International Journal of Modern Physics A* **24** (14): 2601–2627. arXiv:0907.3466. Bibcode:2009IJMPA..24.2601G. doi:10.1142/S0217751X09045431.

[16] B.W. Lee, C. Quigg, H.B. Thacker (1977). "Weak interactions at very high energies: The role of the Higgs-boson mass". *Physical Review D* **16** (5): 1519–1531. Bibcode:1977PhRvD..16.1519L. doi:10.1103/PhysRevD.16.1519.

[17] "Huge $10 billion collider resumes hunt for 'God particle'". CNN. 11 November 2009. Retrieved 2010-05-04.

[18] M. Strassler (10 July 2012). "Higgs Discovery: Is it a Higgs?". Retrieved 2013-08-06.

[19] "CERN experiments observe particle consistent with long-sought Higgs boson". CERN. 4 July 2012. Retrieved 2012-07-04.

[20] "Observation of a New Particle with a Mass of 125 GeV". CERN. 4 July 2012. Retrieved 2012-07-05.

[21] "ATLAS Experiment". ATLAS. 1 January 2006. Retrieved 2012-07-05.

[22] "Confirmed: CERN discovers new particle likely to be the Higgs boson". *YouTube*. Russia Today. 4 July 2012. Retrieved 2013-08-06.

[23] D. Overbye (4 July 2012). "A New Particle Could Be Physics' Holy Grail". *New York Times*. Retrieved 2012-07-04.

[24] "New results indicate that new particle is a Higgs boson". CERN. 14 March 2013. Retrieved 2013-08-06.

[25] S. Braibant, G. Giacomelli, M. Spurio (2009). *Particles and Fundamental Interactions: An Introduction to Particle Physics*. Springer. pp. 313–314. ISBN 978-94-007-2463-1.

[26] http://home.web.cern.ch/about/physics/standard-model Official CERN website

[27] http://www.pha.jhu.edu/~{ }dfehling/particle.gif

[28] "BABAR Data in Tension with the Standard Model". SLAC. 31 May 2012. Retrieved 2013-08-06.

[29] BaBar Collaboration (2012). "Evidence for an excess of B → D$^{(*)}$ τ$^-$ vτ decays". *Physical Review Letters* **109** (10): 101802. arXiv:1205.5442. Bibcode:2012PhRvL.109j1802L. doi:10.1103/PhysRevLett.109.101802.

[30] "BaBar data hint at cracks in the Standard Model". *e! Science News*. 18 June 2012. Retrieved 2013-08-06.

[31] J. Bagdonaite; et al. (2012). "A Stringent Limit on a Drifting Proton-to-Electron Mass Ratio from Alcohol in the Early Universe". *Science* **339** (6115): 46. Bibcode:2013Sci...339...46B. doi:10.1126/science.1224898.

[32] C. Moskowitz (13 December 2012). "Phew! Universe's Constant Has Stayed Constant". Space.com. Retrieved 2012-12-14.

[33] "Particle chameleon caught in the act of changing". CERN. 31 May 2010. Retrieved 2012-07-05.

[34] S. Weinberg (1979). "Baryon and Lepton Nonconserving Processes". *Physical Review Letters* **43** (21): 1566. Bibcode:1979PhW. doi:10.1103/PhysRevLett.43.1566.

[35] P. Minkowski (1977). "μ → e γ at a Rate of One Out of 10^9 Muon Decays?". *Physics Letters B* **67** (4): 421. Bibcode:1977PhLB...67..421M. doi:10.1016/0370-2693(77)90435-X.

[36] R. N. Mohapatra, G. Senjanovic (1980). "Neutrino Mass and Spontaneous Parity Nonconservation". *Physical Review Letters* **44** (14): 912–915. Bibcode:1980PhRvL..44..912M. doi:10.1103/PhysRevLett.44.912.

[37] M. Gell-Mann, P. Ramond and R. Slansky (1979). F. van Nieuwenhuizen and D. Z. Freedman, ed. *Supergravity*. North Holland. pp. 315–321. ISBN 0-444-85438-X.

[38] Salvio, Strumia (2014-03-17). "Agravity". *JHEP 1406 (2014) 080*. arXiv:1403.4226. Bibcode:2014JHEP...06..080S. doi:

3.11   Further reading

- R. Oerter (2006). *The Theory of Almost Everything: The Standard Model, the Unsung Triumph of Modern Physics*. Plume.

- B.A. Schumm (2004). *Deep Down Things: The Breathtaking Beauty of Particle Physics*. Johns Hopkins University Press. ISBN 0-8018-7971-X.

- "The Standard Model of Particle Physics Interactive Graphic".

Introductory textbooks

- I. Aitchison, A. Hey (2003). *Gauge Theories in Particle Physics: A Practical Introduction*. Institute of Physics. ISBN 978-0-585-44550-2.

- W. Greiner, B. Müller (2000). *Gauge Theory of Weak Interactions*. Springer. ISBN 3-540-67672-4.

- G.D. Coughlan, J.E. Dodd, B.M. Gripaios (2006). *The Ideas of Particle Physics: An Introduction for Scientists*. Cambridge University Press.

- D.J. Griffiths (1987). *Introduction to Elementary Particles*. John Wiley & Sons. ISBN 0-471-60386-4.

- G.L. Kane (1987). *Modern Elementary Particle Physics*. Perseus Books. ISBN 0-201-11749-5.

Advanced textbooks

- T.P. Cheng, L.F. Li (2006). *Gauge theory of elementary particle physics*. Oxford University Press. ISBN 0-19-851961-3. Highlights the gauge theory aspects of the Standard Model.

- J.F. Donoghue, E. Golowich, B.R. Holstein (1994). *Dynamics of the Standard Model*. Cambridge University Press. ISBN 978-0-521-47652-2. Highlights dynamical and phenomenological aspects of the Standard Model.

- L. O'Raifeartaigh (1988). *Group structure of gauge theories*. Cambridge University Press. ISBN 0-521-34785-8.

- Nagashima Y. Elementary Particle Physics: Foundations of the Standard Model, Volume 2. (Wiley 2013) 920 pаnyы

- Schwartz, M.D. Quantum Field Theory and the Standard Model (Cambridge University Press 2013) 952 pages

- Langacker P. The standard model and beyond. (CRC Press, 2010) 670 pages Highlights group-theoretical aspects of the Standard Model.

Journal articles

- E.S. Abers, B.W. Lee (1973). "Gauge theories".*Physics Reports* **9**: 1–141. Bibcode:1973PhR.....9....1A.doi:10.1573(73)90027-6.

- M. Baak; et al. (2012). "The Electroweak Fit of the Standard Model after the Discovery of a New Boson at the LHC". *The European Physical Journal C* **72** (11). arXiv:1209.2716. Bibcode:2012EPJC...72.2205B. doi:10.1140/epjc/s10052-012-2205-9.

- Y. Hayato; et al. (1999). "Search for Proton Decay through $p \rightarrow \nu K^+$ in a Large Water Cherenkov Detector". *Physical Review Letters* **83** (8): 1529. arXiv:hep-ex/9904020. Bibcode:1999PhRvL..83.1529H. doi:

- S.F. Novaes (2000). "Standard Model: An Introduction". arXiv:hep-ph/0001283 [hep-ph].

- D.P. Roy (1999). "Basic Constituents of Matter and their Interactions — A Progress Report". arXiv:hep-ph/9912523 [hep-ph].

- F. Wilczek (2004). "The Universe Is A Strange Place". *Nuclear Physics B - Proceedings Supplements* **134**: 3. arXiv:astro-ph/0401347. Bibcode:2004NuPhS.134....3W. doi:10.1016/j.nuclphysbps.2004.08.001.

3.12 External links

- "The Standard Model explained in Detail by CERN's John Ellis" omega tau podcast.

- "LHC sees hint of lightweight Higgs boson" "New Scientist".

- "Standard Model may be found incomplete," *New Scientist*.

- "Observation of the Top Quark" at Fermilab.

- "The Standard Model Lagrangian." After electroweak symmetry breaking, with no explicit Higgs boson.

- "Standard Model Lagrangian" with explicit Higgs terms. PDF, PostScript, and LaTeX versions.

- "The particle adventure." Web tutorial.

- Nobes, Matthew (2002) "Introduction to the Standard Model of Particle Physics" on Kuro5hin: Part 1, Part 2, Part 3a, Part 3b.

- "The Standard Model" The Standard Model on the CERN web site explains how the basic building blocks of matter interact, governed by four fundamental forces.

Chapter 4

Quantum

For other uses, see Quantum (disambiguation).

In physics, a **quantum** (plural: **quanta**) is the minimum amount of any physical entity involved in an interaction. Behind this, one finds the fundamental notion that a physical property may be "quantized," referred to as "the hypothesis of quantization".[1] This means that the magnitude can take on only certain discrete values.

A photon is a single quantum of (visible) light as well as all other forms of electromagnetic radiation and can be referred to as a "light quantum". The energy of an electron bound to an atom is quantized, which results in the stability of atoms, and hence of matter in general.

As incorporated into the theory of quantum mechanics, this is regarded by physicists as part of the fundamental framework for understanding and describing nature.

4.1 Etymology and discovery

The word "quantum" comes from the Latin "quantus", meaning "how much". "Quanta", short for "quanta of electricity" (electrons) was used in a 1902 article on the photoelectric effect by Philipp Lenard, who credited Hermann von Helmholtz for using the word in the area of electricity. However, the word quantum in general was well known before 1900.[2] It was often used by physicians, such as in the term quantum satis. Both Helmholtz and Julius von Mayer were physicians as well as physicists. Helmholtz used "quantum" with reference to heat in his article[3] on Mayer's work, and indeed, the word "quantum" can be found in the formulation of the first law of thermodynamics by Mayer in his letter[4] dated July 24, 1841. Max Planck used "quanta" to mean "quanta of matter and electricity",[5] gas, and heat.[6] In 1905, in response to Planck's work and the experimental work of Lenard (who explained his results by using the term "quanta of electricity"), Albert Einstein suggested that radiation existed in spatially localized packets which he called "quanta of light" ("Lichtquanta").[7]

The concept of quantization of radiation was discovered in 1900 by Max Planck, who had been trying to understand the emission of radiation from heated objects, known as black-body radiation. By assuming that energy can only be absorbed or released in tiny, differential, discrete packets he called "bundles" or "energy elements",[8] Planck accounted for the fact that certain objects change colour when heated.[9] On December 14, 1900, Planck reported his revolutionary findings to the German Physical Society, and introduced the idea of quantization for the first time as a part of his research on black-body radiation.[10] As a result of his experiments, Planck deduced the numerical value of h, known as the Planck constant, and could also report a more precise value for the Avogadro–Loschmidt number, the number of real molecules in a mole and the unit of electrical charge, to the German Physical Society. After his theory was validated, Planck was awarded the Nobel Prize in Physics in 1918 for his discovery.

4.2 Beyond electromagnetic radiation

While quantization was first discovered in electromagnetic radiation, it describes a fundamental aspect of energy not just restricted to photons.[11] In the attempt to bring experiment into agreement with theory, Max Planck postulated that electromagnetic energy is absorbed or emitted in discrete packets, or quanta.[12]

4.3 See also

- Elementary particle

- Graviton

- Introduction to quantum mechanics

- Magnetic flux quantum

- Photon

- Photon polarization

- Quantal analysis

- Quantization (physics)

- Quantum cellular automata

- Quantum channel

- Quantum coherence

- Quantum chromodynamics

- Quantum computer

- Quantum cryptography

- Quantum dot

- Quantum electronics

- Quantum entanglement

- Quantum immortality

- Quantum lithography

- Quantum mechanics

- Quantum number

- Quantum sensor

- Quantum state

- Subatomic particle

4.4 References

[1] Wiener, N. (1966). *Differential Space, Quantum Systems, and Prediction.* Cambridge: The Massachusetts Institute of Technology Press

[2] E. Cobham Brewer 1810–1897. Dictionary of Phrase and Fable. 1898.

[3] E. Helmholtz, Robert Mayer's Priorität (German)

[4] Herrmann,A. Weltreich der Physik, GNT-Verlag (1991) (German)

[5] Planck, M. (1901). "Ueber die Elementarquanta der Materie und der Elektricität". *Annalen der Physik* (in German) **309** (3): 564–566. Bibcode:1901AnP...309..564P. doi:10.1002/andp.19013090311.

[6] Planck, Max (1883). "Ueber das thermodynamische Gleichgewicht von Gasgemengen". *Annalen der Physik* (in German) **255** (6): 358. Bibcode:1883AnP...255..358P. doi:10.1002/andp.18832550612.

[7] Einstein, A. (1905). "Über einen die Erzeugung und Verwandlung des Lichtes betreffenden heuristischen Gesichtspunkt" (PDF). *Annalen der Physik* (in German) **17** (6): 132–148. Bibcode:1905AnP...322..132E. doi:10.1002/andp.19053220607.. A partial English translation is available from Wikisource.

[8] Max Planck (1901). "Ueber das Gesetz der Energieverteilung im Normalspectrum (On the Law of Distribution of Energy in the Normal Spectrum)". *Annalen der Physik* **309** (3): 553. Bibcode:1901AnP...309..553P. doi:10.1002/andp.19013090310. Archived from the original on 2008-04-18.

[9] Brown, T., LeMay, H., Bursten, B. (2008). *Chemistry: The Central Science* Upper Saddle River, NJ: Pearson Education ISBN 0-13-600617-5

[10] Klein, Martin J. (1961). "Max Planck and the beginnings of the quantum theory". *Archive for History of Exact Sciences* **1** (5): 459. doi:10.1007/BF00327765.

[11] Melville, K. (2005, February 11). Real-World Quantum Effects Demonstrated

[12] Modern Applied Physics-Tippens third edition; McGraw-Hill.

4.5 Further reading

- B. Hoffmann, *The Strange Story of the Quantum*, Pelican 1963.

- Lucretius, *On the Nature of the Universe*, transl. from the Latin by R.E. Latham, Penguin Books Ltd., Harmondsworth 1951. There are, of course, many translations, and the translation's title varies. Some put emphasis on how things work, others on what things are found in nature.

- J. Mehra and H. Rechenberg, *The Historical Development of Quantum Theory*, Vol.1, Part 1, Springer-Verlag New York Inc., New York 1982.

- M. Planck, *A Survey of Physical Theory*, transl. by R. Jones and D.H. Williams, Methuen & Co., Ltd., London 1925 (Dover editions 1960 and 1993) including the Nobel lecture.

- Rodney, Brooks (2011) *Fields of Color: The theory that escaped Einstein.* Allegra Print & Imaging.

Chapter 5

Mass generation

In theoretical physics, a **mass generation** mechanism is a theory that describes the origin of mass from the most fundamental laws of physics. Physicists have proposed a number of models that advocate different views of the origin of mass. The problem is complicated because mass is strongly connected to gravitational interaction, and no theory of gravitational interaction reconciles with the currently popular Standard Model of particle physics.

There are two types of mass generation models: gravity-free models and models that involve gravity.

5.1 Gravity-free models

In these theories, as in the Standard Model itself, the gravitational interaction either is not involved or does not play a crucial role.

- The Higgs mechanism is based on a symmetry-breaking scalar field potential, such as the quartic. The Standard Model uses this mechanism as part of the Glashow–Weinberg–Salam model to unify electromagnetic and weak interactions. This model was one of several that predicted the existence of the scalar Higgs boson.

- Technicolor models break electroweak symmetry through new gauge interactions, which were originally modeled on quantum chromodynamics.[1][2]

- Coleman–Weinberg mechanism (spontaneous symmetry breaking through radiative corrections).

- Models of composite W and Z vector bosons.[3]

- Top quark condensate.

- Asymptotically safe weak interactions [4][5] based on some nonlinear sigma models.[6]

- Symmetry breaking driven by non-equilibrium dynamics of quantum fields above the electroweak scale.[7][8]

- Unparticle physics and the unhiggs[9][10] models posit that the Higgs sector and Higgs boson are scaling invariant, also known as unparticle physics.

- UV-Completion by Classicalization, in which the unitarization of the WW scattering happens by creation of classical configurations.[11]

5.2 Models that involve gravity

- Extra-dimensional Higgsless models use the fifth component of the gauge fields in place of the Higgs fields. It is possible to produce electroweak symmetry breaking by imposing certain boundary conditions on the extra dimensional fields, increasing the unitarity breakdown scale up to the energy scale of the extra dimension.[12][13] Through the AdS/QCD correspondence this model can be related to technicolor models and to *UnHiggs* models, in which the Higgs field is of unparticle nature.[14]

- Unitary Weyl gauge. If one adds a suitable gravitational term to the standard model action with gravitational coupling, the theory becomes locally scale-invariant (i.e. Weyl-invariant) in the unitary gauge for the local SU(2). Weyl transformations act multiplicatively on the Higgs field, so one can fix the Weyl gauge by requiring that the Higgs scalar be a constant.[15]

- Preon and models inspired by preons such as the Ribbon model of Standard Model particles by Sundance Bilson-Thompson, based in braid theory and compatible with loop quantum gravity and similar theories.[16] This model not only explains the origin of mass, but also interprets electric charge as a topological quantity (twists carried on the individual ribbons), and colour charge as modes of twisting.

- In the theory of superfluid vacuum, masses of elementary particles arise from interaction with a physical vacuum, similarly to the gap generation mechanism in superfluids.[17] The low-energy limit of this theory suggests an effective potential for the Higgs sector that is different from the Standard Model's, yet it yields the mass generation.[18][19] Under certain conditions, this potential gives rise to an elementary particle with a role and characteristics similar to the Higgs boson.

5.3 See also

- Mass

- Higgs mechanism

- Spontaneous symmetry breaking

5.4 References

[1]Steven Weinberg (1976), "Implications of dynamical symmetry breaking",*Physical Review***D13**(4): 974–996,Bibcode:, doi:10.1103/PhysRevD.13.974.
S. Weinberg (1979), "Implications of dynamical symmetry breaking: An addendum", *Physical Review* **D19** (4): 1277–1280, Bibcode:1979PhRvD..19.1277W, doi:10.1103/PhysRevD.19.1277.

[2] Leonard Susskind (1979), "Dynamics of spontaneous symmetry breaking in the Weinberg-Salam theory", *Physical Review* **D20** (10): 2619–2625, Bibcode:1979PhRvD..20.2619S, doi:10.1103/PhysRevD.20.2619.

[3]Abbott, L. F.; Farhi, E. (1981), "Are the Weak Interactions Strong?",*Physics Letters B***101**(1–2): 69,Bibcode:1981PhLB..101, doi:10.1016/0370-2693(81)90492-5

[4]Calmet, X. (2011), "Asymptotically safe weak interactions",*Mod. Phys. Lett.***A26**: 1571–1576,arXiv:1012.5529,Bibcod1C, doi:10.1142/S0217732311035900

[5] Calmet, X. (2011), "An Alternative view on the electroweak interactions", *Int.J.Mod.Phys.* **A26**: 2855–2864, arXiv:1008.3780, Bibcode:2011IJMPA..26.2855C, doi:10.1142/S0217751X11053699

[6] Codello, A.; Percacci, R. (2008), "Fixed Points of Nonlinear Sigma Models in d>2", *Physics Letters B* **672** (3): 280–283, arXiv:0810.0715, Bibcode:2009PhLB..672..280C, doi:10.1016/j.physletb.2009.01.032

[7]"Bifurcations and pattern formation in particle physics: An introductory study".*EPL (Europhysics Letters)***82**: 11001.BiG. doi:10.1209/0295-5075/82/11001.

[8] http://www.ejtp.com/articles/ejtpv7i24p219.pdf

[9] http://arxiv.org/PS_cache/arxiv/pdf/0807/0807.3961v2.pdf

[10] http://arxiv.org/PS_cache/arxiv/pdf/0901/0901.3777v2.pdf

[11] Dvali, Gia; Giudice, Gian F.; Gomez, Cesar; Kehagias, Alex (2011). "UV-Completion by Classicalization". arXiv:1010.1415. Bibcode:2011JHEP...08..108D. doi:10.1007/JHEP08(2011)108.

[12] Csaki, C.; Grojean, C.; Pilo, L.; Terning, J. (2004), "Towards a realistic model of Higgsless electroweak symmetry breaking", *Physical Review Letters* **92** (10): 101802, arXiv:hep-ph/0308038, Bibcode:2004PhRvL..92j1802C, doi:10.1103/PhysRevL, PMID 15089195

[13] Csaki, C.; Grojean, C.; Murayama, H.; Pilo, L.; Terning, John (2004), "Gauge theories on an interval: Unitarity without a Higgs", *Physical Review D* **69** (5): 055006, arXiv:hep-ph/0305237, Bibcode:2004PhRvD..69e5006C, doi:

[14] Calmet, X.; Deshpande, N. G.; He, X. G.; Hsu, S. D. H. (2008), "Invisible Higgs boson, continuous mass fields and unHiggs mechanism", *Physical Review D* **79** (5): 055021, arXiv:0810.2155, Bibcode:2009PhRvD..79e5021C, doi:10.1103/Phys1

[15] Pawlowski, M.; Raczka, R. (1994), "A Unified Conformal Model for Fundamental Interactions without Dynamical Higgs Field", *Foundations of Physics* **24** (9): 1305–1327, arXiv:hep-th/9407137, Bibcode:1994FoPh...24.1305P, doi:10.1007/BF02148570

[16] Bilson-Thompson, Sundance O.; Markopoulou, Fotini; Smolin, Lee (2007), "Quantum gravity and the standard model", *Class. Quantum Grav.* **24** (16): 3975–3993, arXiv:hep-th/0603022, Bibcode:2007CQGra..24.3975B,doi:10.1088/0264-9381/.

[17] A. V. Avdeenkov and K. G. Zloshchastiev, *Quantum Bose liquids with logarithmic nonlinearity: Self-sustainability and emergence of spatial extent*, J. Phys. B: At. Mol. Opt. Phys. **44** (2011) 195303. ArXiv:1108.0847.

[18] K. G. Zloshchastiev, *Spontaneous symmetry breaking and mass generation as built-in phenomena in logarithmic nonlinear quantum theory*, Acta Phys. Polon. B **42** (2011) 261-292 ArXiv:0912.4139.

[19] V. Dzhunushaliev and K.G. Zloshchastiev (2012). *Singularity-free model of electric charge in physical vacuum: Non-zero spatial extent and mass generation*. ArXiv:1204.6380.

Chapter 6

Symmetry (physics)

For other uses, see Symmetry (disambiguation).

In physics, a **symmetry** of a physical system is a physical or mathematical feature of the system (observed or intrinsic) that is preserved or remains unchanged under some transformation.

A family of particular transformations may be *continuous* (such as rotation of a circle) or *discrete* (e.g., reflection of a bilaterally symmetric figure, or rotation of a regular polygon). Continuous and discrete transformations give rise to corresponding types of symmetries. Continuous symmetries can be described by Lie groups while discrete symmetries are described by finite groups (see Symmetry group).

These two concepts, Lie and finite groups, are the foundation for the fundamental theories of modern physics. Symmetries are frequently amenable to mathematical formulations such as group representations and can, in addition, be exploited to simplify many problems.

Arguably the most important example of a symmetry in physics is that the speed of light has the same value in all frames of reference, which is known in mathematical terms as Poincare group, the symmetry group of special relativity. Another important example is the invariance of the form of physical laws under arbitrary differentiable coordinate transformations, which is an important idea in general relativity.

6.1 Symmetry as invariance

Invariance is specified mathematically by transformations that leave some quantity unchanged. This idea can apply to basic real-world observations. For example, temperature may be constant throughout a room. Since the temperature is independent of position within the room, the temperature is *invariant* under a shift in the measurer's position.

Similarly, a uniform sphere rotated about its center will appear exactly as it did before the rotation. The sphere is said to exhibit spherical symmetry. A rotation about any axis of the sphere will preserve how the sphere "looks".

6.1.1 Invariance in force

The above ideas lead to the useful idea of *invariance* when discussing observed physical symmetry; this can be applied to symmetries in forces as well.

For example, an electric field due to a wire is said to exhibit cylindrical symmetry, because the electric field strength at a given distance r from the electrically charged wire of infinite length will have the same magnitude at each point on the surface of a cylinder (whose axis is the wire) with radius r. Rotating the wire about its own axis does not change its position or charge density, hence it will preserve the field. The field strength at a rotated position is the same. Suppose some configuration of charges (may be non-stationary) produce an electric field in some direction, then rotating the configuration of the charges (without disturbing the internal dynamics that produces the particular field) will lead to a net

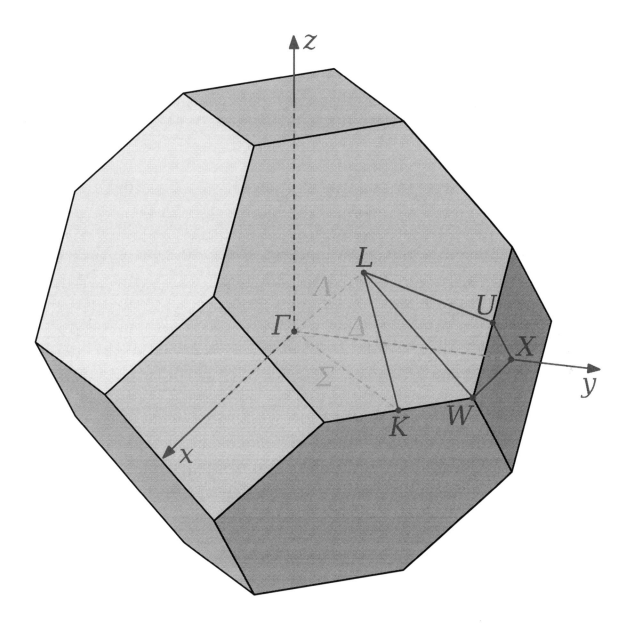

First Brillouin zone of FCC lattice showing symmetry labels

rotation of the direction of the electric field. These two properties are interconnected through the more general property that rotating *any* system of charges causes a corresponding rotation of the electric field.

In Newton's theory of mechanics, given two bodies, each with mass m, starting from rest at the origin and moving along the x-axis in opposite directions, one with speed v_1 and the other with speed v_2 the total kinetic energy of the system (as calculated from an observer at the origin) is $\frac{1}{2}m(v_1{}^2 + v_2{}^2)$ and remains the same if the velocities are interchanged. The total kinetic energy is preserved under a reflection in the y-axis.

The last example above illustrates another way of expressing symmetries, namely through the equations that describe some aspect of the physical system. The above example shows that the total kinetic energy will be the same if v_1 and v_2 are interchanged.

6.2 Local and global symmetries

Main articles: Global symmetry and Local symmetry

Symmetries may be broadly classified as *global* or *local*. A *global symmetry* is one that holds at all points of spacetime, whereas a *local symmetry* is one that has a different symmetry transformation at different points of spacetime; specifically a local symmetry transformation is parameterised by the spacetime co-ordinates. Local symmetries play an important role in physics as they form the basis for gauge theories.

6.3 Continuous symmetries

The two examples of rotational symmetry described above - spherical and cylindrical - are each instances of continuous symmetry. These are characterised by invariance following a continuous change in the geometry of the system. For example, the wire may be rotated through any angle about its axis and the field strength will be the same on a given cylinder. Mathematically, continuous symmetries are described by continuous or smooth functions. An important subclass of continuous symmetries in physics are spacetime symmetries.

6.3.1 Spacetime symmetries

Main article: Spacetime symmetries

Continuous *spacetime symmetries* are symmetries involving transformations of space and time. These may be further classified as *spatial symmetries*, involving only the spatial geometry associated with a physical system; *temporal symmetries*, involving only changes in time; or *spatio-temporal symmetries*, involving changes in both space and time.

- **Time translation**: A physical system may have the same features over a certain interval of time δt ; this is expressed mathematically as invariance under the transformation $t \rightarrow t + a$ for any real numbers t and a in the interval. For example, in classical mechanics, a particle solely acted upon by gravity will have gravitational potential energy mgh when suspended from a height h above the Earth's surface. Assuming no change in the height of the particle, this will be the total gravitational potential energy of the particle at all times. In other words, by considering the state of the particle at some time (in seconds) t_0 and also at $t_0 + 3$, say, the particle's total gravitational potential energy will be preserved.

- **Spatial translation**: These spatial symmetries are represented by transformations of the form $\vec{r} \rightarrow \vec{r} + \vec{a}$ and describe those situations where a property of the system does not change with a continuous change in location. For example, the temperature in a room may be independent of where the thermometer is located in the room.

- **Spatial rotation**: These spatial symmetries are classified as proper rotations and improper rotations. The former are just the 'ordinary' rotations; mathematically, they are represented by square matrices with unit determinant. The latter are represented by square matrices with determinant -1 and consist of a proper rotation combined with a spatial reflection (inversion). For example, a sphere has proper rotational symmetry. Other types of spatial rotations are described in the article *Rotation symmetry*.

- **Poincaré transformations**: These are spatio-temporal symmetries which preserve distances in Minkowski space-time, i.e. they are isometries of Minkowski space. They are studied primarily in special relativity. Those isometries that leave the origin fixed are called Lorentz transformations and give rise to the symmetry known as Lorentz co-variance.

- **Projective symmetries**: These are spatio-temporal symmetries which preserve the geodesic structure of spacetime. They may be defined on any smooth manifold, but find many applications in the study of exact solutions in general relativity.

- *Inversion transformations*: These are spatio-temporal symmetries which generalise Poincaré transformations to include other conformal one-to-one transformations on the space-time coordinates. Lengths are not invariant under inversion transformations but there is a cross-ratio on four points that is invariant.

Mathematically, spacetime symmetries are usually described by smooth vector fields on a smooth manifold. The underlying local diffeomorphisms associated with the vector fields correspond more directly to the physical symmetries, but the vector fields themselves are more often used when classifying the symmetries of the physical system.

Some of the most important vector fields are Killing vector fields which are those spacetime symmetries that preserve the underlying metric structure of a manifold. In rough terms, Killing vector fields preserve the distance between any two points of the manifold and often go by the name of isometries.

6.4 Discrete symmetries

Main article: Discrete symmetry

A **discrete symmetry** is a symmetry that describes non-continuous changes in a system. For example, a square possesses discrete rotational symmetry, as only rotations by multiples of right angles will preserve the square's original appearance. Discrete symmetries sometimes involve some type of 'swapping', these swaps usually being called *reflections* or *interchanges*.

- *Time reversal*: Many laws of physics describe real phenomena when the direction of time is reversed. Mathematically, this is represented by the transformation, $t \rightarrow -t$. For example, Newton's second law of motion still holds if, in the equation $F = m\ddot{r}$, t is replaced by $-t$. This may be illustrated by recording the motion of an object thrown up vertically (neglecting air resistance) and then playing it back. The object will follow the same parabolic trajectory through the air, whether the recording is played normally or in reverse. Thus, position is symmetric with respect to the instant that the object is at its maximum height.

- *Spatial inversion*: These are represented by transformations of the form $\vec{r} \rightarrow -\vec{r}$ and indicate an invariance property of a system when the coordinates are 'inverted'. Said another way, these are symmetries between a certain object and its mirror image.

- *Glide reflection*: These are represented by a composition of a translation and a reflection. These symmetries occur in some crystals and in some planar symmetries, known as wallpaper symmetries.

6.4.1 C, P, and T symmetries

The Standard model of particle physics has three related natural near-symmetries. These state that the actual universe about us is indistinguishable from one where:

- Every particle is replaced with its antiparticle. This is C-symmetry (charge symmetry);

- Everything appears as if reflected in a mirror. This is P-symmetry (parity symmetry);

- The direction of time is reversed. This is T-symmetry (time symmetry).

T-symmetry is counterintuitive (surely the future and the past are not symmetrical) but explained by the fact that the Standard model describes local properties, not global ones like entropy. To properly reverse the direction of time, one would have to put the big bang and the resulting low-entropy state in the "future." Since we perceive the "past" ("future") as having lower (higher) entropy than the present (see perception of time), the inhabitants of this hypothetical time-reversed universe would perceive the future in the same way as we perceive the past.

These symmetries are near-symmetries because each is broken in the present-day universe. However, the Standard Model predicts that the combination of the three (that is, the simultaneous application of all three transformations) must be a symmetry, called CPT symmetry. In <ref name=*qm*>G. Kalmbach H.E.: *Quantum Mathematics: WIGRIS*. RGN Publications, Delhi, 2014.</ref> the 4 dimensional matrix description of P,T is through a diagonal matrix, the negative identity, as well as C. Hence CPT is the identity operator. CP violation, the violation of the combination of C- and P-symmetry, is necessary for the presence of significant amounts of baryonic matter in the universe. CP violation is a fruitful area of current research in particle physics.

6.4.2 Supersymmetry

Main article: Supersymmetry

A type of symmetry known as supersymmetry has been used to try to make theoretical advances in the standard model. Supersymmetry is based on the idea that there is another physical symmetry beyond those already developed in the standard model, specifically a symmetry between bosons and fermions. Supersymmetry asserts that each type of boson has, as a supersymmetric partner, a fermion, called a superpartner, and vice versa. Supersymmetry has not yet been experimentally verified: no known particle has the correct properties to be a superpartner of any other known particle. If superpartners exist they must have masses greater than current particle accelerators can generate.

6.5 Mathematics of physical symmetry

Main article: Symmetry group
See also: Symmetry in quantum mechanics and Symmetries in general relativity

The transformations describing physical symmetries typically form a mathematical group. Group theory is an important area of mathematics for physicists.

Continuous symmetries are specified mathematically by *continuous groups* (called Lie groups). Many physical symmetries are isometries and are specified by symmetry groups. Sometimes this term is used for more general types of symmetries. The set of all proper rotations (about any angle) through any axis of a sphere form a Lie group called the special orthogonal group $SO(3)$. (The 3 refers to the three-dimensional space of an ordinary sphere.) Thus, the symmetry group of the sphere with proper rotations is $SO(3)$. Any rotation preserves distances on the surface of the ball. The set of all Lorentz transformations form a group called the Lorentz group (this may be generalised to the Poincaré group).

Discrete symmetries are described by discrete groups. For example, the symmetries of an equilateral triangle are described by the symmetric group S_3.

An important type of physical theory based on *local* symmetries is called a *gauge* theory and the symmetries natural to such a theory are called gauge symmetries. Gauge symmetries in the Standard model, used to describe three of the fundamental interactions, are based on the SU(3) × SU(2) × U(1) group. (Roughly speaking, the symmetries of the SU(3) group describe the strong force, the SU(2) group describes the weak interaction and the U(1) group describes the electromagnetic force.)

Also, the reduction by symmetry of the energy functional under the action by a group and spontaneous symmetry breaking of transformations of symmetric groups appear to elucidate topics in particle physics (for example, the unification of electromagnetism and the weak force in physical cosmology).

6.5.1 Conservation laws and symmetry

Main article: Noether's theorem

The symmetry properties of a physical system are intimately related to the conservation laws characterizing that system.

Noether's theorem gives a precise description of this relation. The theorem states that each continuous symmetry of a physical system implies that some physical property of that system is conserved. Conversely, each conserved quantity has a corresponding symmetry. For example, the isometry of space gives rise to conservation of (linear) momentum, and isometry of time gives rise to conservation of energy.

The following table summarizes some fundamental symmetries and the associated conserved quantity.

6.6 Mathematics

Continuous symmetries in physics preserve transformations. One can specify a symmetry by showing how a very small transformation affects various particle fields. The commutator of two of these infinitessimal transformations are equivalent to a third infinitessimal transformation of the same kind hence they form a Lie algebra.

A general coordinate transformation (also known as a diffeomorphism) has the infinitessimal effect on a scalar, spinor and vector field for example:

$$\delta\phi(x) = h^\mu(x)\partial_\mu\phi(x)$$

$$\delta\psi^\alpha(x) = h^\mu(x)\partial_\mu\psi^\alpha(x) + \partial_\mu h_\nu(x)\sigma^{\alpha\beta}_{\mu\nu}\psi^\beta(x)$$

$$\delta A_\mu(x) = h^\nu(x)\partial_\nu A_\mu(x) + A_\nu(x)\partial_\mu h^\nu(x)$$

for a general field, $h(x)$. Without gravity only the Poincaré symmetries are preserved which restricts $h(x)$ to be of the form:

$$h^\mu(x) = M^{\mu\nu}x_\nu + P^\mu$$

where **M** is an antisymmetric matrix (giving the Lorentz and rotational symmetries) and **P** is a general vector (giving the translational symmetries). Other symmetries affect multiple fields simultaneously. For example local gauge transformations apply to both a vector and spinor field:

$$\delta\psi^\alpha(x) = \lambda(x).\tau^{\alpha\beta}\psi^\beta(x)$$

$$\delta A_\mu(x) = \partial_\mu\lambda(x)$$

where τ are generators of a particular Lie group. So far the transformations on the right have only included fields of the same type. Supersymmetries are defined according to how the mix fields of *different* types.

Another symmetry which is part of some theories of physics and not in others is scale invariance which involve Weyl transformations of the following kind:

$$\delta\phi(x) = \Omega(x)\phi(x)$$

If the fields have this symmetry then it can be shown that the field theory is almost certainly conformally invariant also. This means that in the absence of gravity h(x) would restricted to the form:

$$h^\mu(x) = M^{\mu\nu}x_\nu + P^\mu + Dx_\mu + K^\mu|x|^2 - 2K^\nu x_\nu x_\mu$$

with **D** generating scale transformations and **K** generating special conformal transformations. For example N=4 super-Yang-Mills theory has this symmetry while General Relativity doesn't although other theories of gravity such as conformal gravity do. The 'action' of a field theory is an invariant under all the symmetries of the theory. Much of modern theoretical physics is to do with speculating on the various symmetries the Universe may have and finding the invariants to construct field theories as models.

In string theories, since a string can be decomposed into an infinite number of particle fields, the symmetries on the string world sheet is equivalent to special transformations which mix an infinite number of fields.

6.7 See also

- Conservation law

- Conserved current

- Coordinate-free

- Covariance and contravariance

- Diffeomorphism

- Fictitious force

- Galilean invariance

- Gauge theory

- General covariance

- Harmonic coordinate condition

- Inertial frame of reference

- Lie group

- List of mathematical topics in relativity

- Lorentz covariance

- Noether's theorem

- Poincaré group

- Special relativity

- Spontaneous symmetry breaking

- Standard model

- Standard model (mathematical formulation)

- Symmetry breaking

- Wheeler–Feynman Time-Symmetric Theory

6.8   References

6.8.1   General readers

- Leon Lederman and Christopher T. Hill (2005) *Symmetry and the Beautiful Universe*. Amherst NY: Prometheus Books.

- Schumm, Bruce (2004) *Deep Down Things*. Johns Hopkins Univ. Press.

- Victor J. Stenger (2000) *Timeless Reality: Symmetry, Simplicity, and Multiple Universes*. Buffalo NY: Prometheus Books. Chpt. 12 is a gentle introduction to symmetry, invariance, and conservation laws.

- Anthony Zee (2007) *Fearful Symmetry: The search for beauty in modern physics,* 2nd ed. Princeton University Press. ISBN 978-0-691-00946-9. 1986 1st ed. published by Macmillan.

6.8.2 Technical readers

- Brading, K., and Castellani, E., eds. (2003) *Symmetries in Physics: Philosophical Reflections.* Cambridge Univ. Press.

- -------- (2007) "Symmetries and Invariances in Classical Physics" in Butterfield, J., and John Earman, eds., *Philosophy of Physic Part B.* North Holland: 1331-68.

- Debs, T. and Redhead, M. (2007) *Objectivity, Invariance, and Convention: Symmetry in Physical Science.* Harvard Univ. Press.

- John Earman (2002) "Laws, Symmetry, and Symmetry Breaking: Invariance, Conservations Principles, and Objectivity." Address to the 2002 meeting of the Philosophy of Science Association.

- G. Kalmbach H.E.: *Quantum Mathematics: WIGRIS.* RGN Publications, Delhi, 2014

- Mainzer, K. (1996) *Symmetries of nature.* Berlin: De Gruyter.

- Mouchet, A. "Reflections on the four facets of symmetry: how physics exemplifies rational thinking". European Physical Journal H 38 (2013) 661 hal.archives-ouvertes.fr:hal-00637572

- Thompson, William J. (1994) *Angular Momentum: An Illustrated Guide to Rotational Symmetries for Physical Systems.* Wiley. ISBN 0-471-55264-X.

- Bas Van Fraassen (1989) *Laws and symmetry.* Oxford Univ. Press.

- Eugene Wigner (1967) *Symmetries and Reflections.* Indiana Univ. Press.

6.9 External links

- Stanford Encyclopedia of Philosophy: "Symmetry"—by K. Brading and E. Castellani.

- Pedagogic Aids to Quantum Field Theory Click on link to Chapter 6: Symmetry, Invariance, and Conservation for a simplified, step-by-step introduction to symmetry in physics.

Chapter 7

Weak interaction

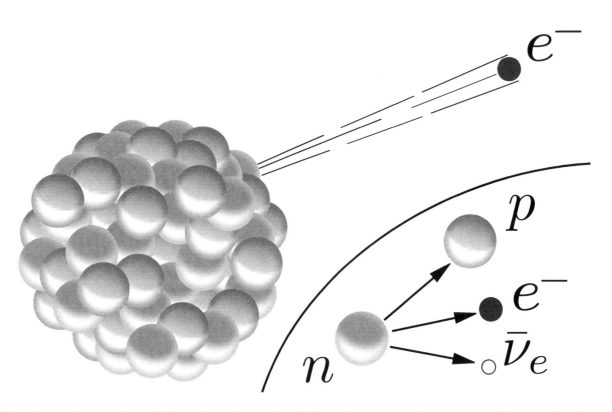

The radioactive beta decay is possible due to the weak interaction, which transforms a neutron into: a proton, an electron, and an electron antineutrino.

In particle physics, the **weak interaction** is the mechanism responsible for the **weak force** or **weak nuclear force**, one of the four known fundamental interactions of nature, alongside the strong interaction, electromagnetism, and gravitation. The weak interaction is responsible for the radioactive decay of subatomic particles, and it plays an essential role in nuclear fission. The theory of the weak interaction is sometimes called **quantum flavordynamics** (**QFD**), in analogy with the terms QCD and QED, but the term is rarely used because the weak force is best understood in terms of electro-weak theory (EWT).[1]

In the Standard Model of particle physics, the weak interaction is caused by the emission or absorption of W and Z bosons. All known fermions interact through the weak interaction. Fermions are particles that have half-integer spin (one

71

of the fundamental properties of particles). A fermion can be an elementary particle, such as the electron, or it can be a composite particle, such as the proton. The masses of W^+, W^-, and Z bosons are each far greater than that of protons or neutrons, consistent with the short range of the weak force. The force is termed *weak* because its field strength over a given distance is typically several orders of magnitude less than that of the strong nuclear force and electromagnetic force.

During the quark epoch, the electroweak force split into the electromagnetic and weak forces. Important examples of weak interaction include beta decay, and the production, from hydrogen, of deuterium needed to power the sun's thermonuclear process. Most fermions will decay by a weak interaction over time. Such decay also makes radiocarbon dating possible, as carbon-14 decays through the weak interaction to nitrogen-14. It can also create radioluminescence, commonly used in tritium illumination, and in the related field of betavoltaics.[2]

Quarks, which make up composite particles like neutrons and protons, come in six "flavours" – up, down, strange, charm, top and bottom – which give those composite particles their properties. The weak interaction is unique in that it allows for quarks to swap their flavour for another. For example, during beta minus decay, a down quark decays into an up quark, converting a neutron to a proton. Also the weak interaction is the only fundamental interaction that breaks parity-symmetry, and similarly, the only one to break CP-symmetry.

7.1   History

In 1933, Enrico Fermi proposed the first theory of the weak interaction, known as Fermi's interaction. He suggested that beta decay could be explained by a four-fermion interaction, involving a contact force with no range.[3][4]

However, it is better described as a non-contact force field having a finite range, albeit very short. In 1968, Sheldon Glashow, Abdus Salam and Steven Weinberg unified the electromagnetic force and the weak interaction by showing them to be two aspects of a single force, now termed the electro-weak force.

The existence of the W and Z bosons was not directly confirmed until 1983.

7.2   Properties

The weak interaction is unique in a number of respects:

1. It is the only interaction capable of changing the flavor of quarks (i.e., of changing one type of quark into another).

2. It is the only interaction that violates **P** or parity-symmetry. It is also the only one that violates **CP** symmetry.

3. It is propagated by carrier particles (known as gauge bosons) that have significant masses, an unusual feature which is explained in the Standard Model by the Higgs mechanism.

Due to their large mass (approximately 90 GeV/c^2[5]) these carrier particles, termed the W and Z bosons, are short-lived: they have a lifetime of under 1×10^{-24} seconds.[6] The weak interaction has a coupling constant (an indicator of interaction strength) of between 10^{-7} and 10^{-6}, compared to the strong interaction's coupling constant of about 1 and the electromagnetic coupling constant of about 10^{-2};[7] consequently the weak interaction is weak in terms of strength.[8] The weak interaction has a very short range (around 10^{-17}–10^{-16} m[8]).[7] At distances around 10^{-18} meters, the weak interaction has a strength of a similar magnitude to the electromagnetic force, but this starts to decrease exponentially with increasing distance. At distances of around 3×10^{-17} m, the weak interaction is 10,000 times weaker than the electromagnetic.[9]

The weak interaction affects all the fermions of the Standard Model, as well as the Higgs boson; neutrinos interact through gravity and the weak interaction only, and neutrinos were the original reason for the name *weak force*.[8] The weak interaction does not produce bound states (nor does it involve binding energy) – something that gravity does on an astronomical scale, that the electromagnetic force does at the atomic level, and that the strong nuclear force does inside nuclei.[10]

Its most noticeable effect is due to its first unique feature: flavor changing. A neutron, for example, is heavier than a proton (its sister nucleon), but it cannot decay into a proton without changing the flavor (type) of one of its two *down*

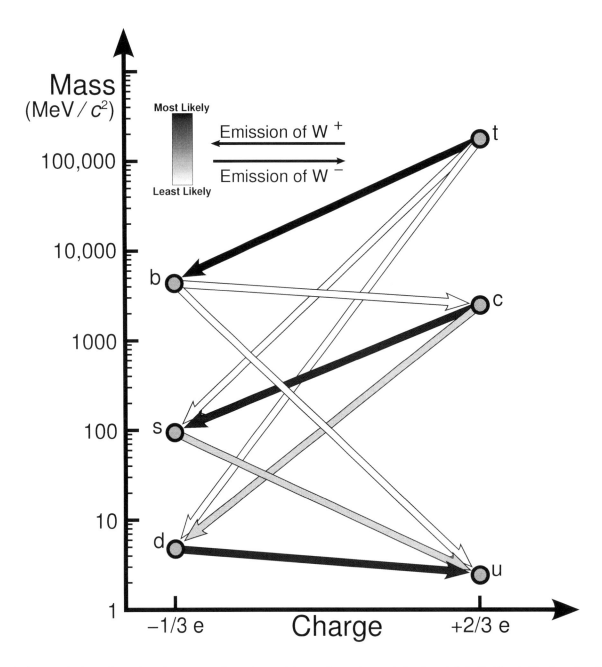

A diagram depicting the various decay routes due to the weak interaction and some indication of their likelihood. The intensity of the lines are given by the CKM parameters.

quarks to *up*. Neither the strong interaction nor electromagnetism permit flavour changing, so this must proceed by **weak decay**; without weak decay, quark properties such as strangeness and charm (associated with the quarks of the same name) would also be conserved across all interactions. All mesons are unstable because of weak decay.[11] In the process known as beta decay, a *down* quark in the neutron can change into an *up* quark by emitting a virtual W− boson which is then converted into an electron and an electron antineutrino.[12] Another example is the electron capture, a common variant of radioactive decay, where a proton (up quark) and an electron within an atom interact, and are changed to a neutron (down quark) and an electron neutrino.

Due to the large mass of a boson, weak decay is much more unlikely than strong or electromagnetic decay, and hence occurs less rapidly. For example, a neutral pion (which decays electromagnetically) has a life of about 10^{-16} seconds, while a charged pion (which decays through the weak interaction) lives about 10^{-8} seconds, a hundred million times longer.[13] In contrast, a free neutron (which also decays through the weak interaction) lives about 15 minutes.[12]

7.2.1   Weak isospin and weak hypercharge

Main article: Weak isospin

All particles have a property called weak isospin (T_3), which serves as a quantum number and governs how that particle interacts in the weak interaction. Weak isospin therefore plays the same role in the weak interaction as electric charge does in electromagnetism, and color charge in the strong interaction. All fermions have a weak isospin value of either $+\frac{1}{2}$ or $-\frac{1}{2}$. For example, the up quark has a T_3 of $+\frac{1}{2}$ and the down quark $-\frac{1}{2}$. A quark never decays through the weak interaction into a quark of the same T_3: quarks with a T_3 of $+\frac{1}{2}$ decay into quarks with a T_3 of $-\frac{1}{2}$ and vice versa.

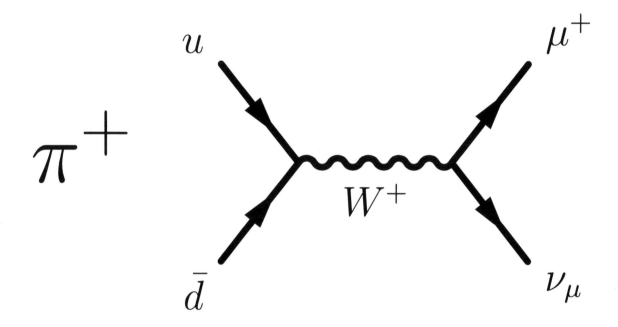

π+ decay through the weak interaction

In any given interaction, weak isospin is conserved: the sum of the weak isospin numbers of the particles entering the interaction equals the sum of the weak isospin numbers of the particles exiting that interaction. For example, a (left-handed) π+, with a weak isospin of 1 normally decays into a v
μ (+1/2) and a μ+ (as a right-handed antiparticle, +1/2).[13]

Following the development of the electroweak theory, another property, weak hypercharge, was developed. It is dependent on a particle's electrical charge and weak isospin, and is defined as:

$$Y_W = 2(Q - T_3)$$

where YW is the weak hypercharge of a given type of particle, Q is its electrical charge (in elementary charge units) and T_3 is its weak isospin. Whereas some particles have a weak isospin of zero, all particles, except gluons, have non-zero weak hypercharge. Weak hypercharge is the generator of the U(1) component of the electroweak gauge group.

7.3 Interaction types

There are two types of weak interaction (called *vertices*). The first type is called the "charged-current interaction" because it is mediated by particles that carry an electric charge (the W+ or W− bosons), and is responsible for the beta decay phenomenon. The second type is called the "neutral-current interaction" because it is mediated by a neutral particle, the Z boson.

7.3.1 Charged-current interaction

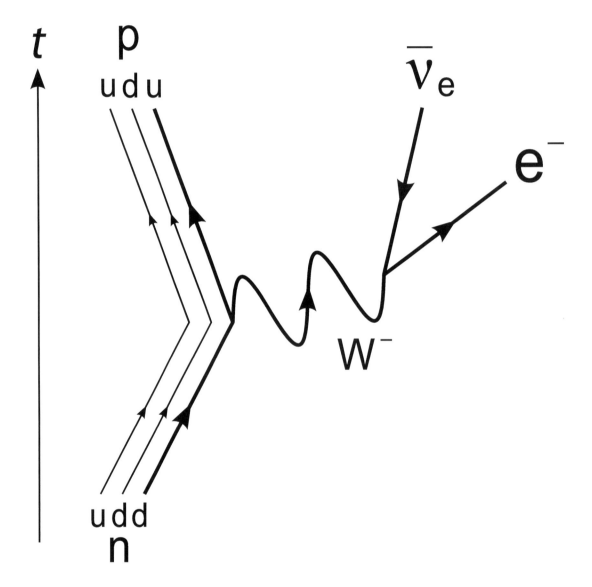

The Feynman diagram for beta-minus decay of a neutron into a proton, electron and electron anti-neutrino, via an intermediate heavy W− boson

In one type of charged current interaction, a charged lepton (such as an electron or a muon, having a charge of −1) can absorb a W+ boson (a particle with a charge of +1) and be thereby converted into a corresponding neutrino (with a charge of 0), where the type ("family") of neutrino (electron, muon or tau) is the same as the type of lepton in the interaction, for example:

$$\mu^- + W^+ \to \nu_\mu$$

Similarly, a down-type quark (d with a charge of $-\frac{1}{3}$) can be converted into an up-type quark (u, with a charge of $+\frac{2}{3}$), by emitting a W− boson or by absorbing a W+ boson. More precisely, the down-type quark becomes a quantum superposition of up-type quarks: that is to say, it has a possibility of becoming any one of the three up-type quarks, with the probabilities given in the CKM matrix tables. Conversely, an up-type quark can emit a W+ boson – or absorb a W− boson – and thereby be converted into a down-type quark, for example:

$$d \to u + W^-$$
$$d + W^+ \to u$$
$$c \to s + W^+$$
$$c + W^- \to s$$

The W boson is unstable so will rapidly decay, with a very short lifetime. For example:

$$W^- \to e^- + \bar{\nu}_e$$
$$W^+ \to e^+ + \nu_e$$

Decay of the W boson to other products can happen, with varying probabilities.[15]

In the so-called beta decay of a neutron (see picture, above), a down quark within the neutron emits a virtual W− boson and is thereby converted into an up quark, converting the neutron into a proton. Because of the energy involved in the process (i.e., the mass difference between the down quark and the up quark), the W− boson can only be converted into an electron and an electron-antineutrino.[16] At the quark level, the process can be represented as:

$$d \to u + e^- + \bar{\nu}_e$$

7.3.2   Neutral-current interaction

In neutral current interactions, a quark or a lepton (e.g., an electron or a muon) emits or absorbs a neutral Z boson. For example:

$$e^- \to e^- + Z^0$$

Like the W boson, the Z boson also decays rapidly,[15] for example:

$$Z^0 \to b + \bar{b}$$

7.4   Electroweak theory

Main article: Electroweak interaction

The Standard Model of particle physics describes the electromagnetic interaction and the weak interaction as two different aspects of a single electroweak interaction, the theory of which was developed around 1968 by Sheldon Glashow, Abdus Salam and Steven Weinberg. They were awarded the 1979 Nobel Prize in Physics for their work.[17] The Higgs mechanism provides an explanation for the presence of three massive gauge bosons (the three carriers of the weak interaction) and the massless photon of the electromagnetic interaction.[18]

According to the electroweak theory, at very high energies, the universe has four massless gauge boson fields similar to the photon and a complex scalar Higgs field doublet. However, at low energies, gauge symmetry is spontaneously broken down to the $U(1)$ symmetry of electromagnetism (one of the Higgs fields acquires a vacuum expectation value). This symmetry breaking would produce three massless bosons, but they become integrated by three photon-like fields (through the Higgs mechanism) giving them mass. These three fields become the W+, W− and Z bosons of the weak interaction, while the fourth gauge field, which remains massless, is the photon of electromagnetism.[18]

This theory has made a number of predictions, including a prediction of the masses of the Z and W bosons before their discovery. On 4 July 2012, the CMS and the ATLAS experimental teams at the Large Hadron Collider independently announced that they had confirmed the formal discovery of a previously unknown boson of mass between 125–127 GeV/c^2, whose behaviour so far was "consistent with" a Higgs boson, while adding a cautious note that further data and analysis were needed before positively identifying the new boson as being a Higgs boson of some type. By 14 March 2013, the Higgs boson was tentatively confirmed to exist .[19]

7.5 Violation of symmetry

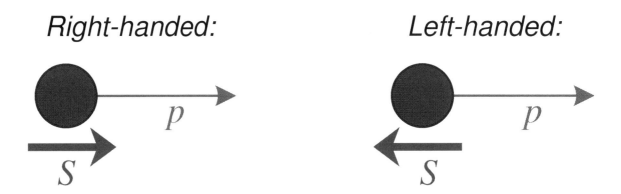

Left- and right-handed particles: p is the particle's momentum and S is its spin. Note the lack of reflective symmetry between the states.

The laws of nature were long thought to remain the same under mirror reflection, the reversal of one spatial axis. The results of an experiment viewed via a mirror were expected to be identical to the results of a mirror-reflected copy of the experimental apparatus. This so-called law of parity conservation was known to be respected by classical gravitation, electromagnetism and the strong interaction; it was assumed to be a universal law.[20] However, in the mid-1950s Chen Ning Yang and Tsung-Dao Lee suggested that the weak interaction might violate this law. Chien Shiung Wu and collaborators in 1957 discovered that the weak interaction violates parity, earning Yang and Lee the 1957 Nobel Prize in Physics.[21]

Although the weak interaction used to be described by Fermi's theory, the discovery of parity violation and renormalization theory suggested that a new approach was needed. In 1957, Robert Marshak and George Sudarshan and, somewhat later, Richard Feynman and Murray Gell-Mann proposed a **V−A** (vector minus axial vector or left-handed) Lagrangian for weak interactions. In this theory, the weak interaction acts only on left-handed particles (and right-handed antiparticles). Since the mirror reflection of a left-handed particle is right-handed, this explains the maximal violation of parity. Interestingly, the **V−A** theory was developed before the discovery of the Z boson, so it did not include the right-handed fields that enter in the neutral current interaction.

However, this theory allowed a compound symmetry **CP** to be conserved. **CP** combines parity **P** (switching left to right) with charge conjugation **C** (switching particles with antiparticles). Physicists were again surprised when in 1964, James

Cronin and Val Fitch provided clear evidence in kaon decays that CP symmetry could be broken too, winning them the 1980 Nobel Prize in Physics.[22] In 1973, Makoto Kobayashi and Toshihide Maskawa showed that CP violation in the weak interaction required more than two generations of particles,[23] effectively predicting the existence of a then unknown third generation. This discovery earned them half of the 2008 Nobel Prize in Physics.[24] Unlike parity violation, CP violation occurs in only a small number of instances, but remains widely held as an answer to the difference between the amount of matter and antimatter in the universe; it thus forms one of Andrei Sakharov's three conditions for baryogenesis.[25]

7.6   See also

- Weakless Universe – the postulate that weak interactions are not anthropically necessary

- Gravity

- Nuclear force

- Electromagnetism

7.7   References

7.7.1   Citations

[1] Griffiths, David (2009). *Introduction to Elementary Particles*. pp. 59–60. ISBN 978-3-527-40601-2.

[2] "The Nobel Prize in Physics 1979: Press Release". *NobelPrize.org*. Nobel Media. Retrieved 22 March 2011.

[3] Fermi, Enrico (1934). "Versuch einer Theorie der β-Strahlen. I". *Zeitschrift für Physik A* **88** (3–4): 161–177. Bibcode: doi:10.1007/BF01351864.

[4] Wilson, Fred L. (December 1968). "Fermi's Theory of Beta Decay". *American Journal of Physics* **36** (12): 1150–1160. Bibcode:1968AmJPh..36.1150W. doi:10.1119/1.1974382.

[5] W.-M. Yao *et al.* (Particle Data Group) (2006). "Review of Particle Physics: Quarks" (PDF). *Journal of Physics G* **33**: 1–1232. arXiv:astro-ph/0601168. Bibcode:2006JPhG...33....1Y. doi:10.1088/0954-3899/33/1/001.

[6] Peter Watkins (1986). *Story of the W and Z*. Cambridge: Cambridge University Press. p. 70. ISBN 978-0-521-31875-4.

[7] "Coupling Constants for the Fundamental Forces". *HyperPhysics*. Georgia State University. Retrieved 2 March 2011.

[8] J. Christman (2001). "The Weak Interaction" (PDF). *Physnet*. Michigan State University.

[9] "Electroweak". *The Particle Adventure*. Particle Data Group. Retrieved 3 March 2011.

[10] Walter Greiner; Berndt Müller (2009). *Gauge Theory of Weak Interactions*. Springer. p. 2. ISBN 978-3-540-87842-1.

[11] Cottingham & Greenwood (1986, 2001), p.29

[12] Cottingham & Greenwood (1986, 2001), p.28

[13] Cottingham & Greenwood (1986, 2001), p.30

[14] Baez, John C.; Huerta, John (2009)."The Algebra of Grand Unified Theories".*Bull.Am.Math.Soc.***0904**: 483–552.arXiv:090. Bibcode:2009arXiv0904.1556B. doi:10.1090/s0273-0979-10-01294-2. Retrieved 15 October 2013.

[15] K. Nakamura *et al.* (Particle Data Group) (2010). "Gauge and Higgs Bosons" (PDF). *Journal of Physics G* **37**. doi:10.1088/0954-3899/37/7a/075021.

[16] K. Nakamura *et al.* (Particle Data Group) (2010). "n" (PDF). *Journal of Physics G* **37**: 7. doi:10.1088/0954-3899/37/7a/075021.

[17] "The Nobel Prize in Physics 1979". *NobelPrize.org*. Nobel Media. Retrieved 26 February 2011.

[18] C. Amsler *et al.* (Particle Data Group) (2008). "Review of Particle Physics – Higgs Bosons: Theory and Searches" (PDF). *Physics Letters B* **667**: 1–6. Bibcode:2008PhLB..667....1P. doi:10.1016/j.physletb.2008.07.018.

[19] "New results indicate that new particle is a Higgs boson | CERN". Home.web.cern.ch. Retrieved 20 September 2013.

[20] Charles W. Carey (2006). "Lee, Tsung-Dao". *American scientists*. Facts on File Inc. p. 225. ISBN 9781438108070.

[21] "The Nobel Prize in Physics 1957". *NobelPrize.org*. Nobel Media. Retrieved 26 February 2011.

[22] "The Nobel Prize in Physics 1980". *NobelPrize.org*. Nobel Media. Retrieved 26 February 2011.

[23] M. Kobayashi, T. Maskawa (1973). "CP-Violation in the Renormalizable Theory of Weak Interaction". *Progress of Theoretical Physics* **49** (2): 652–657. Bibcode:1973PThPh..49..652K. doi:10.1143/PTP.49.652.

[24] "The Nobel Prize in Physics 1980". *NobelPrize.org*. Nobel Media. Retrieved 17 March 2011.

[25] Paul Langacker (2001) [1989]. "Cp Violation and Cosmology". In Cecilia Jarlskog. *CP violation*. London, River Edge: World Scientific Publishing Co. p. 552. ISBN 9789971505615.

7.7.2 General readers

- R. Oerter (2006). *The Theory of Almost Everything: The Standard Model, the Unsung Triumph of Modern Physics*. Plume. ISBN 978-0-13-236678-6.

- B.A. Schumm (2004). *Deep Down Things: The Breathtaking Beauty of Particle Physics*. Johns Hopkins University Press. ISBN 0-8018-7971-X.

7.7.3 Texts

- D.A. Bromley (2000). *Gauge Theory of Weak Interactions*. Springer. ISBN 3-540-67672-4.

- G.D. Coughlan, J.E. Dodd, B.M. Gripaios (2006). *The Ideas of Particle Physics: An Introduction for Scientists* (3rd ed.). Cambridge University Press. ISBN 978-0-521-67775-2.

- W. N. Cottingham; D. A. Greenwood (2001) [1986]. *An introduction to nuclear physics* (2nd ed.). Cambridge University Press. p. 30. ISBN 978-0-521-65733-4.

- D.J. Griffiths (1987). *Introduction to Elementary Particles*. John Wiley & Sons. ISBN 0-471-60386-4.

- G.L. Kane (1987). *Modern Elementary Particle Physics*. Perseus Books. ISBN 0-201-11749-5.

- D.H. Perkins (2000). *Introduction to High Energy Physics*. Cambridge University Press. ISBN 0-521-62196-8.

Chapter 8

Higgs mechanism

In the Standard Model of particle physics, the **Higgs mechanism** is essential to explain the generation mechanism of the property "mass" for gauge bosons. Without the Higgs mechanism, or some other effect like it, all bosons (a type of fundamental particle) would be massless, but measurements show that the W^+, W^-, and Z bosons actually have relatively large masses of around 80 GeV/c^2. The Higgs field resolves this conundrum. The simplest description of the mechanism adds a quantum field (the Higgs field) that permeates all space, to the Standard Model. Below some extremely high temperature, the field causes spontaneous symmetry breaking during interactions. The breaking of symmetry triggers the Higgs mechanism, causing the bosons it interacts with to have mass. In the Standard Model, the phrase "Higgs mechanism" refers specifically to the generation of masses for the $W^\pm$, and Z weak gauge bosons through electroweak symmetry breaking.[1] The Large Hadron Collider at CERN announced results consistent with the Higgs particle on March 14, 2013, making it extremely likely that the field, or one like it, exists, and explaining how the Higgs mechanism takes place in nature.

The mechanism was proposed in 1962 by Philip Warren Anderson,[2] following work in the late 1950s on symmetry breaking in superconductivity and a 1960 paper by Yoichiro Nambu that discussed its application within particle physics. A theory able to finally explain mass generation without "breaking" gauge theory was published almost simultaneously by three independent groups in 1964: by Robert Brout and François Englert;[3] by Peter Higgs;[4] and by Gerald Guralnik, C. R. Hagen, and Tom Kibble.[5][6][7] The Higgs mechanism is therefore also called the **Brout–Englert–Higgs mechanism** or **Englert–Brout–Higgs–Guralnik–Hagen–Kibble mechanism**,[8] **Anderson–Higgs mechanism**,[9] **Anderson–Higgs-Kibble mechanism**,[10] **Higgs–Kibble mechanism** by Abdus Salam[11] and **ABEGHHK'tH mechanism** [for Anderson, Brout, Englert, Guralnik, Hagen, Higgs, Kibble and 't Hooft] by Peter Higgs.[11]

On October 8, 2013, following the discovery at CERN's Large Hadron Collider of a new particle that appeared to be the long-sought Higgs boson predicted by the theory, it was announced that Peter Higgs and François Englert had been awarded the 2013 Nobel Prize in Physics (Englert's co-author Robert Brout had died in 2011 and the Nobel Prize is not usually awarded posthumously).[12]

8.1   Standard model

The Higgs mechanism was incorporated into modern particle physics by Steven Weinberg and Abdus Salam, and is an essential part of the standard model.

In the standard model, at temperatures high enough that electroweak symmetry is unbroken, all elementary particles are massless. At a critical temperature the Higgs field becomes tachyonic, the symmetry is spontaneously broken by condensation, and the W and Z bosons acquire masses. (EWSB, ElectroWeak Symmetry Breaking, is an abbreviation used for this.)

Fermions, such as the leptons and quarks in the Standard Model, can also acquire mass as a result of their interaction with the Higgs field, but not in the same way as the gauge bosons.

8.1.1   Structure of the Higgs field

In the standard model, the Higgs field is an **SU**(2) doublet, a complex scalar with four real components (or equivalently with two complex components). Its (weak hypercharge) **U**(1) charge is 1. That means that it transforms as a spinor under **SU**(2). Under **U**(1) rotations, it is multiplied by a phase, which thus mixes the real and imaginary parts of the complex spinor into each other—so this is *not the same* as two complex spinors mixing under **U**(1) (which would have eight real components between them), but instead is the spinor representation of the group **U**(2).

The Higgs field, through the interactions specified (summarized, represented, or even simulated) by its potential, induces spontaneous breaking of three out of the four generators ("directions") of the gauge group **SU**(2) × **U**(1): three out of its four components would ordinarily amount to Goldstone bosons, if they were not coupled to gauge fields.

However, after symmetry breaking, these three of the four degrees of freedom in the Higgs field mix with the three W and Z bosons (W+, W− and Z), and are only observable as spin components of these weak bosons, which are now massive; while the one remaining degree of freedom becomes the Higgs boson—a new scalar particle.

8.1.2   The photon as the part that remains massless

The gauge group of the electroweak part of the standard model is **SU**(2) × **U**(1). The group **SU**(2) is the group of all 2-by-2 unitary matrices with unit determinant; all the orthonormal changes of coordinates in a complex two dimensional vector space.

Rotating the coordinates so that the second basis vector points in the direction of the Higgs boson makes the vacuum expectation value of H the spinor $(0, v)$. The generators for rotations about the x, y, and z axes are by half the Pauli matrices σx, σy, and σz, so that a rotation of angle θ about the z-axis takes the vacuum to

$$(0, ve^{-i\theta/2}).$$

While the T_x and T_y generators mix up the top and bottom components of the spinor, the T_z rotations only multiply each by opposite phases. This phase can be undone by a **U**(1) rotation of angle $1/2\theta$. Consequently, under both an **SU**(2) T_z-rotation and a **U**(1) rotation by an amount $1/2\theta$, *the vacuum is invariant.*

This combination of generators

$$Q = T_z + \frac{Y}{2}$$

defines the unbroken part of the gauge group, where Q is the electric charge, T_z is the generator of rotations around the z-axis in the **SU**(2) and Y is the hypercharge generator of the **U**(1). This combination of generators (a z rotation in the **SU**(2) and a simultaneous **U**(1) rotation by half the angle) preserves the vacuum, and defines the unbroken gauge group in the standard model, namely *the electric charge* group. The part of the gauge field in this direction stays massless, and amounts to the physical photon.

8.1.3   Consequences for fermions

In spite of the introduction of spontaneous symmetry breaking, the mass terms oppose the chiral gauge invariance. For these fields the mass terms should always be replaced by a gauge-invariant "Higgs" mechanism. One possibility is some kind of "Yukawa coupling" (see below) between the fermion field ψ and the Higgs field Φ, with unknown couplings $G\psi$, which after symmetry breaking (more precisely: after expansion of the Lagrange density around a suitable ground state) again results in the original mass terms, which are now, however (i.e. by introduction of the Higgs field) written in a gauge-invariant way. The Lagrange density for the "Yukawa" interaction of a fermion field ψ and the Higgs field Φ is

$$\mathcal{L}_{\text{Fermion}}(\phi, A, \psi) = \overline{\psi}\gamma^{\mu}D_{\mu}\psi + G_{\psi}\overline{\psi}\phi\psi,$$

where again the gauge field A only enters $D\mu$ (i.e., it is only indirectly visible). The quantities γ^{μ} are the Dirac matrices, and $G\psi$ is the already-mentioned "Yukawa" coupling parameter. Already now the mass-generation follows the same principle as above, namely from the existence of a finite expectation value $|\langle\phi\rangle|$, as described above. Again, this is crucial for the existence of the property "mass".

8.2   History of research

8.2.1   Background

Spontaneous symmetry breaking offered a framework to introduce bosons into relativistic quantum field theories. However, according to Goldstone's theorem, these bosons should be massless.[13] The only observed particles which could be approximately interpreted as Goldstone bosons were the pions, which Yoichiro Nambu related to chiral symmetry breaking.

A similar problem arises with Yang–Mills theory (also known as non-abelian gauge theory), which predicts massless spin-1 gauge bosons. Massless weakly interacting gauge bosons lead to long-range forces, which are only observed for electromagnetism and the corresponding massless photon. Gauge theories of the weak force needed a way to describe massive gauge bosons in order to be consistent.

8.2.2   Discovery

The mechanism was proposed in 1962 by Philip Warren Anderson,[2] who discussed its consequences for particle physics but did not work out an explicit relativistic model. The relativistic model was developed in 1964 by three independent groups – Robert Brout and François Englert;[3] Peter Higgs;[4] and Gerald Guralnik, Carl Richard Hagen, and Tom Kibble.[5][6][7] Slightly later, in 1965, but independently from the other publications[14][15][16][17][18][19] the mechanism was also proposed by Alexander Migdal and Alexander Polyakov,[20] at that time Soviet undergraduate students. However, the paper was delayed by the Editorial Office of JETP, and was published only in 1966.

The mechanism is closely analogous to phenomena previously discovered by Yoichiro Nambu involving the "vacuum structure" of quantum fields in superconductivity.[21] A similar but distinct effect (involving an affine realization of what is now recognized as the Higgs field), known as the Stueckelberg mechanism, had previously been studied by Ernst Stueckelberg.

These physicists discovered that when a gauge theory is combined with an additional field that spontaneously breaks the symmetry group, the gauge bosons can consistently acquire a nonzero mass. In spite of the large values involved (see below) this permits a gauge theory description of the weak force, which was independently developed by Steven Weinberg and Abdus Salam in 1967. Higgs's original article presenting the model was rejected by Physics Letters. When revising the article before resubmitting it to Physical Review Letters, he added a sentence at the end,[22] mentioning that it implies the existence of one or more new, massive scalar bosons, which do not form complete representations of the symmetry group; these are the Higgs bosons.

The three papers by Brout and Englert; Higgs; and Guralnik, Hagen, and Kibble were each recognized as "milestone letters" by *Physical Review Letters* in 2008.[23] While each of these seminal papers took similar approaches, the contributions and differences among the 1964 PRL symmetry breaking papers are noteworthy. All six physicists were jointly awarded the 2010 J. J. Sakurai Prize for Theoretical Particle Physics for this work.[24]

Benjamin W. Lee is often credited with first naming the "Higgs-like" mechanism, although there is debate around when this first occurred.[25][26][27] One of the first times the *Higgs* name appeared in print was in 1972 when Gerardus 't Hooft and Martinus J. G. Veltman referred to it as the "Higgs–Kibble mechanism" in their Nobel winning paper.[28][29]

Philip W. Anderson, the first to propose the mechanism in 1962.

8.3 Examples

The Higgs mechanism occurs whenever a charged field has a vacuum expectation value. In the nonrelativistic context, this is the Landau model of a charged Bose–Einstein condensate, also known as a superconductor. In the relativistic condensate, the condensate is a scalar field, and is relativistically invariant.

Five of the six 2010 APS Sakurai Prize Winners – (L to R) Tom Kibble, Gerald Guralnik, Carl Richard Hagen, François Englert, and Robert Brout

8.3.1   Landau model

The Higgs mechanism is a type of superconductivity which occurs in the vacuum. It occurs when all of space is filled with a sea of particles which are charged, or, in field language, when a charged field has a nonzero vacuum expectation value. Interaction with the quantum fluid filling the space prevents certain forces from propagating over long distances (as it does in a superconducting medium; e.g., in the Ginzburg–Landau theory).

A superconductor expels all magnetic fields from its interior, a phenomenon known as the Meissner effect. This was mysterious for a long time, because it implies that electromagnetic forces somehow become short-range inside the superconductor. Contrast this with the behavior of an ordinary metal. In a metal, the conductivity shields electric fields by rearranging charges on the surface until the total field cancels in the interior. But magnetic fields can penetrate to any distance, and if a magnetic monopole (an isolated magnetic pole) is surrounded by a metal the field can escape without collimating into a string. In a superconductor, however, electric charges move with no dissipation, and this allows for permanent surface currents, not just surface charges. When magnetic fields are introduced at the boundary of a superconductor, they produce surface currents which exactly neutralize them. The Meissner effect is due to currents in a thin surface layer, whose thickness, the London penetration depth, can be calculated from a simple model (the Ginzburg–Landau theory).

This simple model treats superconductivity as a charged Bose–Einstein condensate. Suppose that a superconductor contains bosons with charge q. The wavefunction of the bosons can be described by introducing a quantum field, ψ, which obeys the Schrödinger equation as a field equation (in units where the reduced Planck constant, $\hbar$, is set to 1):

$$i\frac{\partial}{\partial t}\psi = \frac{(\nabla - iqA)^2}{2m}\psi.$$

The operator $\psi(x)$ annihilates a boson at the point x, while its adjoint $\psi^\dagger$ creates a new boson at the same point. The

Number six: Peter Higgs 2009

wavefunction of the Bose–Einstein condensate is then the expectation value ψ of $\psi(x)$, which is a classical function that obeys the same equation. The interpretation of the expectation value is that it is the phase that one should give to a newly created boson so that it will coherently superpose with all the other bosons already in the condensate.

When there is a charged condensate, the electromagnetic interactions are screened. To see this, consider the effect of a gauge transformation on the field. A gauge transformation rotates the phase of the condensate by an amount which changes from point to point, and shifts the vector potential by a gradient:

$$\psi \to e^{iq\phi(x)}\psi$$
$$A \to A + \nabla\phi.$$

When there is no condensate, this transformation only changes the definition of the phase of ψ at every point. But when there is a condensate, the phase of the condensate defines a preferred choice of phase.

The condensate wave function can be written as

$$\psi(x) = \rho(x)\,e^{i\theta(x)},$$

where ρ is real amplitude, which determines the local density of the condensate. If the condensate were neutral, the flow would be along the gradients of θ, the direction in which the phase of the Schrödinger field changes. If the phase θ changes slowly, the flow is slow and has very little energy. But now θ can be made equal to zero just by making a gauge transformation to rotate the phase of the field.

The energy of slow changes of phase can be calculated from the Schrödinger kinetic energy,

$$H = \frac{1}{2m}|(qA + \nabla)\psi|^2,$$

and taking the density of the condensate ρ to be constant,

$$H \approx \frac{\rho^2}{2m}(qA + \nabla\theta)^2.$$

Fixing the choice of gauge so that the condensate has the same phase everywhere, the electromagnetic field energy has an extra term,

$$\frac{q^2\rho^2}{2m}A^2.$$

When this term is present, electromagnetic interactions become short-ranged. Every field mode, no matter how long the wavelength, oscillates with a nonzero frequency. The lowest frequency can be read off from the energy of a long wavelength A mode,

$$E \approx \frac{\dot{A}^2}{2} + \frac{q^2\rho^2}{2m}A^2.$$

This is a harmonic oscillator with frequency

$$\sqrt{\frac{1}{m}q^2\rho^2}.$$

The quantity $|\psi|^2$ ($=\rho^2$) is the density of the condensate of superconducting particles.

In an actual superconductor, the charged particles are electrons, which are fermions not bosons. So in order to have superconductivity, the electrons need to somehow bind into Cooper pairs. The charge of the condensate q is therefore twice the electron charge e. The pairing in a normal superconductor is due to lattice vibrations, and is in fact very weak; this means that the pairs are very loosely bound. The description of a Bose–Einstein condensate of loosely bound pairs is actually more difficult than the description of a condensate of elementary particles, and was only worked out in 1957 by Bardeen, Cooper and Schrieffer in the famous BCS theory.

8.3.2   Abelian Higgs mechanism

Gauge invariance means that certain transformations of the gauge field do not change the energy at all. If an arbitrary gradient is added to A, the energy of the field is exactly the same. This makes it difficult to add a mass term, because a mass term tends to push the field toward the value zero. But the zero value of the vector potential is not a gauge invariant idea. What is zero in one gauge is nonzero in another.

So in order to give mass to a gauge theory, the gauge invariance must be broken by a condensate. The condensate will then define a preferred phase, and the phase of the condensate will define the zero value of the field in a gauge-invariant way. The gauge-invariant definition is that a gauge field is zero when the phase change along any path from parallel transport is equal to the phase difference in the condensate wavefunction.

The condensate value is described by a quantum field with an expectation value, just as in the Ginzburg-Landau model.

In order for the phase of the vacuum to define a gauge, the field must have a phase (also referred to as 'to be charged'). In order for a scalar field Φ to have a phase, it must be complex, or (equivalently) it should contain two fields with a symmetry which rotates them into each other. The vector potential changes the phase of the quanta produced by the field when they move from point to point. In terms of fields, it defines how much to rotate the real and imaginary parts of the fields into each other when comparing field values at nearby points.

The only renormalizable model where a complex scalar field Φ acquires a nonzero value is the Mexican-hat model, where the field energy has a minimum away from zero. The action for this model is

$$S(\phi) = \int \frac{1}{2}|\partial\phi|^2 - \lambda\left(|\phi|^2 - \Phi^2\right)^2,$$

which results in the Hamiltonian

$$H(\phi) = \frac{1}{2}|\dot{\phi}|^2 + |\nabla\phi|^2 + V(|\phi|).$$

The first term is the kinetic energy of the field. The second term is the extra potential energy when the field varies from point to point. The third term is the potential energy when the field has any given magnitude.

This potential energy, $V(z, \Phi) = \lambda(|z|^2 - \Phi^2)^2$,[30] has a graph which looks like a Mexican hat, which gives the model its name. In particular, the minimum energy value is not at $z = 0$, but on the circle of points where the magnitude of z is Φ.

When the field $\Phi(x)$ is not coupled to electromagnetism, the Mexican-hat potential has flat directions. Starting in any one of the circle of vacua and changing the phase of the field from point to point costs very little energy. Mathematically, if

$$\phi(x) = \Phi e^{i\theta(x)}$$

with a constant prefactor, then the action for the field $\theta(x)$, i.e., the "phase" of the Higgs field $\Phi(x)$, has only derivative terms. This is not a surprise. Adding a constant to $\theta(x)$ is a symmetry of the original theory, so different values of $\theta(x)$ cannot have different energies. This is an example of Goldstone's theorem: spontaneously broken continuous symmetries normally produce massless excitations.

The Abelian Higgs model is the Mexican-hat model coupled to electromagnetism:

$$S(\phi, A) = \int -\frac{1}{4}F^{\mu\nu}F_{\mu\nu} + |(\partial - iqA)\phi|^2 - \lambda(|\phi|^2 - \Phi^2)^2.$$

The classical vacuum is again at the minimum of the potential, where the magnitude of the complex field φ is equal to Φ. But now the phase of the field is arbitrary, because gauge transformations change it. This means that the field $\theta(x)$ can be set to zero by a gauge transformation, and does not represent any actual degrees of freedom at all.

Furthermore, choosing a gauge where the phase of the vacuum is fixed, the potential energy for fluctuations of the vector field is nonzero. So in the abelian Higgs model, the gauge field acquires a mass. To calculate the magnitude of the mass, consider a constant value of the vector potential A in the x direction in the gauge where the condensate has constant phase. This is the same as a sinusoidally varying condensate in the gauge where the vector potential is zero. In the gauge where A is zero, the potential energy density in the condensate is the scalar gradient energy:

$$E = \frac{1}{2}\left|\partial\left(\Phi e^{iqAx}\right)\right|^2 = \frac{1}{2}q^2\Phi^2A^2.$$

This energy is the same as a mass term $1/2m^2A^2$ where $m = q\Phi$.

8.3.3 Nonabelian Higgs mechanism

The Nonabelian Higgs model has the following action:

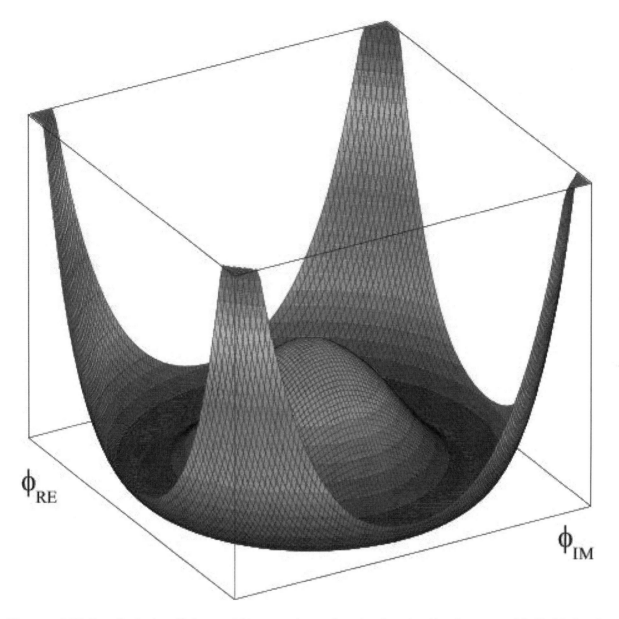

Higgs potential V. *For a fixed value of* λ *the potential is presented upwards against the real and imaginary parts of* Φ. *The* Mexican-hat *or* champagne-bottle profile *at the ground should be noted.*

$$S(\phi, \mathbf{A}) = \int \frac{1}{4g^2} \mathrm{tr}(F^{\mu\nu} F_{\mu\nu}) + |D\phi|^2 + V(|\phi|)$$

where now the nonabelian field $\mathbf{A}$ is contained in D and in the tensor components $F^{\mu\nu}$ and $F_{\mu\nu}$ (the relation between $\mathbf{A}$ and those components is well-known from the Yang–Mills theory).

It is exactly analogous to the Abelian Higgs model. Now the field φ is in a representation of the gauge group, and the gauge covariant derivative is defined by the rate of change of the field minus the rate of change from parallel transport using the gauge field A as a connection.

$$D\phi = \partial\phi - iA^k t_k \phi$$

Again, the expectation value of Φ defines a preferred gauge where the vacuum is constant, and fixing this gauge, fluctuations in the gauge field A come with a nonzero energy cost.

Depending on the representation of the scalar field, not every gauge field acquires a mass. A simple example is in the renormalizable version of an early electroweak model due to Julian Schwinger. In this model, the gauge group is **SO**(3) (or **SU**(2) – there are no spinor representations in the model), and the gauge invariance is broken down to **U**(1) or **SO**(2) at long distances. To make a consistent renormalizable version using the Higgs mechanism, introduce a scalar field φ^a which transforms as a vector (a triplet) of **SO**(3). If this field has a vacuum expectation value, it points in some direction in field space. Without loss of generality, one can choose the *z*-axis in field space to be the direction that φ is pointing, and then the vacuum expectation value of φ is $(0, 0, A)$, where A is a constant with dimensions of mass ($c = \hbar = 1$).

Rotations around the *z*-axis form a **U**(1) subgroup of **SO**(3) which preserves the vacuum expectation value of φ, and this is the unbroken gauge group. Rotations around the *x* and *y*-axis do not preserve the vacuum, and the components of the **SO**(3) gauge field which generate these rotations become massive vector mesons. There are two massive W mesons in the Schwinger model, with a mass set by the mass scale A, and one massless **U**(1) gauge boson, similar to the photon.

The Schwinger model predicts magnetic monopoles at the electroweak unification scale, and does not predict the Z meson. It doesn't break electroweak symmetry properly as in nature. But historically, a model similar to this (but not using the Higgs mechanism) was the first in which the weak force and the electromagnetic force were unified.

8.3.4 Affine Higgs mechanism

Ernst Stueckelberg discovered[31] a version of the Higgs mechanism by analyzing the theory of quantum electrodynamics with a massive photon. Effectively, Stueckelberg's model is a limit of the regular Mexican hat Abelian Higgs model, where the vacuum expectation value H goes to infinity and the charge of the Higgs field goes to zero in such a way that their product stays fixed. The mass of the Higgs boson is proportional to H, so the Higgs boson becomes infinitely massive and decouples, so is not present in the discussion. The vector meson mass, however, equals to the product eH, and stays finite.

The interpretation is that when a **U**(1) gauge field does not require quantized charges, it is possible to keep only the angular part of the Higgs oscillations, and discard the radial part. The angular part of the Higgs field θ has the following gauge transformation law:

$$\theta \to \theta + e\alpha$$

$$A \to A + \alpha.$$

The gauge covariant derivative for the angle (which is actually gauge invariant) is:

$$D\theta = \partial\theta - eA.$$

In order to keep θ fluctuations finite and nonzero in this limit, θ should be rescaled by H, so that its kinetic term in the action stays normalized. The action for the theta field is read off from the Mexican hat action by substituting $\phi = He^{\frac{1}{H}i\theta}$
.

$$S = \int \frac{1}{4}F^2 + \frac{1}{2}(D\theta)^2 = \int \frac{1}{4}F^2 + \frac{1}{2}(\partial\theta - HeA)^2 = \int \frac{1}{4}F^2 + \frac{1}{2}(\partial\theta - mA)^2$$

since eH is the gauge boson mass. By making a gauge transformation to set θ = 0, the gauge freedom in the action is eliminated, and the action becomes that of a massive vector field:

$$S = \int \frac{1}{4}F^2 + \frac{1}{2}m^2A^2.$$

To have arbitrarily small charges requires that the $\mathbf{U}(1)$ is not the circle of unit complex numbers under multiplication, but the real numbers $\mathbf{R}$ under addition, which is only different in the global topology. Such a $\mathbf{U}(1)$ group is *non-compact*. The field θ transforms as an affine representation of the gauge group. Among the allowed gauge groups, only non-compact $\mathbf{U}(1)$ admits affine representations, and the $\mathbf{U}(1)$ of electromagnetism is experimentally known to be compact, since charge quantization holds to extremely high accuracy.

The Higgs condensate in this model has infinitesimal charge, so interactions with the Higgs boson do not violate charge conservation. The theory of quantum electrodynamics with a massive photon is still a renormalizable theory, one in which electric charge is still conserved, but magnetic monopoles are not allowed. For nonabelian gauge theory, there is no affine limit, and the Higgs oscillations cannot be too much more massive than the vectors.

8.4   See also

- Electromagnetic mass

- Higgs bundle

- Mass generation

- QCD vacuum

- Quantum triviality

- Top quark condensate

- Yang–Mills–Higgs equations

8.5   References

[1] G. Bernardi, M. Carena, and T. Junk: "Higgs bosons: theory and searches", Reviews of Particle Data Group: Hypothetical particles and Concepts, 2007, http://pdg.lbl.gov/2008/reviews/higgs_s055.pdf

[2] P. W. Anderson (1962). "Plasmons, Gauge Invariance, and Mass". *Physical Review* **130**(1): 439–442. Bibcode:1963PhRv..1. doi:10.1103/PhysRev.130.439.

[3] F. Englert and R. Brout (1964). "Broken Symmetry and the Mass of Gauge Vector Mesons". *Physical Review Letters* **13** (9): 321–323. Bibcode:1964PhRvL..13..321E. doi:10.1103/PhysRevLett.13.321.

[4] Peter W. Higgs (1964). "Broken Symmetries and the Masses of Gauge Bosons". *Physical Review Letters* **13** (16): 508–509. Bibcode:1964PhRvL..13..508H. doi:10.1103/PhysRevLett.13.508.

[5] G. S. Guralnik, C. R. Hagen, and T. W. B. Kibble (1964). "Global Conservation Laws and Massless Particles". *Physical Review Letters* **13** (20): 585–587. Bibcode:1964PhRvL..13..585G. doi:10.1103/PhysRevLett.13.585.

[6] Gerald S. Guralnik (2009). "The History of the Guralnik, Hagen and Kibble development of the Theory of Spontaneous Symmetry Breaking and Gauge Particles". *International Journal of Modern Physics* **A24** (14): 2601–2627. arXiv:0907.3466. Bibcode:2009IJMPA..24.2601G. doi:10.1142/S0217751X09045431.

[7] History of Englert–Brout–Higgs–Guralnik–Hagen–Kibble Mechanism. Scholarpedia.

[8] "Englert–Brout–Higgs–Guralnik–Hagen–Kibble Mechanism". Scholarpedia. Retrieved 2012-06-16.

[9] Liu, G. Z.; Cheng, G. (2002). "Extension of the Anderson-Higgs mechanism". *Physical Review B* **65** (13): 132513. arXiv:cond-mat/0106070. Bibcode:2002PhRvB..65m2513L. doi:10.1103/PhysRevB.65.132513.

[10] Matsumoto, H.; Papastamatiou, N. J.; Umezawa, H.; Vitiello, G. (1975). "Dynamical rearrangement in the Anderson-Higgs-Kibble mechanism". *Nuclear Physics B* **97**: 61. doi:10.1016/0550-3213(75)90215-1.

[11] Close, Frank (2011). *The Infinity Puzzle: Quantum Field Theory and the Hunt for an Orderly Universe*. Oxford: Oxford University Press. ISBN 978-0-19-959350-7.

[12] "Press release from Royal Swedish Academy of Sciences" (PDF). 8 October 2013. Retrieved 8 October 2013.

[13] "Guralnik, G S; Hagen, C R and Kibble, T W B (1967). Broken Symmetries and the Goldstone Theorem. Advances in Physics, vol. 2" (PDF).

[14] A.M. Polyakov, A View From The Island, 1992

[15] Farhi, E., & Jackiw, R. W. (1982). *Dynamical Gauge Symmetry Breaking: A Collection Of Reprints.* Singapore: World Scientific Pub. Co.

[16] Frank Close. "The Infinity Puzzle." 2011, p.158

[17] Norman Dombey, "Higgs Boson: Credit Where It's Due". The Guardian, July 6, 2012

[18] Cern Courier, Mar 1, 2006

[19] Sean Carrol, "The Particle At The End Of The Universe: The Hunt For The Higgs And The Discovery Of A New World", 2012, p.228

[20] A. A. Migdal and A. M. Polyakov, "Spontaneous Breakdown of Strong Interaction Symmetry and Absence of Massless Particles", *JETP* **51**, 135, July 1966 (English translation: *Soviet Physics JETP*, **24**, 1, January 1967)

[21] Nambu, Y (1960). "Quasiparticles and Gauge Invariance in the Theory of Superconductivity". *Physical Review* **117** (3): 648–663. Bibcode:1960PhRv..117..648N. doi:10.1103/PhysRev.117.648.

[22] Higgs, Peter (2007). "Prehistory of the Higgs boson". *Comptes Rendus Physique* **8** (9): 970–972. Bibcode:2007CRPhy...8..970H. doi:10.1016/j.crhy.2006.12.006.

[23] "Physical Review Letters – 50th Anniversary Milestone Papers". Prl.aps.org. Retrieved 2012-06-16.

[24] "American Physical Society – J. J. Sakurai Prize Winners". Aps.org. Retrieved 2012-06-16.

[25] Department of Physics and Astronomy. "Rochester's Hagen Sakurai Prize Announcement". Pas.rochester.edu. Retrieved 2012-06-16.

[26] FermiFred (2010-02-15). "C.R. Hagen discusses naming of Higgs Boson in 2010 Sakurai Prize Talk". Youtube.com. Retrieved 2012-06-16.

[27] Sample, Ian (2009-05-29). "Anything but the God particle by Ian Sample". Guardian. Retrieved 2012-06-16.

[28] G. 't Hooft and M. Veltman (1972). "Regularization and Renormalization of Gauge Fields". *Nuclear Physics B* **44** (1): 189–219. Bibcode:1972NuPhB..44..189T. doi:10.1016/0550-3213(72)90279-9.

[29] "Regularization and Renormalization of Gauge Fields by t'Hooft and Veltman (PDF)" (PDF). Retrieved 2012-06-16.

[30] Goldstone, J. (1961). "Field theories with " Superconductor " solutions". *Il Nuovo Cimento* **19**: 154–164. doi:10.1007/BF.

[31] Stueckelberg, E. C. G. (1938), "Die Wechselwirkungskräfte in der Elektrodynamik und in der Feldtheorie der Kräfte", *Helv. Phys. Acta.* **11:** 225

8.6 Further reading

- Schumm, Bruce A. (2004) *Deep Down Things.* Johns Hopkins Univ. Press. Chpt. 9.

- Englert-Brout-Higgs-Guralnik-Hagen-Kibble mechanism Tom W B Kibble Scholarpedia, 4(1):6441. doi:10.4249/scholarpedia.6441

8.7   External links

- Guralnik, G.S.; Hagen, C.R.; Kibble, T.W.B. (1964). "Global Conservation Laws and Massless Particles". *Physical Review Letters* **13** (20): 585–87. Bibcode:1964PhRvL..13..585G. doi:10.1103/PhysRevLett.13.585.

- Mark D. Roberts (1999) "A Generalized Higgs Model"

- 2010 Sakurai Prize - All Events - YouTube

- From BCS to the LHC - CERN Courier Jan 21, 2008, Steven Weinberg, University of Texas at Austin.

- Higgs, dark matter and supersymmetry: What the Large Hadron Collider will tell us (Steven Weinberg) - YouTube on YouTube 06-11-2009

- Gerry Guralnik speaks at Brown University about the 1964 PRL papers

- Guralnik, Gerald (2013). "Heretical Ideas that Provided the Cornerstone for the Standard Model of Particle Physics". SPG MITTEILUNGEN March 2013, No. 39, (p. 14)

- Steven Weinberg Praises Teams for Higgs Boson Theory

- Physical Review Letters – 50th Anniversary Milestone Papers

- Imperial College London on PRL 50th Anniversary Milestone Papers

- Englert–Brout–Higgs–Guralnik–Hagen–Kibble Mechanism on Scholarpedia

- History of Englert–Brout–Higgs–Guralnik–Hagen–Kibble Mechanism on Scholarpedia

- The Hunt for the Higgs at Tevatron

- The Mystery of Empty Space on YouTube. A lecture with UCSD physicist Kim Griest (43 minutes)

Chapter 9

Branching fraction

In particle physics and nuclear physics, the **branching fraction** for a decay is the fraction of particles which decay by an individual decay mode with respect to the total number of particles which decay.[1] It is equal to the ratio of the **partial decay constant** to the overall decay constant. Sometimes a **partial half-life** is given, but this term is misleading; due to competing modes it is not true that half of the particles will decay through a particular decay mode after its partial half-life. The partial half-life is merely an alternate way to specify the partial decay constant λ, the two being related through:

$$t_{1/2} = \frac{\ln 2}{\lambda}.$$

For example, for spontaneous decays of ^{132}Cs, 98.1% are ε or β^+ decays, and 1.9% are β^- decays. The partial decay constants can be calculated from the branching fraction and the half-life of ^{132}Cs (6.479 d), they are: 0.10 d^{-1} ($\varepsilon + \beta^+$) and .0020 d^{-1} (β^-). The partial half-lives are 6.60 d ($\varepsilon + \beta^+$) and 341 d (β^-). Here the problem with the term partial half-life is evident: after (341+6.60) days almost all the nuclei will have decayed, not only half as one may initially think.

Isotopes with significant branching of decay modes include copper-64, arsenic-74, rhodium-102, indium-112, iodine-126 and holmium-164.

9.1 References

[1] IUPAC, *Compendium of Chemical Terminology*, 2nd ed. (the "Gold Book") (1997). Online corrected version: (2006–) "branching fraction".

9.2 External links

- NUCLEONICA Nuclear Science Portal

- NUCLEONICA wiki: Decay Engine

- LBNL Isotopes Project

- Particle Data Group (listings for particle physics)

- **Nuclear Structure and Decay Data - IAEA** for nuclear decays

Chapter 10

Excited state

"Excited" redirects here. For other uses, see Excited (disambiguation).

Excitation is an elevation in energy level above an arbitrary baseline energy state. In physics there is a specific technical

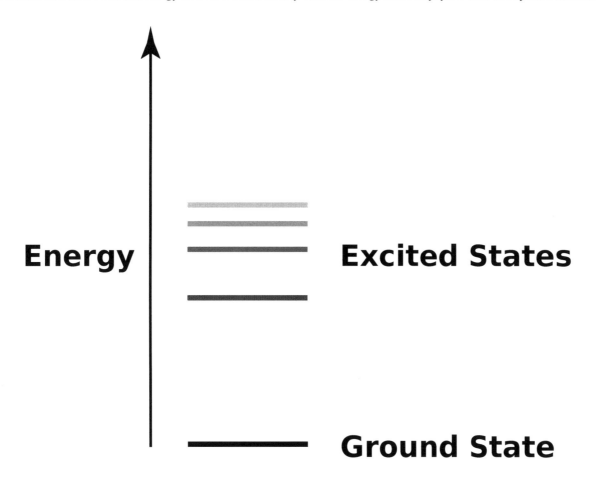

After absorbing energy, an electron may jump from the ground state to a higher energy excited state.

definition for energy level which is often associated with an atom being raised to an excited state.

In quantum mechanics an **excited state** of a system (such as an atom, molecule or nucleus) is any quantum state of the system that has a higher energy than the ground state (that is, more energy than the absolute minimum). The temperature

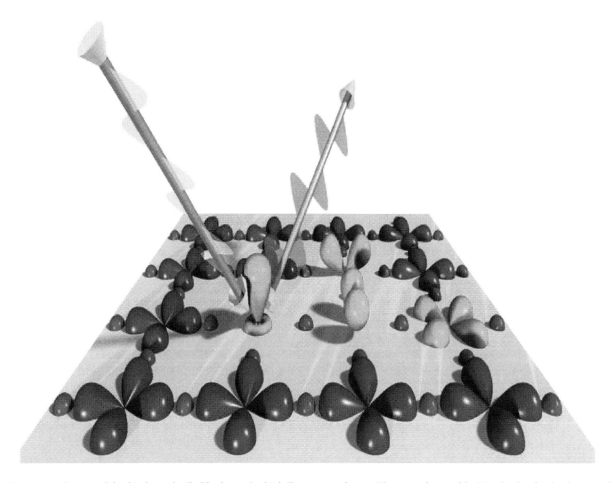

Excitations of copper 3d orbitals on the CuO2-plane of a high Tc superconductor; The ground state (blue) is x2-y2 orbitals; the excited orbitals are in green; the arrows illustrate inelastic x-ray spectroscopy

of a group of particles is indicative of the level of excitation (with the notable exception of systems that exhibit Negative temperature).

The lifetime of a system in an excited state is usually short: spontaneous or induced emission of a quantum of energy (such as a photon or a phonon) usually occurs shortly after the system is promoted to the excited state, returning the system to a state with lower energy (a less excited state or the ground state). This return to a lower energy level is often loosely described as decay and is the inverse of excitation.

Long-lived excited states are often called metastable. Long-lived nuclear isomers and singlet oxygen are two examples of this.

10.1 Atomic excitation

A simple example of this concept comes by considering the hydrogen atom.

The ground state of the hydrogen atom corresponds to having the atom's single electron in the lowest possible orbit (that is, the spherically symmetric "1s" wavefunction, which has the lowest possible quantum numbers). By giving the atom additional energy (for example, by the absorption of a photon of an appropriate energy), the electron is able to move into an excited state (one with one or more quantum numbers greater than the minimum possible). If the photon has too much energy, the electron will cease to be bound to the atom, and the atom will become ionised.

After excitation the atom may return to the ground state or a lower excited state, by emitting a photon with a characteristic energy. Emission of photons from atoms in various excited states leads to an electromagnetic spectrum showing a series of characteristic emission lines (including, in the case of the hydrogen atom, the Lyman, Balmer, Paschen and Brackett series.)

An atom in a high excited state is termed Rydberg atom. A system of highly excited atoms can form a long-lived condensed excited state e.g. a condensed phase made completely of excited atoms: Rydberg matter. Hydrogen can also be excited by heat or electricity.

10.2   Perturbed gas excitation

A collection of molecules forming a gas can be considered in an excited state if one or more molecules are elevated to kinetic energy levels such that the resulting velocity distribution departs from the equilibrium Boltzmann distribution. This phenomenon has been studied in the case of a two-dimensional gas in some detail, analyzing the time taken to relax to equilibrium.

10.3   Calculation of excited states

Excited states are often calculated using Coupled cluster, Møller–Plesset perturbation theory, Multi-configurational self-consistent field, Configuration interaction,[1] and Time-dependent density functional theory. These calculations are more difficult than non-excited state calculations.[2][3][4][5][6]

10.4   Reaction

A further consequence is reaction of the atom in the excited state, as in photochemistry. Excited states give rise to chemical reaction.

10.5   See also

- Rydberg formula

- Stationary state

- Repulsive state

10.6   References

[1] Hehre, Warren J. (2003). *A Guide to Molecular Mechanics and Quantum Chemical Calculations* (PDF). Irvine, California: Wavefunction, Inc. ISBN 1-890661-06-6.

[2] Glaesemann, Kurt R.; Govind, Niranjan; Krishnamoorthy, Sriram; Kowalski, Karol (2010). "EOMCC, MRPT, and TDDFT Studies of Charge Transfer Processes in Mixed-Valence Compounds: Application to the Spiro Molecule". *The Journal of Physical Chemistry A* **114** (33): 8764–8771. doi:10.1021/jp101761d. PMID 20540550.

[3] Dreuw, Andreas; Head-Gordon, Martin (2005). "Single-Reference ab Initio Methods for the Calculation of Excited States of Large Molecules". *Chemical Reviews* **105** (11): 4009–37. doi:10.1021/cr0505627. PMID 16277369.

[4] Knowles, Peter J.; Werner, Hans-Joachim (1992). "Internally contracted multiconfiguration-reference configuration interaction calculations for excited states". *Theoretica Chimica Acta* **84**: 95. doi:10.1007/BF01117405.

[5] Foresman, James B.; Head-Gordon, Martin; Pople, John A.; Frisch, Michael J. (1992). "Toward a systematic molecular orbital theory for excited states". *The Journal of Physical Chemistry* **96**: 135. doi:10.1021/j100180a030.

[6] Glaesemann, Kurt R.; Gordon, Mark S.; Nakano, Haruyuki (1999). "A study of FeCO+ with correlated wavefunctions". *Physical Chemistry Chemical Physics* **1** (6): 967–975. Bibcode:1999PCCP....1..967G. doi:10.1039/a808518h.

10.7 External links

- NASA background information on ground and excited states

Chapter 11

Scalar boson

A **scalar boson** is a boson whose spin equals zero. *Boson* means that it has an integer-valued spin; the *scalar* fixes this value to 0.

The name "scalar boson" arises from quantum field theory. It refers to the particular transformation properties under Lorentz transformation.

11.1 Examples

- Various known composite particles are scalar bosons, e.g. the alpha particle and the pi meson. Among the scalar mesons, one distinguishes between the scalar and pseudoscalar mesons, which refers to their transformation property under parity.

- The only fundamental scalar boson in the standard model of elementary particle physics is the Higgs boson, whose existence was confirmed on 14 March 2013 at the Large Hadron Collider. As a result of this confirmation, the 2013 Nobel Prize in physics was awarded to Peter Higgs and François Englert.

- One very popular quantum field theory, which uses scalar bosonic fields and is introduced in many introductory books to quantum field theories[1] for pedagogical reasons, is the so-called φ^4-theory. It usually serves as a toy model to introduce the basic concepts of the field.

11.2 See also

- Scalar meson

- Scalar field theory

- Vector boson

- Higgs Boson

11.3 References

[1] Michael E. Peskin and Daniel V. Schroeder (1995). *An Introduction to Quantum Field Theory*. Westview Press. ISBN 0-201-50397-2.

Chapter 12

Spin (physics)

This article is about spin in quantum mechanics. For rotation in classical mechanics, see angular momentum.

In quantum mechanics and particle physics, **spin** is an intrinsic form of angular momentum carried by elementary particles, composite particles (hadrons), and atomic nuclei.[1][2]

Spin is one of two types of angular momentum in quantum mechanics, the other being *orbital angular momentum*. The orbital angular momentum operator is the quantum-mechanical counterpart to the classical notion of angular momentum: it arises when a particle executes a rotating or twisting trajectory (such as when an electron orbits a nucleus).[3][4] The existence of spin angular momentum is inferred from experiments, such as the Stern–Gerlach experiment, in which particles are observed to possess angular momentum that cannot be accounted for by orbital angular momentum alone.[5]

In some ways, spin is like a vector quantity; it has a definite magnitude, and it has a "direction" (but quantization makes this "direction" different from the direction of an ordinary vector). All elementary particles of a given kind have the same magnitude of spin angular momentum, which is indicated by assigning the particle a *spin quantum number*.[2]

The SI unit of spin is the joule-second, just as with classical angular momentum. In practice, however, it is written as a multiple of the reduced Planck constant $\hbar$, usually in natural units, where the $\hbar$ is omitted, resulting in a unitless number. Spin quantum numbers are unitless numbers by definition.

When combined with the spin-statistics theorem, the spin of electrons results in the Pauli exclusion principle, which in turn underlies the periodic table of chemical elements.

Wolfgang Pauli was the first to propose the concept of spin, but he did not name it. In 1925, Ralph Kronig, George Uhlenbeck and Samuel Goudsmit at Leiden University suggested a physical interpretation of particles spinning around their own axis. The mathematical theory was worked out in depth by Pauli in 1927. When Paul Dirac derived his relativistic quantum mechanics in 1928, electron spin was an essential part of it.

12.1 Quantum number

Main article: Spin quantum number

As the name suggests, spin was originally conceived as the rotation of a particle around some axis. This picture is correct so far as spin obeys the same mathematical laws as quantized angular momenta do. On the other hand, spin has some peculiar properties that distinguish it from orbital angular momenta:

- Spin quantum numbers may take half-integer values.

- Although the direction of its spin can be changed, an elementary particle cannot be made to spin faster or slower.

- The spin of a charged particle is associated with a magnetic dipole moment with a g-factor differing from 1. This could only occur classically if the internal charge of the particle were distributed differently from its mass.

The conventional definition of the **spin quantum number**, s, is $s = n/2$, where n can be any non-negative integer. Hence the allowed values of s are 0, 1/2, 1, 3/2, 2, etc. The value of s for an elementary particle depends only on the type of particle, and cannot be altered in any known way (in contrast to the *spin direction* described below). The spin angular momentum, S, of any physical system is quantized. The allowed values of S are:

$$S = \frac{h}{2\pi}\sqrt{s(s+1)} = \frac{h}{4\pi}\sqrt{n(n+2)},$$

where h is the Planck constant. In contrast, orbital angular momentum can only take on integer values of s; i.e., even-numbered values of n.

12.1.1   Fermions and bosons

Those particles with half-integer spins, such as 1/2, 3/2, 5/2, are known as fermions, while those particles with integer spins, such as 0, 1, 2, are known as bosons. The two families of particles obey different rules and *broadly* have different roles in the world around us. A key distinction between the two families is that fermions obey the Pauli exclusion principle; that is, there cannot be two identical fermions simultaneously having the same quantum numbers (meaning, roughly, having the same position, velocity and spin direction). In contrast, bosons obey the rules of Bose–Einstein statistics and have no such restriction, so they may "bunch together" even if in identical states. Also, composite particles can have spins different from the particles which comprise them. For example, a helium atom can have spin 0 and therefore can behave like a boson even though the quarks and electrons which make it up are all fermions.

This has profound practical applications:

- Quarks and leptons (including electrons and neutrinos), which make up what is classically known as matter, are all fermions with spin 1/2. The common idea that "matter takes up space" actually comes from the Pauli exclusion principle acting on these particles to prevent the fermions that make up matter from being in the same quantum state. Further compaction would require electrons to occupy the same energy states, and therefore a kind of pressure (sometimes known as degeneracy pressure of electrons) acts to resist the fermions being overly close. It is also this pressure which prevents stars collapsing inwardly, and which, when it finally gives way under immense gravitational pressure in a dying massive star, triggers inward collapse and the dramatic explosion into a supernova.

  Elementary fermions with other spins (3/2, 5/2 etc.) are not known to exist, as of 2014.

- Elementary particles which are thought of as carrying forces are all bosons with spin 1. They include the photon which carries the electromagnetic force, the gluon (strong force), and the W and Z bosons (weak force). The ability of bosons to occupy the same quantum state is used in the laser, which aligns many photons having the same quantum number (the same direction and frequency), superfluid liquid helium resulting from helium-4 atoms being bosons, and superconductivity where pairs of electrons (which individually are fermions) act as single composite bosons.

  Elementary bosons with other spins (0, 2, 3 etc.) were not historically known to exist, although they have received considerable theoretical treatment and are well established within their respective mainstream theories. In particular theoreticians have proposed the graviton (predicted to exist by some quantum gravity theories) with spin 2, and the Higgs boson (explaining electroweak symmetry breaking) with spin 0. Since 2013 the Higgs boson with spin 0 has been considered proven to exist. It is the first scalar particle (spin 0) known to exist in nature.

Theoretical and experimental studies have shown that the spin possessed by elementary particles cannot be explained by postulating that they are made up of even smaller particles rotating about a common center of mass analogous to a classical electron radius; as far as can be presently determined, these elementary particles have no inner structure. The spin of an elementary particle is therefore seen as a truly intrinsic physical property, akin to the particle's electric charge and rest mass.

12.1.2 Spin-statistics theorem

The proof that particles with half-integer spin (fermions) obey Fermi–Dirac statistics and the Pauli Exclusion Principle, and particles with integer spin (bosons) obey Bose–Einstein statistics, occupy "symmetric states", and thus can share quantum states, is known as the spin-statistics theorem. The theorem relies on both quantum mechanics and the theory of special relativity, and this connection between spin and statistics has been called "one of the most important applications of the special relativity theory".[6]

12.2 Magnetic moments

Main article: Spin magnetic moment

Particles with spin can possess a magnetic dipole moment, just like a rotating electrically charged body in classical electrodynamics. These magnetic moments can be experimentally observed in several ways, e.g. by the deflection of particles by inhomogeneous magnetic fields in a Stern–Gerlach experiment, or by measuring the magnetic fields generated by the particles themselves.

The intrinsic magnetic moment $\boldsymbol{\mu}$ of a spin-1/2 particle with charge q, mass m, and spin angular momentum $\mathbf{S}$, is[7]

$$\boldsymbol{\mu} = \frac{g_s q}{2m} \mathbf{S}$$

where the dimensionless quantity g_s is called the spin g-factor. For exclusively orbital rotations it would be 1 (assuming that the mass and the charge occupy spheres of equal radius).

The electron, being a charged elementary particle, possesses a nonzero magnetic moment. One of the triumphs of the theory of quantum electrodynamics is its accurate prediction of the electron g-factor, which has been experimentally determined to have the value −2.0023193043622(15), with the digits in parentheses denoting measurement uncertainty in the last two digits at one standard deviation.[8] The value of 2 arises from the Dirac equation, a fundamental equation connecting the electron's spin with its electromagnetic properties, and the correction of 0.002319304... arises from the electron's interaction with the surrounding electromagnetic field, including its own field.[9] Composite particles also possess magnetic moments associated with their spin. In particular, the neutron possesses a non-zero magnetic moment despite being electrically neutral. This fact was an early indication that the neutron is not an elementary particle. In fact, it is made up of quarks, which are electrically charged particles. The magnetic moment of the neutron comes from the spins of the individual quarks and their orbital motions.

Neutrinos are both elementary and electrically neutral. The minimally extended Standard Model that takes into account non-zero neutrino masses predicts neutrino magnetic moments of:[10][11][12]

$$\mu_\nu \approx 3 \times 10^{-19} \mu_B \frac{m_\nu}{\text{eV}}$$

where the $\mu\nu$ are the neutrino magnetic moments, $m\nu$ are the neutrino masses, and μ_B is the Bohr magneton. New physics above the electroweak scale could, however, lead to significantly higher neutrino magnetic moments. It can be shown in a model independent way that neutrino magnetic moments larger than about 10^{-14} μB are unnatural, because they would also lead to large radiative contributions to the neutrino mass. Since the neutrino masses cannot exceed about 1 eV, these radiative corrections must then be assumed to be fine tuned to cancel out to a large degree.[13]

The measurement of neutrino magnetic moments is an active area of research. As of 2001, the latest experimental results have put the neutrino magnetic moment at less than 1.2×10^{-10} times the electron's magnetic moment.

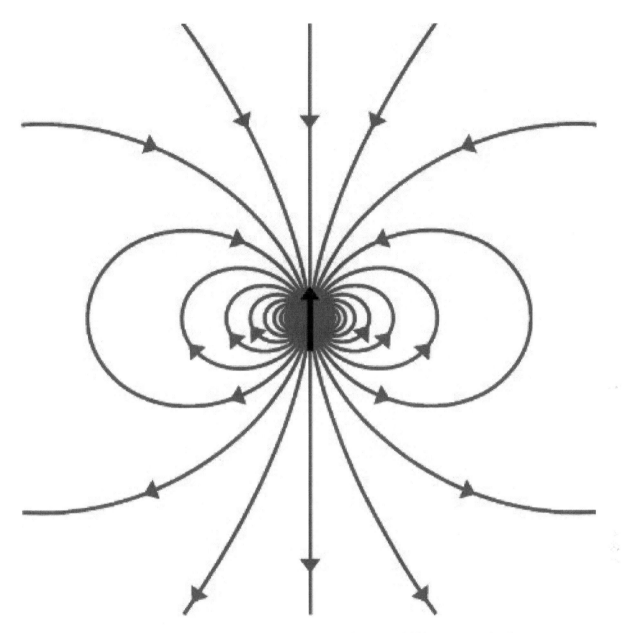

Schematic diagram depicting the spin of the neutron as the black arrow and magnetic field lines associated with the neutron magnetic moment. The neutron has a negative magnetic moment. While the spin of the neutron is upward in this diagram, the magnetic field lines at the center of the dipole are downward.

In ordinary materials, the magnetic dipole moments of individual atoms produce magnetic fields that cancel one another, because each dipole points in a random direction. Ferromagnetic materials below their Curie temperature, however, exhibit magnetic domains in which the atomic dipole moments are locally aligned, producing a macroscopic, non-zero magnetic field from the domain. These are the ordinary "magnets" with which we are all familiar.

In paramagnetic materials, the magnetic dipole moments of individual atoms spontaneously align with an externally applied magnetic field. In diamagnetic materials, on the other hand, the magnetic dipole moments of individual atoms spontaneously align oppositely to any externally applied magnetic field, even if it requires energy to do so.

The study of the behavior of such "spin models" is a thriving area of research in condensed matter physics. For instance, the Ising model describes spins (dipoles) that have only two possible states, up and down, whereas in the Heisenberg model the spin vector is allowed to point in any direction. These models have many interesting properties, which have led to interesting results in the theory of phase transitions.

12.3 Direction

Further information: Angular momentum operator

12.3.1 Spin projection quantum number and multiplicity

In classical mechanics, the angular momentum of a particle possesses not only a magnitude (how fast the body is rotating), but also a direction (either up or down on the axis of rotation of the particle). Quantum mechanical spin also contains information about direction, but in a more subtle form. Quantum mechanics states that the component of angular momentum measured along any direction can only take on the values [14]

$$S_i = \hbar s_i, \quad s_i \in \{-s, -(s-1), \dots, s-1, s\}$$

where S_i is the spin component along the i-axis (either x, y, or z), s_i is the spin projection quantum number along the i-axis, and s is the principal spin quantum number (discussed in the previous section). Conventionally the direction chosen is the z-axis:

$$S_z = \hbar s_z, \quad s_z \in \{-s, -(s-1), \dots, s-1, s\}$$

where S_z is the spin component along the z-axis, s_z is the spin projection quantum number along the z-axis.

One can see that there are $2s+1$ possible values of s_z. The number "$2s + 1$" is the multiplicity of the spin system. For example, there are only two possible values for a spin-1/2 particle: $s_z = +1/2$ and $s_z = -1/2$. These correspond to quantum states in which the spin is pointing in the +z or −z directions respectively, and are often referred to as "spin up" and "spin down". For a spin-3/2 particle, like a delta baryon, the possible values are +3/2, +1/2, −1/2, −3/2.

12.3.2 Vector

For a given quantum state, one could think of a spin vector $\langle S \rangle$ whose components are the expectation values of the spin components along each axis, i.e., $\langle S \rangle = [\langle S_x \rangle, \langle S_y \rangle, \langle S_z \rangle]$. This vector then would describe the "direction" in which the spin is pointing, corresponding to the classical concept of the axis of rotation. It turns out that the spin vector is not very useful in actual quantum mechanical calculations, because it cannot be measured directly: s_x, s_y and s_z cannot possess simultaneous definite values, because of a quantum uncertainty relation between them. However, for statistically large collections of particles that have been placed in the same pure quantum state, such as through the use of a Stern–Gerlach apparatus, the spin vector does have a well-defined experimental meaning: It specifies the direction in ordinary space in which a subsequent detector must be oriented in order to achieve the maximum possible probability (100%) of detecting every particle in the collection. For spin-1/2 particles, this maximum probability drops off smoothly as the angle between the spin vector and the detector increases, until at an angle of 180 degrees—that is, for detectors oriented in the opposite direction to the spin vector—the expectation of detecting particles from the collection reaches a minimum of 0%.

As a qualitative concept, the spin vector is often handy because it is easy to picture classically. For instance, quantum mechanical spin can exhibit phenomena analogous to classical gyroscopic effects. For example, one can exert a kind of "torque" on an electron by putting it in a magnetic field (the field acts upon the electron's intrinsic magnetic dipole moment—see the following section). The result is that the spin vector undergoes precession, just like a classical gyroscope. This phenomenon is known as electron spin resonance (ESR). The equivalent behaviour of protons in atomic nuclei is used in nuclear magnetic resonance (NMR) spectroscopy and imaging.

Mathematically, quantum mechanical spin states are described by vector-like objects known as spinors. There are subtle differences between the behavior of spinors and vectors under coordinate rotations. For example, rotating a spin-1/2

A single point in space can spin continuously without becoming tangled. Notice that after a 360 degree rotation, the spiral flips between clockwise and counterclockwise orientations. It returns to its original configuration after spinning a full 720 degrees.

particle by 360 degrees does not bring it back to the same quantum state, but to the state with the opposite quantum phase; this is detectable, in principle, with interference experiments. To return the particle to its exact original state, one needs a 720 degree rotation. A spin-zero particle can only have a single quantum state, even after torque is applied. Rotating a spin-2 particle 180 degrees can bring it back to the same quantum state and a spin-4 particle should be rotated 90 degrees to bring it back to the same quantum state. The spin 2 particle can be analogous to a straight stick that looks the same even after it is rotated 180 degrees and a spin 0 particle can be imagined as sphere which looks the same after whatever angle it is turned through.

12.4   Mathematical formulation

12.4.1 Operator

Spin obeys commutation relations analogous to those of the orbital angular momentum:

$$[S_i, S_j] = i\hbar\epsilon_{ijk}S_k$$

where ϵ_{ijk} is the Levi-Civita symbol. It follows (as with angular momentum) that the eigenvectors of S^2 and S_z (expressed as kets in the total S basis) are:

$$S^2|s, m\rangle = \hbar^2 s(s + 1)|s, m\rangle$$
$$S_z|s, m\rangle = \hbar m|s, m\rangle.$$

The spin raising and lowering operators acting on these eigenvectors give:

$$S_\pm|s, m\rangle = \hbar\sqrt{s(s + 1) - m(m \pm 1)}|s, m \pm 1\rangle \text{ , where } S_\pm = S_x \pm iS_y.$$

But unlike orbital angular momentum the eigenvectors are not spherical harmonics. They are not functions of θ and φ. There is also no reason to exclude half-integer values of s and m.

In addition to their other properties, all quantum mechanical particles possess an intrinsic spin (though it may have the intrinsic spin 0, too). The spin is quantized in units of the reduced Planck constant, such that the state function of the particle is, say, not $\psi = \psi(\mathbf{r})$, but $\psi = \psi(\mathbf{r}, \sigma)$ where σ is out of the following discrete set of values:

$$\sigma \in \{-s\hbar, -(s - 1)\hbar, \cdots, +(s - 1)\hbar, +s\hbar\}.$$

One distinguishes bosons (integer spin) and fermions (half-integer spin). The total angular momentum conserved in interaction processes is then the *sum* of the orbital angular momentum and the spin.

12.4.2 Pauli matrices

The quantum mechanical operators associated with spin-$\frac{1}{2}$ observables are:

$$\hat{\mathbf{S}} = \frac{\hbar}{2}\sigma$$

where in Cartesian components:

$$S_x = \frac{\hbar}{2}\sigma_x, \quad S_y = \frac{\hbar}{2}\sigma_y, \quad S_z = \frac{\hbar}{2}\sigma_z.$$

For the special case of spin-1/2 particles, σx, σy and σz are the three Pauli matrices, given by:

$$\sigma_x = \begin{pmatrix} 0 & 1 \\ 1 & 0 \end{pmatrix} \quad \sigma_y = \begin{pmatrix} 0 & -i \\ i & 0 \end{pmatrix} \quad \sigma_z = \begin{pmatrix} 1 & 0 \\ 0 & -1 \end{pmatrix}.$$

12.4.3 Pauli exclusion principle

For systems of N identical particles this is related to the Pauli exclusion principle, which states that by interchanges of any two of the N particles one must have

$$\psi(\cdots \mathbf{r}_i, \sigma_i \cdots \mathbf{r}_j, \sigma_j \cdots) = (-1)^{2s} \psi(\cdots \mathbf{r}_j, \sigma_j \cdots \mathbf{r}_i, \sigma_i \cdots).$$

Thus, for bosons the prefactor $(-1)^{2s}$ will reduce to $+1$, for fermions to -1. In quantum mechanics all particles are either bosons or fermions. In some speculative relativistic quantum field theories "supersymmetric" particles also exist, where linear combinations of bosonic and fermionic components appear. In two dimensions, the prefactor $(-1)^{2s}$ can be replaced by any complex number of magnitude 1 such as in the Anyon.

The above permutation postulate for N-particle state functions has most-important consequences in daily life, e.g. the periodic table of the chemists or biologists.

12.4.4   Rotations

See also: symmetries in quantum mechanics

As described above, quantum mechanics states that components of angular momentum measured along any direction can only take a number of discrete values. The most convenient quantum mechanical description of particle's spin is therefore with a set of complex numbers corresponding to amplitudes of finding a given value of projection of its intrinsic angular momentum on a given axis. For instance, for a spin 1/2 particle, we would need two numbers $a_{\pm 1/2}$, giving amplitudes of finding it with projection of angular momentum equal to $\hbar/2$ and $-\hbar/2$, satisfying the requirement

$$\left| a_{\frac{1}{2}} \right|^2 + \left| a_{-\frac{1}{2}} \right|^2 = 1.$$

For a generic particle with spin s, we would need $2s + 1$ such parameters. Since these numbers depend on the choice of the axis, they transform into each other non-trivially when this axis is rotated. It's clear that the transformation law must be linear, so we can represent it by associating a matrix with each rotation, and the product of two transformation matrices corresponding to rotations A and B must be equal (up to phase) to the matrix representing rotation AB. Further, rotations preserve the quantum mechanical inner product, and so should our transformation matrices:

$$\sum_{m=-j}^{j} a_m^* b_m = \sum_{m=-j}^{j} \left(\sum_{n=-j}^{j} U_{nm} a_n \right)^* \left(\sum_{k=-j}^{j} U_{km} b_k \right)$$

$$\sum_{n=-j}^{j} \sum_{k=-j}^{j} U_{np}^* U_{kq} = \delta_{pq}.$$

Mathematically speaking, these matrices furnish a unitary projective representation of the rotation group SO(3). Each such representation corresponds to a representation of the covering group of SO(3), which is SU(2).[15] There is one n-dimensional irreducible representation of SU(2) for each dimension, though this representation is n-dimensional real for odd n and n-dimensional complex for even n (hence of real dimension $2n$). For a rotation by angle θ in the plane with normal vector $\hat{\boldsymbol{\theta}}$, U can be written

$$U = e^{-\frac{i}{\hbar} \boldsymbol{\theta} \cdot \mathbf{S}},$$

where $\boldsymbol{\theta} = \theta \hat{\boldsymbol{\theta}}$ is a and $\mathbf{S}$ is the vector of spin operators.

(Click "show" at right to see a proof or "hide" to hide it.)

Working in the coordinate system where $\hat{\theta} = \hat{z}$, we would like to show that S_x and S_y are rotated into each other by the angle θ. Starting with S_x. Using units where $\hbar = 1$:

$$S_x \to U^\dagger S_x U = e^{i\theta S_z} S_x e^{-i\theta S_z}$$

$$= S_x + (i\theta)[S_z, S_x] + \left(\frac{1}{2!}\right)(i\theta)^2[S_z,[S_z,S_x]] + \left(\frac{1}{3!}\right)(i\theta)^3[S_z,[S_z,[S_z,S_x]]] + \cdots$$

Using the spin operator commutation relations, we see that the commutators evaluate to iS_y for the odd terms in the series, and to S_x for all of the even terms. Thus:

$$U^\dagger S_x U = S_x \left[1 - \frac{\theta^2}{2!} + \ldots\right] - S_y \left[\theta - \frac{\theta^3}{3!}\cdots\right]$$

$$= S_x \cos\theta - S_y \sin\theta$$

as expected. Note that since we only relied on the spin operator commutation relations, this proof holds for any dimension (i.e., for any principal spin quantum number s).[16]

A generic rotation in 3-dimensional space can be built by compounding operators of this type using Euler angles:

$$\mathcal{R}(\alpha, \beta, \gamma) = e^{-i\alpha S_x} e^{-i\beta S_y} e^{-i\gamma S_z}$$

An irreducible representation of this group of operators is furnished by the Wigner D-matrix:

$$D^s_{m'm}(\alpha, \beta, \gamma) \equiv \langle sm'|\mathcal{R}(\alpha, \beta, \gamma)|sm\rangle = e^{-im'\alpha} d^s_{m'm}(\beta) e^{-im\gamma},$$

where

$$d^s_{m'm}(\beta) = \langle sm'|e^{-i\beta s_y}|sm\rangle$$

is Wigner's small d-matrix. Note that for $\gamma = 2\pi$ and $\alpha = \beta = 0$; i.e., a full rotation about the z-axis, the Wigner D-matrix elements become

$$D^s_{m'm}(0, 0, 2\pi) = d^s_{m'm}(0)e^{-im2\pi} = \delta_{m'm}(-1)^{2m}.$$

Recalling that a generic spin state can be written as a superposition of states with definite m, we see that if s is an integer, the values of m are all integers, and this matrix corresponds to the identity operator. However, if s is a half-integer, the values of m are also all half-integers, giving $(-1)^{2m} = -1$ for all m, and hence upon rotation by 2π the state picks up a minus sign. This fact is a crucial element of the proof of the spin-statistics theorem.

12.4.5 Lorentz transformations

We could try the same approach to determine the behavior of spin under general Lorentz transformations, but we would immediately discover a major obstacle. Unlike SO(3), the group of Lorentz transformations SO(3,1) is non-compact and therefore does not have any faithful, unitary, finite-dimensional representations.

In case of spin 1/2 particles, it is possible to find a construction that includes both a finite-dimensional representation and a scalar product that is preserved by this representation. We associate a 4-component Dirac spinor ψ with each particle. These spinors transform under Lorentz transformations according to the law

$$\psi' = \exp\left(\frac{1}{8}\omega_{\mu\nu}[\gamma_\mu, \gamma_\nu]\right)\psi$$

where γ_μ are gamma matrices and $\omega_{\mu\nu}$ is an antisymmetric 4×4 matrix parametrizing the transformation. It can be shown that the scalar product

$$\langle\psi|\phi\rangle = \bar{\psi}\phi = \psi^\dagger\gamma_0\phi$$

is preserved. It is not, however, positive definite, so the representation is not unitary.

12.4.6   Metrology along the *x*, *y*, and *z* axes

Each of the (Hermitian) Pauli matrices has two eigenvalues, +1 and −1. The corresponding normalized eigenvectors are:

$$\psi_{x+} = \frac{1}{\sqrt{2}}\begin{pmatrix}1\\1\end{pmatrix}, \quad \psi_{x-} = \frac{1}{\sqrt{2}}\begin{pmatrix}1\\-1\end{pmatrix},$$
$$\psi_{y+} = \frac{1}{\sqrt{2}}\begin{pmatrix}1\\i\end{pmatrix}, \quad \psi_{y-} = \frac{1}{\sqrt{2}}\begin{pmatrix}1\\-i\end{pmatrix},$$
$$\psi_{z+} = \begin{pmatrix}1\\0\end{pmatrix}, \quad \psi_{z-} = \begin{pmatrix}0\\1\end{pmatrix}.$$

By the postulates of quantum mechanics, an experiment designed to measure the electron spin on the *x*, *y* or *z* axis can only yield an eigenvalue of the corresponding spin operator (*Sx*, *Sy* or *Sz*) on that axis, i.e. $\hbar/2$ or $-\hbar/2$. The quantum state of a particle (with respect to spin), can be represented by a two component spinor:

$$\psi = \begin{pmatrix}a + bi\\c + di\end{pmatrix}.$$

When the spin of this particle is measured with respect to a given axis (in this example, the *x*-axis), the probability that its spin will be measured as $\hbar/2$ is just $|\langle\psi_{x+}|\psi\rangle|^2$. Correspondingly, the probability that its spin will be measured as $-\hbar/2$ is just $|\langle\psi_{x-}|\psi\rangle|^2$. Following the measurement, the spin state of the particle will collapse into the corresponding eigenstate. As a result, if the particle's spin along a given axis has been measured to have a given eigenvalue, all measurements will yield the same eigenvalue (since $|\langle\psi_{x+}|\psi_{x+}\rangle|^2 = 1$, etc), provided that no measurements of the spin are made along other axes.

12.4.7   Metrology along an arbitrary axis

The operator to measure spin along an arbitrary axis direction is easily obtained from the Pauli spin matrices. Let *u* = (*ux*, *uy*, *uz*) be an arbitrary unit vector. Then the operator for spin in this direction is simply

$$S_u = \frac{\hbar}{2}(u_x\sigma_x + u_y\sigma_y + u_z\sigma_z)$$

The operator *Su* has eigenvalues of $\pm\hbar/2$, just like the usual spin matrices. This method of finding the operator for spin in an arbitrary direction generalizes to higher spin states, one takes the dot product of the direction with a vector of the three operators for the three *x*, *y*, *z* axis directions.

A normalized spinor for spin-1/2 in the (*ux*, *uy*, *uz*) direction (which works for all spin states except spin down where it will give 0/0), is:

$$\frac{1}{\sqrt{2 + 2u_z}}\begin{pmatrix}1 + u_z\\u_x + iu_y\end{pmatrix}.$$

The above spinor is obtained in the usual way by diagonalizing the σ_u matrix and finding the eigenstates corresponding to the eigenvalues. In quantum mechanics, vectors are termed "normalized" when multiplied by a normalizing factor, which results in the vector having a length of unity.

12.4.8 Compatibility of metrology

Since the Pauli matrices do not commute, measurements of spin along the different axes are incompatible. This means that if, for example, we know the spin along the x-axis, and we then measure the spin along the y-axis, we have invalidated our previous knowledge of the x-axis spin. This can be seen from the property of the eigenvectors (i.e. eigenstates) of the Pauli matrices that:

$$| \langle \psi_{x\pm} | \psi_{y\pm} \rangle |^2 = | \langle \psi_{x\pm} | \psi_{z\pm} \rangle |^2 = | \langle \psi_{y\pm} | \psi_{z\pm} \rangle |^2 = \frac{1}{2}.$$

So when physicists measure the spin of a particle along the x-axis as, for example, $\hbar/2$, the particle's spin state collapses into the eigenstate $| \psi_{x+} \rangle$. When we then subsequently measure the particle's spin along the y-axis, the spin state will now collapse into either $| \psi_{y+} \rangle$ or $| \psi_{y-} \rangle$, each with probability 1/2. Let us say, in our example, that we measure $-\hbar/2$. When we now return to measure the particle's spin along the x-axis again, the probabilities that we will measure $\hbar/2$ or $-\hbar/2$ are each 1/2 (i.e. they are $| \langle \psi_{x+} | \psi_{y-} \rangle |^2$ and $| \langle \psi_{x-} | \psi_{y-} \rangle |^2$ respectively). This implies that the original measurement of the spin along the x-axis is no longer valid, since the spin along the x-axis will now be measured to have either eigenvalue with equal probability.

12.4.9 Higher spins

The spin-1/2 operator $\mathbf{S} = \hbar/2\boldsymbol{\sigma}$ form the fundamental representation of SU(2). By taking Kronecker products of this representation with itself repeatedly, one may construct all higher irreducible representations. That is, the resulting spin operators for higher spin systems in three spatial dimensions, for arbitrarily large s, can be calculated using this spin operator and ladder operators.

The resulting spin matrices for spin 1 are:

$$S_x = \frac{\hbar}{\sqrt{2}} \begin{pmatrix} 0 & 1 & 0 \\ 1 & 0 & 1 \\ 0 & 1 & 0 \end{pmatrix}$$

$$S_y = \frac{\hbar}{\sqrt{2}} \begin{pmatrix} 0 & -i & 0 \\ i & 0 & -i \\ 0 & i & 0 \end{pmatrix}$$

$$S_z = \hbar \begin{pmatrix} 1 & 0 & 0 \\ 0 & 0 & 0 \\ 0 & 0 & -1 \end{pmatrix}$$

for spin 3/2 they are

$$S_x = \frac{\hbar}{2} \begin{pmatrix} 0 & \sqrt{3} & 0 & 0 \\ \sqrt{3} & 0 & 2 & 0 \\ 0 & 2 & 0 & \sqrt{3} \\ 0 & 0 & \sqrt{3} & 0 \end{pmatrix}$$

$$S_y = \frac{\hbar}{2} \begin{pmatrix} 0 & -i\sqrt{3} & 0 & 0 \\ i\sqrt{3} & 0 & -2i & 0 \\ 0 & 2i & 0 & -i\sqrt{3} \\ 0 & 0 & i\sqrt{3} & 0 \end{pmatrix}$$

$$S_z = \frac{\hbar}{2} \begin{pmatrix} 3 & 0 & 0 & 0 \\ 0 & 1 & 0 & 0 \\ 0 & 0 & -1 & 0 \\ 0 & 0 & 0 & -3 \end{pmatrix}$$

and for spin 5/2 they are

$$S_x = \frac{\hbar}{2}\begin{pmatrix} 0 & \sqrt{5} & 0 & 0 & 0 & 0 \\ \sqrt{5} & 0 & 2\sqrt{2} & 0 & 0 & 0 \\ 0 & 2\sqrt{2} & 0 & 3 & 0 & 0 \\ 0 & 0 & 3 & 0 & 2\sqrt{2} & 0 \\ 0 & 0 & 0 & 2\sqrt{2} & 0 & \sqrt{5} \\ 0 & 0 & 0 & 0 & \sqrt{5} & 0 \end{pmatrix}$$

$$S_y = \frac{\hbar}{2}\begin{pmatrix} 0 & -i\sqrt{5} & 0 & 0 & 0 & 0 \\ i\sqrt{5} & 0 & -2i\sqrt{2} & 0 & 0 & 0 \\ 0 & 2i\sqrt{2} & 0 & -3i & 0 & 0 \\ 0 & 0 & 3i & 0 & -2i\sqrt{2} & 0 \\ 0 & 0 & 0 & 2i\sqrt{2} & 0 & -i\sqrt{5} \\ 0 & 0 & 0 & 0 & i\sqrt{5} & 0 \end{pmatrix}$$

$$S_z = \frac{\hbar}{2}\begin{pmatrix} 5 & 0 & 0 & 0 & 0 & 0 \\ 0 & 3 & 0 & 0 & 0 & 0 \\ 0 & 0 & 1 & 0 & 0 & 0 \\ 0 & 0 & 0 & -1 & 0 & 0 \\ 0 & 0 & 0 & 0 & -3 & 0 \\ 0 & 0 & 0 & 0 & 0 & -5 \end{pmatrix}.$$

The generalization of these matrices for arbitrary s is

$$(S_x)_{ab} = \frac{\hbar}{2}(\delta_{a,b+1} + \delta_{a+1,b})\sqrt{(s+1)(a+b-1)-ab}$$
$$(S_y)_{ab} = \frac{\hbar}{2i}(\delta_{a,b+1} - \delta_{a+1,b})\sqrt{(s+1)(a+b-1)-ab} \quad 1 \le a,b \le 2s+1$$
$$(S_z)_{ab} = \hbar(s+1-a)\delta_{a,b} = \hbar(s+1-b)\delta_{a,b}.$$

Also useful in the quantum mechanics of multiparticle systems, the general Pauli group Gn is defined to consist of all n-fold tensor products of Pauli matrices.

The analog formula of Euler's formula in terms of the Pauli matrices:

$$e^{i\theta(\hat{\mathbf{n}}\cdot\boldsymbol{\sigma})} = I\cos\theta + i(\hat{\mathbf{n}}\cdot\boldsymbol{\sigma})\sin\theta$$

for higher spins is tractable, but less simple.[17]

12.5 Parity

In tables of the spin quantum number s for nuclei or particles, the spin is often followed by a "+" or "−". This refers to the parity with "+" for even parity (wave function unchanged by spatial inversion) and "−" for odd parity (wave function negated by spatial inversion). For example, see the isotopes of bismuth.

12.6 Applications

Spin has important theoretical implications and practical applications. Well-established *direct* applications of spin include:

- Nuclear magnetic resonance (NMR) spectroscopy in chemistry;

- Electron spin resonance spectroscopy in chemistry and physics;

- Magnetic resonance imaging (MRI) in medicine, a type of applied NMR, which relies on proton spin density;

- Giant magnetoresistive (GMR) drive head technology in modern hard disks.

Electron spin plays an important role in magnetism, with applications for instance in computer memories. The manipulation of *nuclear spin* by radiofrequency waves (nuclear magnetic resonance) is important in chemical spectroscopy and medical imaging.

Spin-orbit coupling leads to the fine structure of atomic spectra, which is used in atomic clocks and in the modern definition of the second. Precise measurements of the g-factor of the electron have played an important role in the development and verification of quantum electrodynamics. *Photon spin* is associated with the polarization of light.

An emerging application of spin is as a binary information carrier in spin transistors. The original concept, proposed in 1990, is known as Datta-Das spin transistor.[18] Electronics based on spin transistors are referred to as spintronics. The manipulation of spin in dilute magnetic semiconductor materials, such as metal-doped ZnO or TiO_2 imparts a further degree of freedom and has the potential to facilitate the fabrication of more efficient electronics.[19]

There are many *indirect* applications and manifestations of spin and the associated Pauli exclusion principle, starting with the periodic table of chemistry.

12.7 History

Spin was first discovered in the context of the emission spectrum of alkali metals. In 1924 Wolfgang Pauli introduced what he called a "two-valued quantum degree of freedom" associated with the electron in the outermost shell. This allowed him to formulate the Pauli exclusion principle, stating that no two electrons can share the same quantum state at the same time.

The physical interpretation of Pauli's "degree of freedom" was initially unknown. Ralph Kronig, one of Landé's assistants, suggested in early 1925 that it was produced by the self-rotation of the electron. When Pauli heard about the idea, he criticized it severely, noting that the electron's hypothetical surface would have to be moving faster than the speed of light in order for it to rotate quickly enough to produce the necessary angular momentum. This would violate the theory of relativity. Largely due to Pauli's criticism, Kronig decided not to publish his idea.

In the autumn of 1925, the same thought came to two Dutch physicists, George Uhlenbeck and Samuel Goudsmit at Leiden University. Under the advice of Paul Ehrenfest, they published their results. It met a favorable response, especially after Llewellyn Thomas managed to resolve a factor-of-two discrepancy between experimental results and Uhlenbeck and Goudsmit's calculations (and Kronig's unpublished results). This discrepancy was due to the orientation of the electron's tangent frame, in addition to its position.

Mathematically speaking, a fiber bundle description is needed. The tangent bundle effect is additive and relativistic; that is, it vanishes if *c* goes to infinity. It is one half of the value obtained without regard for the tangent space orientation, but with opposite sign. Thus the combined effect differs from the latter by a factor two (Thomas precession).

Despite his initial objections, Pauli formalized the theory of spin in 1927, using the modern theory of quantum mechanics invented by Schrödinger and Heisenberg. He pioneered the use of Pauli matrices as a representation of the spin operators, and introduced a two-component spinor wave-function.

Pauli's theory of spin was non-relativistic. However, in 1928, Paul Dirac published the Dirac equation, which described the relativistic electron. In the Dirac equation, a four-component spinor (known as a "Dirac spinor") was used for the electron wave-function. In 1940, Pauli proved the *spin-statistics theorem*, which states that fermions have half-integer spin and bosons integer spin.

In retrospect, the first direct experimental evidence of the electron spin was the Stern–Gerlach experiment of 1922. However, the correct explanation of this experiment was only given in 1927.[20]

Wolfgang Pauli

12.8   See also

- Einstein–de Haas effect

- Spin-orbital

- Chirality (physics)

- Dynamic nuclear polarisation

- Helicity (particle physics)

- Pauli equation

- Pauli–Lubanski pseudovector

- Rarita–Schwinger equation

- Representation theory of SU(2)

- Spin-½

- Spin-flip

- Spin isomers of hydrogen

- Spin tensor

- Spin wave

- Spin engineering

- Yrast

- Zitterbewegung

12.9 References

[1] Merzbacher, Eugen (1998). *Quantum Mechanics* (3rd ed.). pp. 372–3.

[2] Griffiths, David (2005). *Introduction to Quantum Mechanics* (2nd ed.). pp. 183–4.

[3] "Angular Momentum Operator Algebra", class notes by Michael Fowler

[4] *A modern approach to quantum mechanics*, by Townsend, p. 31 and p. 80

[5]Eisberg, Robert; Resnick, Robert (1985).*Quantum Physics of Atoms, Molecules, Solids, Nuclei, and Particles*(2nd ed.) 272–3.

[6]Pauli, Wolfgang(1940)."The Connection Between Spin and Statistics"(PDF).*Phys. Rev***58**(8): 716–722.Bibcode:194P. doi:10.1103/PhysRev.58.716.

[7] Physics of Atoms and Molecules, B.H. Bransden, C.J.Joachain, Longman, 1983, ISBN 0-582-44401-2

[8]"CODATA Value: electron g factor".*The NIST Reference on Constants, Units, and Uncertainty*.NIST. 2006 11-15.

[9]R.P. Feynman(1985). "Electrons and Their Interactions".*QED: The Strange Theory of Light and Matter*.Princeton, Jersey: Princeton University Press. p. 115. ISBN 0-691-08388-6.

> "After some years, it was discovered that this value [−g/2] was not exactly 1, but slightly more—something like 1.00116. This correction was worked out for the first time in 1948 by Schwinger as j*j divided by 2 pi [*sic*] [where *j* is the square root of the fine-structure constant], and was due to an alternative way the electron can go from place to place: instead of going directly from one point to another, the electron goes along for a while and suddenly emits a photon; then (horrors!) it absorbs its own photon."

[10]W.J. Marciano, A.I. Sanda (1977). "Exotic decays of the muon and heavy leptons in gauge theories".*Physics Letters*: 303–305. Bibcode:1977PhLB...67..303M. doi:10.1016/0370-2693(77)90377-X.

[11] B.W. Lee, R.E. Shrock (1977). "Natural suppression of symmetry violation in gauge theories: Muon- and electron-lepton-number nonconservation". *Physical Review* **D16** (5): 1444–1473. Bibcode:1977PhRvD..16.1444L. doi:

[12] K. Fujikawa, R. E. Shrock (1980). "Magnetic Moment of a Massive Neutrino and Neutrino-Spin Rotation". *Physical Review Letters* **45** (12): 963–966. Bibcode:1980PhRvL..45..963F. doi:10.1103/PhysRevLett.45.963.

[13] N.F. Bell; Cirigliano, V.; Ramsey-Musolf, M.; Vogel, P.; Wise, Mark; et al. (2005). "How Magnetic is the Dirac Neutrino?". *Physical Review Letters* **95** (15): 151802. arXiv:hep-ph/0504134. Bibcode:2005PhRvL..95o1802B. doi: PMID 16241715.

[14] Quanta: A handbook of concepts, P.W. Atkins, Oxford University Press, 1974, ISBN 0-19-855493-1

[15] B.C. Hall (2013). *Quantum Theory for Mathematicians*. Springer. pp. 354–358.

[16] *Modern Quantum Mechanics*, by J. J. Sakurai, p159

[17] Curtright, T L; Fairlie, D B; Zachos, C K (2014). "A compact formula for rotations as spin matrix polynomials". *SIGMA* **10**: 084. doi:10.3842/SIGMA.2014.084.

[18] Datta. S and B. Das (1990). "Electronic analog of the electrooptic modulator". *Applied Physics Letters* **56** (7): 665–667. Bibcode:1990ApPhL..56..665D. doi:10.1063/1.102730.

[19] Assadi, M.H.N; Hanaor, D.A.H (2013). "Theoretical study on copper's energetics and magnetism in TiO$_2$ polymorphs" (PDF). *Journal of Applied Physics* **113** (23): 233913. doi:10.1063/1.4811539.

[20] B. Friedrich, D. Herschbach (2003). "Stern and Gerlach: How a Bad Cigar Helped Reorient Atomic Physics". *Physics Today* **56** (12): 53. Bibcode:2003PhT....56l..53F. doi:10.1063/1.1650229.

12.10   Further reading

- Cohen-Tannoudji, Claude; Diu, Bernard; Laloë, Franck (2006). *Quantum Mechanics* (2 volume set ed.). John Wiley & Sons. ISBN 978-0-471-56952-7.

- Condon, E. U.; Shortley, G. H. (1935). "Especially Chapter 3". *The Theory of Atomic Spectra*. Cambridge University Press. ISBN 0-521-09209-4.

- Hipple, J. A.; Sommer, H.; Thomas, H.A. (1949). *A precise method of determining the faraday by magnetic resonance.* doi:10.1103/PhysRev.76.1877.2.https://www.academia.edu/6483539/John_A._Hipple_1911-1985_as_knowledge

- Edmonds, A. R. (1957). *Angular Momentum in Quantum Mechanics*. Princeton University Press. ISBN 0-691-07912-9.

- Jackson, John David (1998). *Classical Electrodynamics* (3rd ed.). John Wiley & Sons. ISBN 978-0-471-30932-1.

- Serway, Raymond A.; Jewett, John W. (2004). *Physics for Scientists and Engineers* (6th ed.). Brooks/Cole. ISBN 0-534-40842-7.

- Thompson, William J. (1994). *Angular Momentum: An Illustrated Guide to Rotational Symmetries for Physical Systems*. Wiley. ISBN 0-471-55264-X.

- Tipler, Paul (2004). *Physics for Scientists and Engineers: Mechanics, Oscillations and Waves, Thermodynamics* (5th ed.). W. H. Freeman. ISBN 0-7167-0809-4.

- Sin-Itiro Tomonaga, The Story of Spin, 1997

12.11 External links

- "Spintronics. Feature Article" in *Scientific American*, June 2002.

- Goudsmit on the discovery of electron spin.

- *Nature*: "Milestones in 'spin' since 1896."

- ECE 495N Lecture 36: Spin Online lecture by S. Datta

Chapter 13

Electric charge

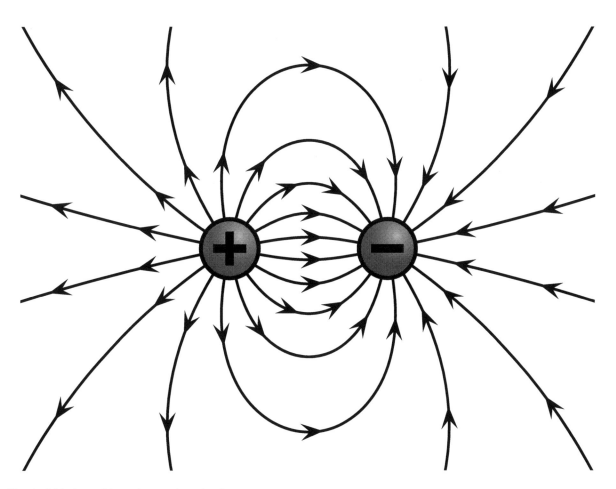

Electric field of a positive and a negative point charge.

Electric charge is the physical property of matter that causes it to experience a force when placed in an electromagnetic field. There are two types of electric charges: positive and negative. Positively charged substances are repelled from other positively charged substances, but attracted to negatively charged substances; negatively charged substances are repelled from negative and attracted to positive. An object is negatively charged if it has an excess of electrons, and is otherwise positively charged or uncharged. The SI derived unit of electric charge is the coulomb (C), although in electrical engineering it is also common to use the ampere-hour (Ah), and in chemistry it is common to use the elementary charge (e) as a unit. The symbol Q is often used to denote charge. The early knowledge of how charged substances interact is now called classical electrodynamics, and is still very accurate if quantum effects do not need to be considered.

The *electric charge* is a fundamental conserved property of some subatomic particles, which determines their electromagnetic interaction. Electrically charged matter is influenced by, and produces, electromagnetic fields. The interaction between a moving charge and an electromagnetic field is the source of the electromagnetic force, which is one of the four fundamental forces (See also: magnetic field).

Twentieth-century experiments demonstrated that electric charge is *quantized*; that is, it comes in integer multiples of individual small units called the elementary charge, e, approximately equal to 1.602×10^{-19} coulombs (except for particles called quarks, which have charges that are integer multiples of $e/3$). The proton has a charge of $+e$, and the electron has a charge of $-e$. The study of charged particles, and how their interactions are mediated by photons, is called quantum electrodynamics.

13.1 Overview

Charge is the fundamental property of forms of matter that exhibit electrostatic attraction or repulsion in the presence of other matter. Electric charge is a characteristic property of many subatomic particles. The charges of free-standing particles are integer multiples of the elementary charge e; we say that electric charge is *quantized*. Michael Faraday, in his electrolysis experiments, was the first to note the discrete nature of electric charge. Robert Millikan's oil-drop experiment demonstrated this fact directly, and measured the elementary charge.

By convention, the charge of an electron is -1, while that of a proton is $+1$. Charged particles whose charges have the same sign repel one another, and particles whose charges have different signs attract. Coulomb's law quantifies the electrostatic force between two particles by asserting that the force is proportional to the product of their charges, and inversely proportional to the square of the distance between them.

The charge of an antiparticle equals that of the corresponding particle, but with opposite sign. Quarks have fractional charges of either $-\frac{1}{3}$ or $+\frac{2}{3}$, but free-standing quarks have never been observed (the theoretical reason for this fact is asymptotic freedom).

The electric charge of a macroscopic object is the sum of the electric charges of the particles that make it up. This charge is often small, because matter is made of atoms, and atoms typically have equal numbers of protons and electrons, in which case their charges cancel out, yielding a net charge of zero, thus making the atom neutral.

An *ion* is an atom (or group of atoms) that has lost one or more electrons, giving it a net positive charge (cation), or that has gained one or more electrons, giving it a net negative charge (anion). *Monatomic ions* are formed from single atoms, while *polyatomic ions* are formed from two or more atoms that have been bonded together, in each case yielding an ion with a positive or negative net charge.

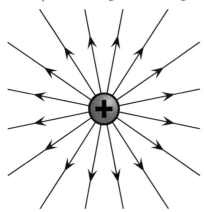

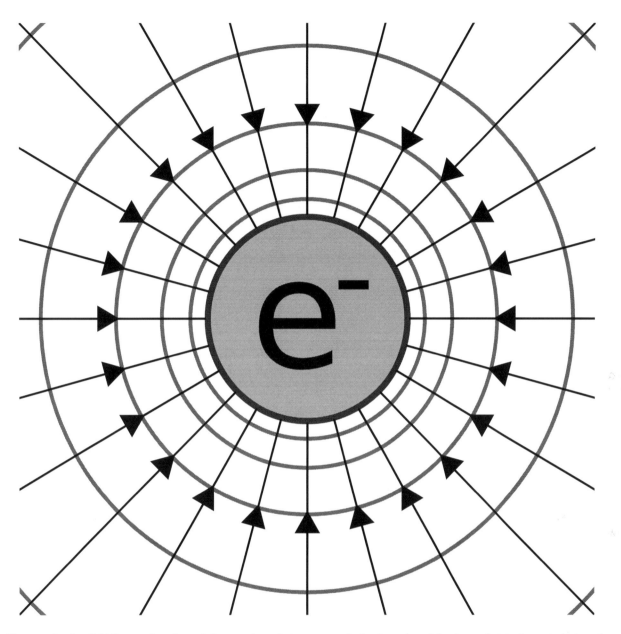

Diagram showing field lines and equipotentials around an electron, a negatively charged particle. In an electrically neutral atom, the number of electrons is equal to the number of protons (which are positively charged), resulting in a net zero overall charge

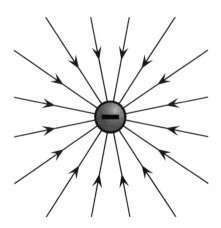

Electric field induced by a positive electric charge (left) and a field induced by a negative electric charge (right).

During formation of macroscopic objects, constituent atoms and ions usually combine to form structures composed of neutral *ionic compounds* electrically bound to neutral atoms. Thus macroscopic objects tend toward being neutral overall, but macroscopic objects are rarely perfectly net neutral.

Sometimes macroscopic objects contain ions distributed throughout the material, rigidly bound in place, giving an overall net positive or negative charge to the object. Also, macroscopic objects made of conductive elements, can more or less easily (depending on the element) take on or give off electrons, and then maintain a net negative or positive charge indefinitely. When the net electric charge of an object is non-zero and motionless, the phenomenon is known as static electricity. This can easily be produced by rubbing two dissimilar materials together, such as rubbing amber with fur or glass with silk. In this way non-conductive materials can be charged to a significant degree, either positively or negatively. Charge taken from one material is moved to the other material, leaving an opposite charge of the same magnitude behind. The law of *conservation of charge* always applies, giving the object from which a negative charge has been taken a positive charge of the same magnitude, and vice versa.

Even when an object's net charge is zero, charge can be distributed non-uniformly in the object (e.g., due to an external electromagnetic field, or bound polar molecules). In such cases the object is said to be polarized. The charge due to polarization is known as bound charge, while charge on an object produced by electrons gained or lost from outside the object is called *free charge*. The motion of electrons in conductive metals in a specific direction is known as electric current.

13.2 Units

The SI unit of quantity of electric charge is the coulomb, which is equivalent to about $6.242{\times}10^{18}$ e (e is the charge of a proton). Hence, the charge of an electron is approximately $-1.602{\times}10^{-19}$ C. The coulomb is defined as the quantity of charge that has passed through the cross section of an electrical conductor carrying one ampere within one second. The symbol Q is often used to denote a quantity of electricity or charge. The quantity of electric charge can be directly measured with an electrometer, or indirectly measured with a ballistic galvanometer.

After finding the quantized character of charge, in 1891 George Stoney proposed the unit 'electron' for this fundamental unit of electrical charge. This was before the discovery of the particle by J.J. Thomson in 1897. The unit is today treated as nameless, referred to as "elementary charge", "fundamental unit of charge", or simply as "e". A measure of charge should be a multiple of the elementary charge e, even if at large scales charge seems to behave as a real quantity. In some contexts it is meaningful to speak of fractions of a charge; for example in the charging of a capacitor, or in the fractional quantum Hall effect.

In systems of units other than SI such as cgs, electric charge is expressed as combination of only three fundamental quantities such as length, mass and time and not four as in SI where electric charge is a combination of length, mass, time and electric current.

13.3 History

As reported by the ancient Greek mathematician Thales of Miletus around 600 BC, charge (or *electricity*) could be accumulated by rubbing fur on various substances, such as amber. The Greeks noted that the charged amber buttons could attract light objects such as hair. They also noted that if they rubbed the amber for long enough, they could even get an electric spark to jump. This property derives from the triboelectric effect.

In 1600, the English scientist William Gilbert returned to the subject in *De Magnete*, and coined the New Latin word *electricus* from ηλεκτρον (*elektron*), the Greek word for *amber*, which soon gave rise to the English words "electric" and "electricity." He was followed in 1660 by Otto von Guericke, who invented what was probably the first electrostatic generator. Other European pioneers were Robert Boyle, who in 1675 stated that electric attraction and repulsion can act across a vacuum; Stephen Gray, who in 1729 classified materials as conductors and insulators; and C. F. du Fay, who

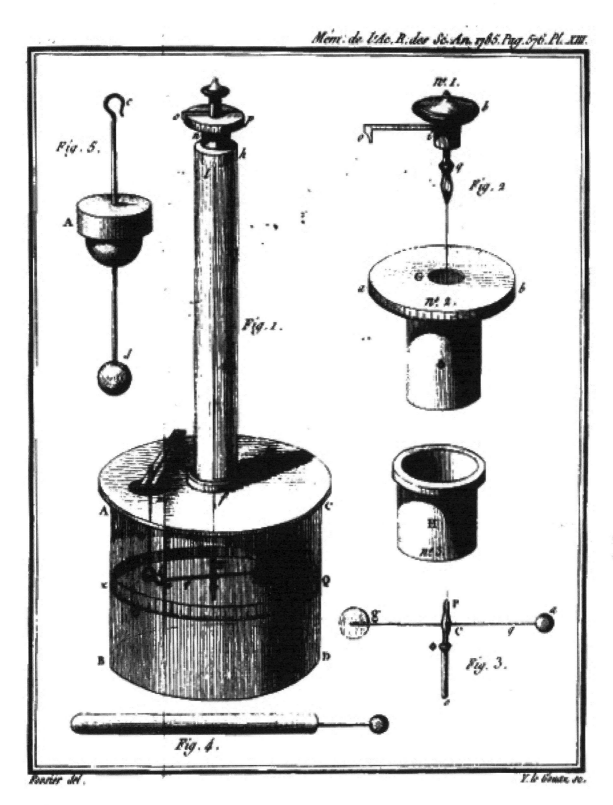

Coulomb's torsion balance

proposed in 1733[1] that electricity comes in two varieties that cancel each other, and expressed this in terms of a two-fluid theory. When glass was rubbed with silk, du Fay said that the glass was charged with *vitreous electricity*, and, when amber was rubbed with fur, the amber was said to be charged with *resinous electricity*. In 1839, Michael Faraday showed that the apparent division between static electricity, current electricity, and bioelectricity was incorrect, and all were a consequence of the behavior of a single kind of electricity appearing in opposite polarities. It is arbitrary which polarity is called positive and which is called negative. Positive charge can be defined as the charge left on a glass rod after being rubbed with silk.[2]

One of the foremost experts on electricity in the 18th century was Benjamin Franklin, who argued in favour of a one-fluid theory of electricity. Franklin imagined electricity as being a type of invisible fluid present in all matter; for example, he believed that it was the glass in a Leyden jar that held the accumulated charge. He posited that rubbing insulating surfaces together caused this fluid to change location, and that a flow of this fluid constitutes an electric current. He also posited that when matter contained too little of the fluid it was "negatively" charged, and when it had an excess it was "positively" charged. For a reason that was not recorded, he identified the term "positive" with vitreous electricity and "negative" with resinous electricity. William Watson arrived at the same explanation at about the same time.

13.4   Static electricity and electric current

Static electricity and electric current are two separate phenomena. They both involve electric charge, and may occur simultaneously in the same object. Static electricity refers to the electric charge of an object and the related electrostatic discharge when two objects are brought together that are not at equilibrium. An electrostatic discharge creates a change in the charge of each of the two objects. In contrast, electric current is the flow of electric charge through an object, which produces no net loss or gain of electric charge.

13.4.1   Electrification by friction

Further information: triboelectric effect

When a piece of glass and a piece of resin—neither of which exhibit any electrical properties—are rubbed together and left with the rubbed surfaces in contact, they still exhibit no electrical properties. When separated, they attract each other.

A second piece of glass rubbed with a second piece of resin, then separated and suspended near the former pieces of glass and resin causes these phenomena:

- The two pieces of glass repel each other.
- Each piece of glass attracts each piece of resin.
- The two pieces of resin repel each other.

This attraction and repulsion is an *electrical phenomena,* and the bodies that exhibit them are said to be *electrified*, or *electrically charged.* Bodies may be electrified in many other ways, as well as by friction. The electrical properties of the two pieces of glass are similar to each other but opposite to those of the two pieces of resin: The glass attracts what the resin repels and repels what the resin attracts.

If a body electrified in any manner whatsoever behaves as the glass does, that is, if it repels the glass and attracts the resin, the body is said to be 'vitreously' electrified, and if it attracts the glass and repels the resin it is said to be 'resinously' electrified. All electrified bodies are found to be either vitreously or resinously electrified.

It is the established convention of the scientific community to define the vitreous electrification as positive, and the resinous electrification as negative. The exactly opposite properties of the two kinds of electrification justify our indicating them by opposite signs, but the application of the positive sign to one rather than to the other kind must be considered as a matter of arbitrary convention, just as it is a matter of convention in mathematical diagram to reckon positive distances towards the right hand.

No force, either of attraction or of repulsion, can be observed between an electrified body and a body not electrified.[3]

Actually, all bodies are electrified, but may appear not to be so by the relative similar charge of neighboring objects in the environment. An object further electrified + or − creates an equivalent or opposite charge by default in neighboring objects, until those charges can equalize. The effects of attraction can be observed in high-voltage experiments, while lower voltage effects are merely weaker and therefore less obvious. The attraction and repulsion forces are codified by Coulomb's Law (attraction falls off at the square of the distance, which has a corollary for acceleration in a gravitational field, suggesting that gravitation may be merely electrostatic phenomenon between relatively weak charges in terms of scale). See also the Casimir effect.

It is now known that the Franklin/Watson model was fundamentally correct. There is only one kind of electrical charge, and only one variable is required to keep track of the amount of charge.[4] On the other hand, just knowing the charge is not a complete description of the situation. Matter is composed of several kinds of electrically charged particles, and these particles have many properties, not just charge.

The most common charge carriers are the positively charged proton and the negatively charged electron. The movement of any of these charged particles constitutes an electric current. In many situations, it suffices to speak of the *conventional current* without regard to whether it is carried by positive charges moving in the direction of the conventional current or by negative charges moving in the opposite direction. This macroscopic viewpoint is an approximation that simplifies electromagnetic concepts and calculations.

At the opposite extreme, if one looks at the microscopic situation, one sees there are many ways of carrying an electric current, including: a flow of electrons; a flow of electron "holes" that act like positive particles; and both negative and positive particles (ions or other charged particles) flowing in opposite directions in an electrolytic solution or a plasma.

Beware that, in the common and important case of metallic wires, the direction of the conventional current is opposite to the drift velocity of the actual charge carriers, i.e., the electrons. This is a source of confusion for beginners.

13.5   Properties

Aside from the properties described in articles about electromagnetism, charge is a relativistic invariant. This means that any particle that has charge Q, no matter how fast it goes, always has charge Q. This property has been experimentally verified by showing that the charge of *one* helium nucleus (two protons and two neutrons bound together in a nucleus and moving around at high speeds) is the same as *two* deuterium nuclei (one proton and one neutron bound together, but moving much more slowly than they would if they were in a helium nucleus).

13.6   Conservation of electric charge

Main article: Charge conservation

The total electric charge of an isolated system remains constant regardless of changes within the system itself. This law is inherent to all processes known to physics and can be derived in a local form from gauge invariance of the wave function. The conservation of charge results in the charge-current continuity equation. More generally, the net change in charge density ϱ within a volume of integration V is equal to the area integral over the current density $\mathbf{J}$ through the closed surface $S = \partial V$, which is in turn equal to the net current I:

$$-\frac{d}{dt} \int_V \rho \, dV = \oiint_{\partial V} \mathbf{J} \cdot d\mathbf{S} = \int J dS \cos\theta = I.$$

Thus, the conservation of electric charge, as expressed by the continuity equation, gives the result:

$$I = \frac{dQ}{dt}.$$

The charge transferred between times t_i and t_f is obtained by integrating both sides:

$$Q = \int_{t_i}^{t_f} I \, dt$$

where I is the net outward current through a closed surface and Q is the electric charge contained within the volume defined by the surface.

13.7 See also

- Quantity of electricity
- SI electromagnetism units

13.8 References

[1] Two Kinds of Electrical Fluid: Vitreous and Resinous – 1733

[2] Electromagnetic Fields (2nd Edition), Roald K. Wangsness, Wiley, 1986. ISBN 0-471-81186-6 (intermediate level textbook)

[3] James Clerk Maxwell *A Treatise on Electricity and Magnetism*, pp. 32-33, Dover Publications Inc., 1954 ASIN: B000HFDK0K, 3rd ed. of 1891

[4] One Kind of Charge

13.9 External links

- How fast does a charge decay?
- Science Aid: Electrostatic charge Easy-to-understand page on electrostatic charge.
- History of the electrical units.

Chapter 14

Color charge

Color charge is a property of quarks and gluons that is related to the particles' strong interactions in the theory of quantum chromodynamics (QCD). The color charge of quarks and gluons is completely unrelated to visual perception of color,[1] because it is a property that has almost no manifestation at distances above the size of an atomic nucleus. The term *color* was chosen because the charge responsible for the strong force between particles can be analogized to the three primary colors of human vision: red, green, and blue.[2] Another color scheme is "red, yellow, and blue",[3] using paint, rather than light as the perceptible analogy.

Particles have corresponding antiparticles. A particle with red, green, or blue charge has a corresponding antiparticle in which the color charge must be the anticolor of red, green, and blue, respectively, for the color charge to be conserved in particle-antiparticle creation and annihilation. Particle physicists call these antired, antigreen, and antiblue. All three colors mixed together, or any one of these colors and its complement (or negative), is "colorless" or "white" and has a net color charge of zero. Free particles have a color charge of zero: baryons are composed of three quarks, but the individual quarks can have red, green, or blue charges, or negatives; mesons are made from a quark and antiquark, the quark can be any color, and the antiquark will have the negative of that color. This color charge differs from electromagnetic charges since electromagnetic charges have only one kind of value. Positive and negative electrical charges are the same kind of charge as they only differ by the sign.

Shortly after the existence of quarks was first proposed in 1964, Oscar W. Greenberg introduced the notion of color charge to explain how quarks could coexist inside some hadrons in otherwise identical quantum states without violating the Pauli exclusion principle. The theory of quantum chromodynamics has been under development since the 1970s and constitutes an important component of the Standard Model of particle physics.

14.1 Red, green, and blue

In QCD, a quark's color can take one of three values or charges, red, green, and blue. An antiquark can take one of three anticolors, called antired, antigreen, and antiblue (represented as cyan, magenta and yellow, respectively). Gluons are mixtures of two colors, such as red and antigreen, which constitutes their color charge. QCD considers eight gluons of the possible nine color–anticolor combinations to be unique; see *eight gluon colors* for an explanation.

The following illustrates the coupling constants for color-charged particles:

- The quark colors (red, green, blue) combine to be colorless

- The quark anticolors (antired, antigreen, antiblue) also combine to be colorless

- A hadron with 3 quarks (red, green, blue) before a color change

- Blue quark emits a blue-antigreen gluon

- Green quark has absorbed the blue-antigreen gluon and is now blue; color remains conserved

- An animation of the interaction inside a neutron. The gluons are represented as circles with the color charge in the center and the anti-color charge on the outside.

14.1.1   Field lines from color charges

Main article: Field (physics)

Analogous to an electric field and electric charges, the strong force acting between color charges can be depicted using field lines. However, the color field lines do not arc outwards from one charge to another as much, because they are pulled together tightly by gluons (within 1 fm).[4] This effect confines quarks within hadrons.

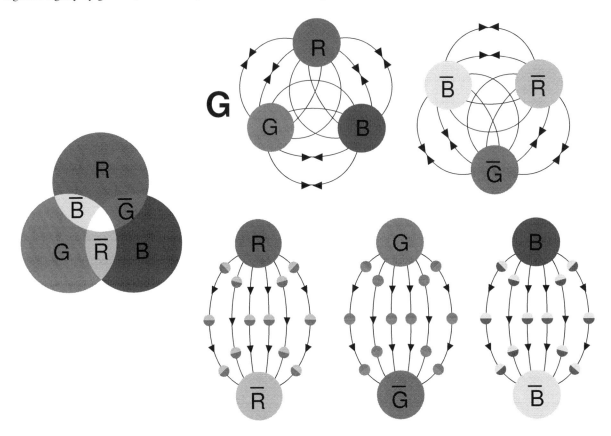

*Fields due to color charges, as in quarks (**G** is the gluon field strength tensor). These are "colorless" combinations. **Top:** Color charge has "ternary neutral states" as well as binary neutrality (analogous to electric charge). **Bottom:** Quark/antiquark combinations.[5][6]*

14.2   Coupling constant and charge

In a quantum field theory, a coupling constant and a charge are different but related notions. The coupling constant sets the magnitude of the force of interaction; for example, in quantum electrodynamics, the fine-structure constant is a coupling constant. The charge in a gauge theory has to do with the way a particle transforms under the gauge symmetry; i.e., its representation under the gauge group. For example, the electron has charge −1 and the positron has charge +1, implying that the gauge transformation has opposite effects on them in some sense. Specifically, if a local gauge transformation $\phi(x)$ is applied in electrodynamics, then one finds (using tensor index notation):

$$A_\mu \to A_\mu + \partial_\mu \phi(x) \, , \, \psi \to \exp[iQ\phi(x)]\psi \text{ and } \overline{\psi} \to \exp[-iQ\phi(x)]\overline{\psi}$$

where A_μ is the photon field, and ψ is the electron field with $Q = -1$ (a bar over ψ denotes its antiparticle — the positron). Since QCD is a non-abelian theory, the representations, and hence the color charges, are more complicated. They are dealt with in the next section.

14.3   Quark and gluon fields and color charges

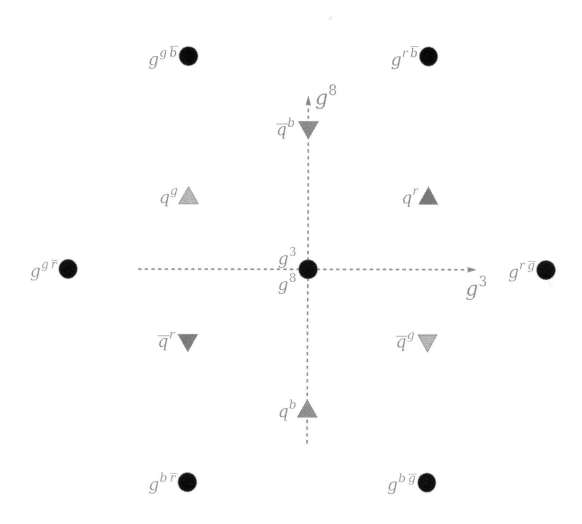

The pattern of strong charges for the three colors of quark, three antiquarks, and eight gluons (with two of zero charge overlapping).

In QCD the gauge group is the non-abelian group SU(3). The *running coupling* is usually denoted by α_s. Each flavor of quark belongs to the fundamental representation (**3**) and contains a triplet of fields together denoted by ψ. The antiquark field belongs to the complex conjugate representation (**3***) and also contains a triplet of fields. We can write

$$\psi = \begin{pmatrix} \psi_1 \\ \psi_2 \\ \psi_3 \end{pmatrix} \text{ and } \overline{\psi} = \begin{pmatrix} \overline{\psi}_1^* \\ \overline{\psi}_2^* \\ \overline{\psi}_3^* \end{pmatrix}.$$

The gluon contains an octet of fields (see gluon field), and belongs to the adjoint representation (**8**), and can be written using the Gell-Mann matrices as

$$\mathbf{A}_\mu = A_\mu^a \lambda_a.$$

(there is an implied summation over $a = 1, 2, \dots 8$). All other particles belong to the trivial representation (**1**) of color SU(3). The **color charge** of each of these fields is fully specified by the representations. Quarks have a color charge of red, green or blue and antiquarks have a color charge of antired, antigreen or antiblue. Gluons have a combination of two color charges (one of red, green or blue and one of antired, antigreen and antiblue) in a superposition of states which are given by the Gell-Mann matrices. All other particles have zero color charge. Mathematically speaking, the color charge of a particle is the value of a certain quadratic Casimir operator in the representation of the particle.

In the simple language introduced previously, the three indices "1", "2" and "3" in the quark triplet above are usually identified with the three colors. The colorful language misses the following point. A gauge transformation in color SU(3) can be written as $\psi \to U \psi$, where U is a 3×3 matrix which belongs to the group SU(3). Thus, after gauge transformation, the new colors are linear combinations of the old colors. In short, the simplified language introduced before is not gauge invariant.

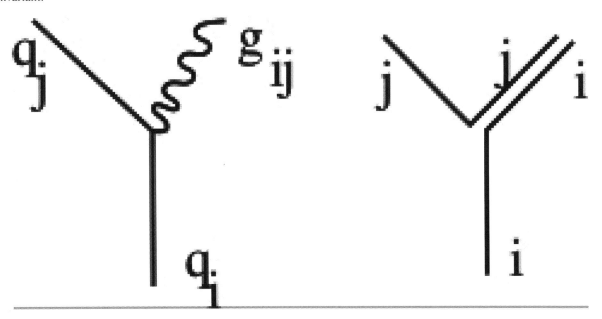

Color-line representation of QCD vertex

Color charge is conserved, but the book-keeping involved in this is more complicated than just adding up the charges, as is done in quantum electrodynamics. One simple way of doing this is to look at the interaction vertex in QCD and replace it by a color-line representation. The meaning is the following. Let ψ_i represent the i-th component of a quark field (loosely called the i-th color). The *color* of a gluon is similarly given by **A** which corresponds to the particular Gell-Mann matrix it is associated with. This matrix has indices i and j. These are the *color labels* on the gluon. At the interaction vertex one has qi → gi j + qj. The **color-line** representation tracks these indices. Color charge conservation means that the ends of these color-lines must be either in the initial or final state, equivalently, that no lines break in the middle of a diagram.

Since gluons carry color charge, two gluons can also interact. A typical interaction vertex (called the three gluon vertex) for gluons involves g + g → g. This is shown here, along with its color-line representation. The color-line diagrams can be

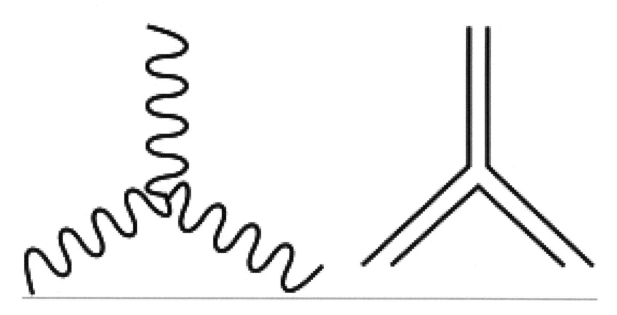

Color-line representation of 3-gluon vertex

restated in terms of conservation laws of color; however, as noted before, this is not a gauge invariant language. Note that in a typical non-abelian gauge theory the gauge boson carries the charge of the theory, and hence has interactions of this kind; for example, the W boson in the electroweak theory. In the electroweak theory, the W also carries electric charge, and hence interacts with a photon.

14.4   See also

- Color confinement

- Gluon field strength tensor

14.5   References

[1] Feynman, Richard (1985), *QED: The Strange Theory of Light and Matter*, Princeton University Press, p. 136, ISBN 0-691-08388-6, The idiot physicists, unable to come up with any wonderful Greek words anymore, call this type of polarization by the unfortunate name of 'color,' which has nothing to do with color in the normal sense.

[2] Close (2007)

[3] R. Penrose (2005). *The Road to Reality*. Vintage books. p. 648. ISBN 978-00994-40680.

[4] R. Resnick, R. Eisberg (1985), *Quantum Physics of Atoms, Molecules, Solids, Nuclei and Particles* (2nd ed.), John Wiley & Sons, p. 684, ISBN 978-0-471-87373-0

[5] Parker, C.B. (1994), *McGraw Hill Encyclopaedia of Physics* (2nd ed.), Mc Graw Hill, ISBN 0-07-051400-3

[6] M. Mansfield, C. O'Sullivan (2011), *Understanding Physics* (4th ed.), John Wiley & Sons, ISBN 978-0-47-0746370

14.6   Further reading

- Georgi, Howard (1999), *Lie algebras in particle physics*, Perseus Books Group, ISBN 0-7382-0233-9.

- Griffiths, David J. (1987), *Introduction to Elementary Particles*, New York: John Wiley & Sons, ISBN 0-471-60386-4.

- Christman, J. Richard (2001), "Colour and Charm" (PDF), *Project PHYSNET document MISN-0-283* External link in |work= (help).

- Hawking, Stephen (1998), *A Brief History of Time*, Bantam Dell Publishing Group, ISBN 978-0-553-10953-5.

- Close, Frank (2007), *The New Cosmic Onion*, Taylor & Francis, ISBN 1-58488-798-2.

Chapter 15

Doublet state

In quantum mechanics, a **doublet** is a quantum state of a system with a spin of 1/2, such that there are two allowed values of the spin component, −1/2 and +1/2. Quantum systems with two possible states are sometimes called two-level systems. Essentially all occurrences of doublets in nature arise from rotational symmetry; spin 1/2 is associated with the fundamental representation of the Lie group SU(2), the group that defines rotational symmetry in three-dimensional space.

15.1 History and applications

The term "doublet" dates back to the 19th century, when it was observed that certain spectral lines of an ionized, excited gas would split into two under the influence of a strong magnetic field, in an effect known as the anomalous Zeeman effect. Such spectral lines were observed not only in the laboratory, but also in astronomical spectroscopy observations, allowing astronomers to deduce the existence of, and measure the strength of magnetic fields around the sun, stars and galaxies. Conversely, it was the observation of doublets in spectroscopy that allowed physicists to deduce that the electron had a spin, and that furthermore, the magnitude of the spin had to be 1/2. See the history section of the article on spin (physics) for greater detail.

Doublets continue to play an important role in physics. For example, the healthcare technology of magnetic resonance imaging is based on nuclear magnetic resonance. In this technology, a spectroscopic doublet occurs in a spin-1/2 atomic nucleus, whose doublet splitting is in the radio-frequency range. By applying both a magnetic field and carefully tuning a radio-frequency transmitter, the nuclear spins will flip and re-emit radiation, in an effect known as the Rabi cycle. The strength and frequency of the emitted radio waves allows the concentration of such nuclei to be measured.

15.2 References

for verification : IUPAC. Compendium of Chemical Terminology, 2nd ed. (the "Gold Book"). Compiled by A. D. McNaught and A. Wilkinson. Blackwell Scientific Publications, Oxford (1997). XML on-line corrected version: (2006-) created by M. Nic, J. Jirat, B. Kosata; updates compiled by A. Jenkins. ISBN 0-9678550-9-8. doi:10.1351/goldbook

15.3 See also

- Singlet state
- Triplet state
- Spin multiplicity

Chapter 16

Weak isospin

In particle physics, **weak isospin** is a quantum number relating to the weak interaction, and parallels the idea of isospin under the strong interaction. Weak isospin is usually given the symbol T or I with the third component written as Tz, T_3, I_z or I_3.[1] Weak isospin is a complement of the weak hypercharge, which unifies weak interactions with electromagnetic interactions. It can be understood as the eigenvalue of a charge operator.

The **weak isospin conservation law** relates the conservation of T_3; all weak interactions must preserve T_3. It is also conserved by the other interactions and is therefore a conserved quantity in general. For this reason T_3 is more important than T and often the term "weak isospin" refers to the "3rd component of weak isospin".

16.1 Relation with chirality

Fermions with negative chirality (also called left-handed fermions) have $T = \frac{1}{2}$ and can be grouped into doublets with $T_3 = \pm\frac{1}{2}$ that behave the same way under the weak interaction. For example, up-type quarks (u, c, t) have $T_3 = +\frac{1}{2}$ and always transform into down-type quarks (d, s, b), which have $T_3 = -\frac{1}{2}$, and vice versa. On the other hand, a quark never decays weakly into a quark of the same T_3. Something similar happens with left-handed leptons, which exist as doublets containing a charged lepton (e−, μ−, τ−) with $T_3 = -\frac{1}{2}$ and a neutrino (ν

e, ν

μ, ν

τ) with $T_3 = \frac{1}{2}$.

Fermions with positive chirality (also called right-handed fermions) have $T = 0$ and form singlets that do not undergo weak interactions.

Electric charge, Q, is related to weak isospin, T_3, and weak hypercharge, YW, by

$$Q = T_3 + \frac{Y_{\mathrm{W}}}{2}.$$

16.2 Weak isospin and the W bosons

The symmetry associated with spin is SU(2). This requires gauge bosons to transform between weak isospin charges: bosons W+, W− and W0. This implies that W bosons have a $T = 1$, with three different values of T_3.

- W+ boson ($T_3 = +1$) is emitted in transitions $\{(T_3 = +\frac{1}{2}) \rightarrow (T_3 = -\frac{1}{2})\}$,

- W− boson ($T_3 = -1$) is emitted in transitions $\{(T_3 = -\frac{1}{2}) \rightarrow (T_3 = +\frac{1}{2})\}$.

- W0 boson ($T_3 = 0$) would be emitted in reactions where T_3 does not change. However, under electroweak unification, the W0 boson mixes with the weak hypercharge gauge boson B, resulting in the observed Z0 boson and the photon of Quantum Electrodynamics.

16.3   See also

- Field theoretical formulation of standard model

- Weak hypercharge

16.4   References

[1] Ambiguities: I is also used as sign for the 'normal' isospin, same for the third component I_3 aka I_z. T is also used as the sign for Topness. This article uses T and T_3.

Chapter 17

Tachyonic field

A **tachyonic field**, or simply **tachyon**, is a quantum field with an imaginary mass.[1] Although tachyons (particles that move faster than light) are a purely hypothetical concept, fields with imaginary mass have come to play an important role in modern physics[2][3][4] and are discussed in popular books on physics.[1][5] Under no circumstances do any excitations ever propagate faster than light in such theories — the presence or absence of a tachyonic mass has no effect whatsoever on the maximum velocity of signals (there is no violation of causality).[6]

The term "tachyon" was coined by Gerald Feinberg in a 1967 paper[7] that studied quantum fields with imaginary mass. Feinberg believed such fields permitted faster than light propagation, but it was soon realized that Feinberg's model in fact did not allow for superluminal speeds.[6] Instead, the imaginary mass creates an instability in the configuration:- any configuration in which one or more field excitations are tachyonic will spontaneously decay, and the resulting configuration contains no physical tachyons. This process is known as tachyon condensation. A famous example is the condensation of the Higgs boson in the Standard Model of particle physics.

Although the notion of a tachyonic imaginary mass might seem troubling because there is no classical interpretation of an imaginary mass, the mass is not quantized. Rather, the scalar field is; even for tachyonic quantum fields, the field operators at spacelike separated points still commute (or anticommute), thus preserving causality. Therefore information still does not propagate faster than light,[8] and solutions grow exponentially, but not superluminally (there is no violation of causality). Tachyon condensation drives a physical system that has reached a local limit and might naively be expected to produce physical tachyons, to an alternate stable state where no physical tachyons exist. Once the tachyonic field reaches the minimum of the potential, its quanta are not tachyons any more but rather are ordinary particles with a positive mass-squared.[9]

In modern physics, all fundamental particles are regarded as localized excitations of fields. Tachyons are unusual because the instability prevents any such localized excitations from existing. Any localized perturbation, no matter how small, starts an exponentially growing cascade that strongly affects physics everywhere inside the future light cone of the perturbation.[6]

17.1 Interpretation

17.1.1 Overview of tachyonic condensation

Main article: Tachyon condensation

Although the notion of a tachyonic imaginary mass might seem troubling because there is no classical interpretation of an imaginary mass, the mass is not quantized. Rather, the scalar field is; even for tachyonic quantum fields, the field operators at spacelike separated points still commute (or anticommute), thus preserving causality. Therefore information still does not propagate faster than light,[8] and solutions grow exponentially, but not superluminally (there is no violation of causality).

The "imaginary mass" really means that the system becomes unstable. The zero value field is at a local maximum rather than a local minimum of its potential energy, much like a ball at the top of a hill. A very small impulse (which will always happen due to quantum fluctuations) will lead the field to roll down with exponentially increasing amplitudes toward the local minimum. In this way, tachyon condensation drives a physical system that has reached a local limit and might naively be expected to produce physical tachyons, to an alternate stable state where no physical tachyons exist. Once the tachyonic field reaches the minimum of the potential, its quanta are not tachyons any more but rather are ordinary particles with a positive mass-squared, such as the Higgs boson.[9]

17.1.2   Physical interpretation of a tachyonic field and signal propagation

There is a simple mechanical analogy that illustrates that tachyonic fields do not propagate faster than light, why they represent instabilities, and helps explain the meaning of imaginary mass (negative squared mass).[6]

Consider a long line of pendulums, all pointing straight down. The mass on the end of each pendulum is connected to the masses of its two neighbors by springs. Wiggling one of the pendulums will create two ripples that propagate in both directions down the line. As the ripple passes, each pendulum in its turn oscillates a few times about the straight down position. The speed of propagation of these ripples is determined in a simple way by the tension of the springs and the inertial mass of the pendulum weights. Formally, these parameters can be chosen so that the propagation speed is the speed of light. In the limit of an infinite density of closely spaced pendulums, this model becomes identical to a relativistic field theory, where the ripples are the analog of particles. Displacing the pendulums from pointing straight down requires positive energy, which indicates that the squared mass of those particles is positive.

Now consider an initial condition where at time t=0, all the pendulums are pointing straight up. Clearly this is unstable, but at least in classical physics one can imagine that they are so carefully balanced they will remain pointing straight up indefinitely so long as they are not perturbed. Wiggling one of the upside-down pendulums will have a very different effect from before. The speed of propagation of the effects of the wiggle is identical to what it was before, since neither the spring tension nor the inertial mass have changed. However, the effects on the pendulums affected by the perturbation are dramatically different. Those pendulums that feel the effects of the perturbation will begin to topple over, and will pick up speed exponentially. Indeed, it is easy to show that any localized perturbation kicks off an exponentially growing instability that affects everything within its future "ripple cone" (a region of size equal to time multiplied by the ripple propagation speed). In the limit of infinite pendulum density, this model is a tachyonic field theory.

17.2   Importance in physics

Tachyonic fields play a very important role in modern physics. Perhaps the most famous example of a tachyon is the Higgs boson of the Standard model of particle physics. In its uncondensed phase, the Higgs field has a negative mass squared and is, therefore, a tachyon.

The phenomenon of spontaneous symmetry breaking, which is closely related to tachyon condensation, plays a central part in many aspects of theoretical physics, including the Ginzburg–Landau and BCS theories of superconductivity.

Other examples include the inflaton field in certain models of cosmic inflation (such as new inflation[10][11]), and the tachyon of bosonic string theory.[5][12][13]

17.3   Condensation

Main article: Tachyon condensation

In quantum field theory, a tachyon is a quantum of a field—usually a scalar field—whose squared mass is negative, and is used to describe spontaneous symmetry breaking: The existence of such a field implies the instability of the field vacuum; the field is at a local maximum rather than a local minimum of its potential energy, much like a ball at the top of a hill. A very small impulse (which will always happen due to quantum fluctuations) will lead the field (ball) to roll down with exponentially increasing amplitudes: it will induce tachyon condensation. It is important to realize that once the tachyonic field reaches the minimum of the potential, its quanta are not tachyons any more but rather have a positive mass-squared. The Higgs boson of the standard model of particle physics is an example.[9]

Technically, the squared mass is the second derivative of the effective potential. For a tachyonic field the second derivative is negative, meaning that the effective potential is at a local maximum rather than a local minimum. Therefore this situation is unstable and the field will roll down the potential.

Because a tachyon's squared mass is negative, it formally has an imaginary mass. This is a special case of the general rule, where unstable massive particles are formally described as having a complex mass, with the real part being their mass in usual sense, and the imaginary part being the decay rate in natural units.[9]

However, in quantum field theory, a particle (a "one-particle state") is roughly defined as a state which is constant over time; i.e., an eigenvalue of the Hamiltonian. An unstable particle is a state which is only approximately constant over time; If it exists long enough to be measured, it can be formally described as having a complex mass, with the real part of the mass greater than its imaginary part. If both parts are of the same magnitude, this is interpreted as a resonance appearing in a scattering process rather than particle, as it is considered not to exist long enough to be measured independently of the scattering process. In the case of a tachyon the real part of the mass is zero, and hence no concept of a particle can be attributed to it.

Even for tachyonic quantum fields, the field operators at space-like separated points still commute (or anticommute), thus preserving the principle of causality. For closely related reasons, the maximum velocity of signals sent with a tachyonic field is strictly bounded from above by the speed of light.[6] Therefore information never moves faster than light regardless of the presence or absence of tachyonic fields.

Examples for tachyonic fields are all cases of spontaneous symmetry breaking. In condensed matter physics a notable example is ferromagnetism; in particle physics the best known example is the Higgs mechanism in the standard model.

17.4 Tachyons in string theory

In string theory, tachyons have the same interpretation as in quantum field theory. However, string theory can, at least, in principle, not only describe the physics of tachyonic fields, but also predict whether such fields appear.

Tachyonic fields indeed arise in many versions of string theory. In general, string theory states that what we see as "particles" (electrons, photons, gravitons and so forth) are actually different vibrational states of the same underlying string. The mass of the particle can be deduced from the vibrations which the string exhibits; roughly speaking, the mass depends upon the "note" which the string sounds. Tachyons frequently appear in the spectrum of permissible string states, in the sense that some states have negative mass-squared, and therefore, imaginary mass. If the tachyon appears as a vibrational mode of an open string, this signals an instability of the underlying D-brane system to which the string is attached.[14] The system will then decay to a state of closed strings and/or stable D-branes. If the tachyon is a closed string vibrational mode, this indicates an instability in spacetime itself. Generally, it is not known (or theorized) what this system will decay to. However, if the closed string tachyon is localized around a spacetime singularity, the endpoint of the decay process will often have the singularity resolved.

17.5 See also

- D-brane
- Standard-Model Extension
- Poincaré group
- Quantum tunnelling
- Tachyonic antitelephone

17.6 References

[1] Lisa Randall, *Warped Passages: Unraveling the Mysteries of the Universe's Hidden Dimensions*, p.286: "People initially thought of tachyons as particles travelling faster than the speed of light...But we now know that a tachyon indicates an instability in a theory that contains it. Regrettably for science fiction fans, tachyons are not real physical particles that appear in nature."

[2] Sen, Ashoke (April 2002). "Rolling Tachyon". *J. High Energy Phys.* **2002** (0204): 048. arXiv:hep-th/0203211.Bibcode:20. doi:10.1088/1126-6708/2002/04/048.

[3] G. W. Gibbons, "Cosmological evolution of the rolling tachyon," Phys. Lett. B **537**, 1 (2002)

[4] Kutasov, David; Marino, Marcos & Moore, Gregory W. (2000). "Some exact results on tachyon condensation in string field theory". *JHEP* **0010**: 045.

[5] Brian Greene, *The Elegant Universe*, Vintage Books (2000)

[6] Aharonov, Y.; Komar, A.; Susskind, L. (1969). "Superluminal Behavior, Causality, and Instability". *Phys. Rev.* (American Physical Society) **182** (5): 1400–1403. Bibcode:1969PhRv..182.1400A. doi:10.1103/PhysRev.182.1400.

[7] Feinberg, G. (1967). "Possibility of Faster-Than-Light Particles".*Physical Review***159**(5): 1089–1105.Bibcode:1967PhRv.. doi:10.1103/PhysRev.159.1089.

[8] Feinberg, Gerald(1967). "Possibility of Faster-Than-Light Particles".*Physical Review***159**(5): 1089–1105.Bibcode:1967. doi:10.1103/PhysRev.159.1089.

[9] Michael E. Peskin and Daniel V. Schroeder (1995). *An Introduction to Quantum Field Theory*, Perseus books publishing.

[10] Linde, A (1982). "A new inflationary universe scenario: A possible solution of the horizon, flatness, homogeneity, isotropy and primordial monopole problems". *Physics Letters B* **108** (6): 389–393. Bibcode:1982PhLB..108..389L. doi:10.1016/0370-2693(82)91219-9.

[11] Albrecht, Andreas; Steinhardt, Paul (1982). "Cosmology for Grand Unified Theories with Radiatively Induced Symmetry Breaking" (PDF). *Physical Review Letters* **48** (17): 1220–1223. Bibcode:1982PhRvL..48.1220A. doi:

[12] J. Polchinski, *String Theory*, Cambridge University Press, Cambridge, UK (1998)

[13] NOVA, "The Elegant Universe", PBS television special, http://www.pbs.org/wgbh/nova/elegant/

[14] Sen, A. (1998). "Tachyon condensation on the brane antibrane system". *Journal of High Energy Physics* **8** (8): 12. arXiv:hep-th/9805170. Bibcode:1998JHEP...08..012S. doi:10.1088/1126-6708/1998/08/012.

17.7 External links

- The Faster Than Light (FTL) FAQ (from the Internet Archive)

- Weisstein, Eric W., Tachyon from ScienceWorld.

- Tachyon entry from the *Physics FAQ*

Chapter 18

Tachyon condensation

Tachyon condensation is a process in particle physics in which the system can lower its energy by spontaneously producing particles. The end result is a "condensate" of particles that fills the volume of the system. Tachyon condensation is closely related to second-order phase transitions.

18.1 Technical overview

Tachyon condensation is a process in which a tachyonic field—usually a scalar field—with a complex mass acquires a vacuum expectation value and reaches the minimum of the potential energy. While the field is tachyonic and unstable near the local maximum of the potential, the field gets a non-negative squared mass and becomes stable near the minimum.

The appearance of tachyons is a potentially serious problem for any theory; examples of tachyonic fields amenable to condensation are all cases of spontaneous symmetry breaking. In condensed matter physics a notable example is ferromagnetism; in particle physics the best known example is the Higgs mechanism in the standard model that breaks the electroweak symmetry.

18.2 Condensation evolution

Although the notion of a tachyonic imaginary mass might seem troubling because there is no classical interpretation of an imaginary mass, the mass is not quantized. Rather, the scalar field is; even for tachyonic quantum fields, the field operators at spacelike separated points still commute (or anticommute), thus preserving causality. Therefore information still does not propagate faster than light,[1] and solutions grow exponentially, but not superluminally (there is no violation of causality).

The "imaginary mass" really means that the system becomes unstable. The zero value field is at a local maximum rather than a local minimum of its potential energy, much like a ball at the top of a hill. A very small impulse (which will always happen due to quantum fluctuations) will lead the field to roll down with exponentially increasing amplitudes toward the local minimum. In this way, tachyon condensation drives a physical system that has reached a local limit and might naively be expected to produce physical tachyons, to an alternate stable state where no physical tachyons exist. Once the tachyonic field reaches the minimum of the potential, its quanta are not tachyons any more but rather are ordinary particles with a positive mass-squared, such as the Higgs boson.[2]

18.3 Tachyon condensation in string theory

In the late 1990s, Ashoke Sen conjectured[3] that the tachyons carried by open strings attached to D-branes in string theory reflect the instability of the D-branes with respect to their complete annihilation. The total energy carried by these

137

tachyons has been calculated in string field theory; it agrees with the total energy of the D-branes, and all other tests have confirmed Sen's conjecture as well. Tachyons therefore became an active area of interest in the early 2000s.

The character of closed-string tachyon condensation is more subtle, though the first steps towards our understanding of their fate have been made by Adams, Polchinski, and Silverstein, in the case of twisted closed string tachyons, and by Simeon Hellerman and Ian Swanson, in a wider array of cases. The fate of the closed string tachyon in the 26-dimensional bosonic string theory remains unknown, though recent progress has revealed interesting new developments.

18.4   See also

- Bose–Einstein condensation — a condensation process that was experimentally observed 70 years after it was theoretically proposed.

18.5   References

[1] Feinberg, Gerald (1967). "Possibility of Faster-Than-Light Particles". *Physical Review* **159** (5): 1089–1105. Bibcode:1967PhRv. doi:10.1103/PhysRev.159.1089.

[2] Michael E. Peskin and Daniel V. Schroeder (1995). *An Introduction to Quantum Field Theory*, Perseus books publishing.

[3] Sen, Ashoke (1998). "Tachyon condensation on the brane antibrane system". *JHEP* **8** (8): 012–012. arXiv:hep-th/9805170. Bibcode:1998JHEP...08..012S. doi:10.1088/1126-6708/1998/08/012.

18.6   External links

- Tachyon condensation on arxiv.org

Chapter 19

Gauge boson

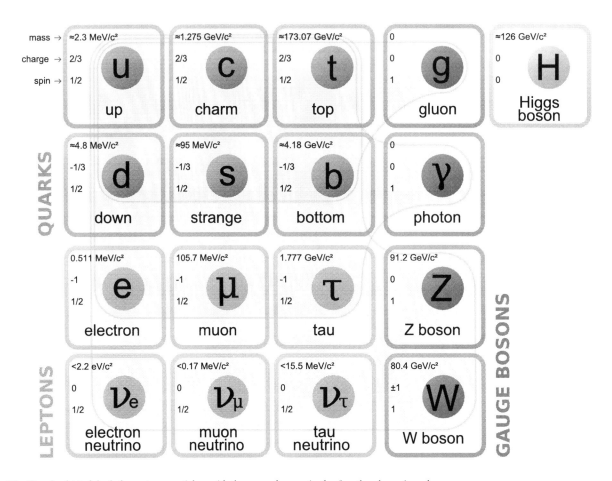

The Standard Model of elementary particles, with the gauge bosons in the fourth column in red

In particle physics, a **gauge boson** is a force carrier, a bosonic particle that carries any of the fundamental interactions of nature.[1][2] Elementary particles, whose interactions are described by a gauge theory, interact with each other by the exchange of gauge bosons—usually as virtual particles.

19.1   Gauge bosons in the Standard Model

The Standard Model of particle physics recognizes four kinds of gauge bosons: photons, which carry the electromagnetic interaction; W and Z bosons, which carry the weak interaction; and gluons, which carry the strong interaction.[3]

Isolated gluons do not occur at low energies because they are color-charged, and subject to color confinement.

19.1.1   Multiplicity of gauge bosons

In a quantized gauge theory, gauge bosons are quanta of the gauge fields. Consequently, there are as many gauge bosons as there are generators of the gauge field. In quantum electrodynamics, the gauge group is $U(1)$; in this simple case, there is only one gauge boson. In quantum chromodynamics, the more complicated group $SU(3)$ has eight generators, corresponding to the eight gluons. The three W and Z bosons correspond (roughly) to the three generators of $SU(2)$ in GWS theory.

19.1.2   Massive gauge bosons

For technical reasons involving gauge invariance, gauge bosons are described mathematically by field equations for massless particles. Therefore, at a naïve theoretical level all gauge bosons are required to be massless, and the forces that they describe are required to be long-ranged. The conflict between this idea and experimental evidence that the weak and strong interactions have a very short range requires further theoretical insight.

According to the Standard Model, the W and Z bosons gain mass via the Higgs mechanism. In the Higgs mechanism, the four gauge bosons (of $SU(2){\times}U(1)$ symmetry) of the unified electroweak interaction couple to a Higgs field. This field undergoes spontaneous symmetry breaking due to the shape of its interaction potential. As a result, the universe is permeated by a nonzero Higgs vacuum expectation value (VEV). This VEV couples to three of the electroweak gauge bosons (the Ws and Z), giving them mass; the remaining gauge boson remains massless (the photon). This theory also predicts the existence of a scalar Higgs boson, which has been observed in experiments that were reported on 4 July 2012.[4]

19.2   Beyond the Standard Model

19.2.1   Grand unification theories

A grand unified theory predicts additional gauge bosons named X and Y bosons. The hypothetical X and Y bosons direct interactions between quarks and leptons, hence violating conservation of baryon number and causing proton decay. Such bosons would be even more massive than W and Z bosons due to symmetry breaking. Analysis of data collected from such sources as the Super-Kamiokande neutrino detector has yielded no evidence of X and Y bosons.

19.2.2   Gravitons

The fourth fundamental interaction, gravity, may also be carried by a boson, called the graviton. In the absence of experimental evidence and a mathematically coherent theory of quantum gravity, it is unknown whether this would be a gauge boson or not. The role of gauge invariance in general relativity is played by a similar symmetry: diffeomorphism invariance.

19.2.3   W' and Z' bosons

Main article: W' and Z' bosons

W' and Z' bosons refer to hypothetical new gauge bosons (named in analogy with the Standard Model W and Z bosons).

19.3 See also

- 1964 PRL symmetry breaking papers

- Boson

- Glueball

- Quantum chromodynamics

- Quantum electrodynamics

19.4 References

[1] Gribbin, John (2000). *Q is for Quantum – An Encyclopedia of Particle Physics*. Simon & Schuster. ISBN 0-684-85578-X.

[2] Clark, John, E.O. (2004). *The Essential Dictionary of Science*. Barnes & Noble. ISBN 0-7607-4616-8.

[3] Veltman, Martinus (2003). *Facts and Mysteries in Elementary Particle Physics*. World Scientific. ISBN 981-238-149-X.

[4] "CERN experiments observe particle consistent with long-sought Higgs boson". CERN. Retrieved 4 July 2012.

19.5 External links

- Explanation of gauge boson and gauge fields by Christopher T. Hill

Chapter 20

W and Z bosons

The **W and Z bosons** (together known as the **weak bosons** or, less specifically, the **intermediate vector bosons**) are the elementary particles that mediate the weak interaction; their symbols are W+, W−, and Z. The W bosons have a positive and negative electric charge of 1 elementary charge respectively and are each other's antiparticles. The Z boson is electrically neutral and is its own antiparticle. The three particles have a spin of 1, and the W bosons have a magnetic moment, while the Z has none. All three of these particles are very short-lived, with a half-life of about 3×10^{-25} s. Their discovery was a major success for what is now called the Standard Model of particle physics.

The W bosons are named after the *w*eak force. The physicist Steven Weinberg named the additional particle the "Z particle",[3] later giving the explanation that it was the last additional particle needed by the model – the W bosons had already been named – and that it has *z*ero electric charge.[4]

The two **W bosons** are best known as mediators of neutrino absorption and emission, where their charge is associated with electron or positron emission or absorption, always causing nuclear transmutation. The Z boson is not involved in the absorption or emission of electrons and positrons.

The **Z boson** mediates the transfer of momentum, spin, and energy when neutrinos scatter *elastically* from matter, something that must happen without the production or absorption of new, charged particles. Such behaviour (which is almost as common as inelastic neutrino interactions) is seen in bubble chambers irradiated with neutrino beams. Whenever an electron simply "appears" in such a chamber as a new free particle suddenly moving with kinetic energy, and moves in the direction of the neutrinos as the apparent result of a new impulse, and this behavior happens more often when the neutrino beam is present, it is inferred to be a result of a neutrino interacting directly with the electron. Here, the neutrino simply strikes the electron and scatters away from it, transferring some of the neutrino's momentum to the electron. Since (i) neither neutrinos nor electrons are affected by the strong force, (ii) neutrinos are electrically neutral (therefore don't interact electromagnetically), and (iii) the incredibly small masses of these particles make any gravitational force between them negligible, such an interaction can only happen via the weak force. Since such an electron is not created from a nucleon, and is unchanged except for the new force impulse imparted by the neutrino, this weak force interaction between the neutrino and the electron must be mediated by a weak-force boson particle with no charge. Thus, this interaction requires a Z boson.

20.1 Basic properties

These bosons are among the heavyweights of the elementary particles. With masses of 80.4 GeV/c^2 and 91.2 GeV/c^2, respectively, the W and Z bosons are almost 100 times as large as the proton – heavier, even, than entire atoms of iron. The masses of these bosons are significant because they act as the force carriers of a quite short-range fundamental force: their high masses thus limit the range of the weak nuclear force. By way of contrast, the electromagnetic force has an infinite range, because its force carrier, the photon, has zero mass, and the same is supposed of the hypothetical graviton.

All three bosons have particle spin $s = 1$. The emission of a W+ or W− boson either raises or lowers the electric charge of the emitting particle by one unit, and also alters the spin by one unit. At the same time, the emission or absorption of a W boson can change the type of the particle – for example changing a strange quark into an up quark. The neutral Z boson cannot change the electric charge of any particle, nor can it change any other of the so-called "charges" (such as strangeness, baryon number, charm, etc.). The emission or absorption of a Z boson can only change the spin, momentum, and energy of the other particle. (See also *weak neutral current*.)

20.2 Weak nuclear force

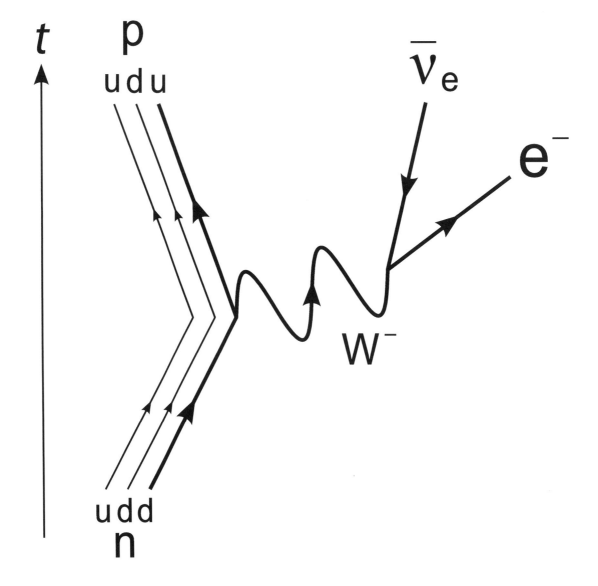

The Feynman diagram for beta decay of a neutron into a proton, electron, and electron antineutrino via an intermediate heavy W boson

The W and Z bosons are carrier particles that mediate the weak nuclear force, much as the photon is the carrier particle for the electromagnetic force.

20.2.1   W bosons

The W bosons are best known for their role in nuclear decay. Consider, for example, the beta decay of cobalt-60.

$$^{60}_{27}\text{Co} \rightarrow {}^{60}_{28}\text{Ni}^+ + e- + \nu_e$$

This reaction does not involve the whole cobalt-60 nucleus, but affects only one of its 33 neutrons. The neutron is converted into a proton while also emitting an electron (called a beta particle in this context) and an electron antineutrino:

$$n0 \rightarrow p+ + e- + \nu_e$$

Again, the neutron is not an elementary particle but a composite of an up quark and two down quarks (udd). It is in fact one of the down quarks that interacts in beta decay, turning into an up quark to form a proton (uud). At the most fundamental level, then, the weak force changes the flavour of a single quark:

$$d \rightarrow u + W-$$

which is immediately followed by decay of the W− itself:

$$W- \rightarrow e- + \nu_e$$

20.2.2   Z boson

The Z boson is its own antiparticle. Thus, all of its flavour quantum numbers and charges are zero. The exchange of a Z boson between particles, called a neutral current interaction, therefore leaves the interacting particles unaffected, except for a transfer of momentum. Z boson interactions involving neutrinos have distinctive signatures: They provide the only known mechanism for elastic scattering of neutrinos in matter; neutrinos are almost as likely to scatter elastically (via Z boson exchange) as inelastically (via W boson exchange). The first prediction of Z bosons was made by Brazilian physicist José Leite Lopes in 1958,[5] by devising an equation which showed the analogy of the weak nuclear interactions with electromagnetism. Steve Weinberg, Sheldon Glashow and Abdus Salam used later these results to develop the electroweak unification,[6] in 1973. Weak neutral currents via Z boson exchange were confirmed shortly thereafter in 1974, in a neutrino experiment in the Gargamelle bubble chamber at CERN.

20.3   Predicting the W and Z

Following the spectacular success of quantum electrodynamics in the 1950s, attempts were undertaken to formulate a similar theory of the weak nuclear force. This culminated around 1968 in a unified theory of electromagnetism and weak interactions by Sheldon Glashow, Steven Weinberg, and Abdus Salam, for which they shared the 1979 Nobel Prize in Physics.[7] Their electroweak theory postulated not only the W bosons necessary to explain beta decay, but also a new Z boson that had never been observed.

The fact that the W and Z bosons have mass while photons are massless was a major obstacle in developing electroweak theory. These particles are accurately described by an SU(2) gauge theory, but the bosons in a gauge theory must be massless. As a case in point, the photon is massless because electromagnetism is described by a U(1) gauge theory. Some mechanism is required to break the SU(2) symmetry, giving mass to the W and Z in the process. One explanation, the Higgs mechanism, was forwarded by the 1964 PRL symmetry breaking papers. It predicts the existence of yet another

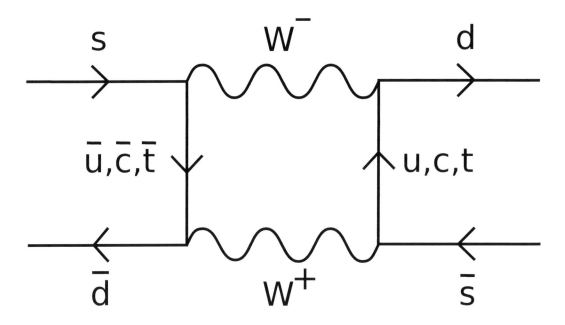

A Feynman diagram showing the exchange of a pair of W bosons. This is one of the leading terms contributing to neutral Kaon oscillation.

new particle; the Higgs boson. Of the four components of a Goldstone boson created by the Higgs field, three are "eaten" by the W^+, Z^0, and W^- bosons to form their longitudinal components and the remainder appears as the spin 0 Higgs boson.

The combination of the SU(2) gauge theory of the weak interaction, the electromagnetic interaction, and the Higgs mechanism is known as the Glashow-Weinberg-Salam model. These days it is widely accepted as one of the pillars of the Standard Model of particle physics. As of 13 December 2011, intensive search for the Higgs boson carried out at CERN has indicated that if the particle is to be found, it seems likely to be found around 125 GeV. On 4 July 2012, the CMS and the ATLAS experimental collaborations at CERN announced the discovery of a new particle with a mass of 125.3 ± 0.6 GeV that appears consistent with a Higgs boson.

20.4   Discovery

Unlike beta decay, the observation of neutral current interactions that involve particles *other than neutrinos* requires huge investments in particle accelerators and detectors, such as are available in only a few high-energy physics laboratories in the world (and then only after 1983). This is because Z-bosons behave in somewhat the same manner as photons, but do not become important until the energy of the interaction is comparable with the relatively huge mass of the Z boson.

The discovery of the W and Z bosons was considered a major success for CERN. First, in 1973, came the observation of neutral current interactions as predicted by electroweak theory. The huge Gargamelle bubble chamber photographed the tracks of a few electrons suddenly starting to move, seemingly of their own accord. This is interpreted as a neutrino interacting with the electron by the exchange of an unseen Z boson. The neutrino is otherwise undetectable, so the only observable effect is the momentum imparted to the electron by the interaction.

The discovery of the W and Z bosons themselves had to wait for the construction of a particle accelerator powerful enough to produce them. The first such machine that became available was the Super Proton Synchrotron, where unambiguous signals of W bosons were seen in January 1983 during a series of experiments made possible by Carlo Rubbia and Simon van der Meer. The actual experiments were called UA1 (led by Rubbia) and UA2 (led by Pierre Darriulat),[8] and were the collaborative effort of many people. Van der Meer was the driving force on the accelerator end (stochastic cooling). UA1 and UA2 found the Z boson a few months later, in May 1983. Rubbia and van der Meer were promptly awarded

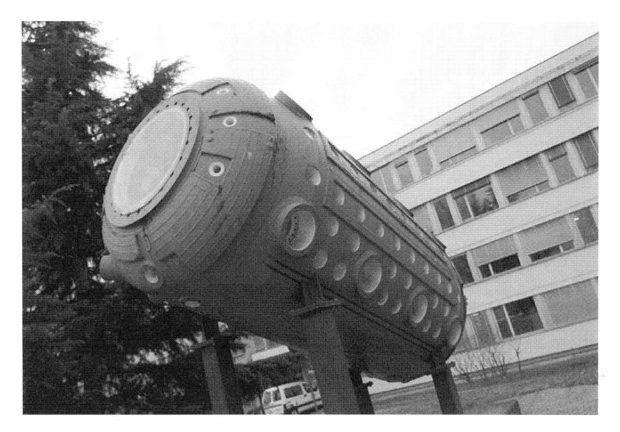

The Gargamelle bubble chamber, now exhibited at CERN

the 1984 Nobel Prize in Physics, a most unusual step for the conservative Nobel Foundation.[9]

The W+, W−, and Z0 bosons, together with the photon (γ), comprise the four gauge bosons of the electroweak interaction.

20.5   Decay

The W and Z bosons decay to fermion–antifermion pairs but neither the W nor the Z bosons can decay into the higher-mass top quark. Neglecting phase space effects and higher order corrections, simple estimates of their branching fractions can be calculated from the coupling constants.

20.5.1   W bosons

W bosons can decay to a lepton and neutrino or to an up-type quark and a down-type quark. The decay width of the W boson to a quark–antiquark pair is proportional to the corresponding squared CKM matrix element and the number of quark colours, $NC = 3$. The decay widths for the W bosons are then proportional to:

Here, e+, μ+, τ+ denote the three flavours of leptons (more exactly, the positive charged antileptons). ν

e, ν

μ, ν

τ denote the three flavours of neutrinos. The other particles, starting with u and d, all denote quarks and antiquarks (factor NC is applied). The various Vij denote the corresponding CKM matrix coefficients.

Unitarity of the CKM matrix implies that $|V_{ud}|^2 + |V_{us}|^2 + |V_{ub}|^2 = |V_{cd}|^2 + |V_{cs}|^2 + |V_{cb}|^2 = 1$. Therefore the leptonic branching ratios of the W boson are approximately B(e+ν

e) = B(μ+ν

μ) = $B(\tau + \nu$

$\tau) = \frac{1}{9}$. The hadronic branching ratio is dominated by the CKM-favored ud and cs final states. The sum of the hadronic branching ratios has been measured experimentally to be 67.60±0.27%, with $B(l^+\nu_l)$ = 10.80±0.09%.[1]

20.5.2 Z bosons

Z bosons decay into a fermion and its antiparticle. As the Z-boson is a mixture of the pre-symmetry-breaking W^0 and B^0 bosons (see weak mixing angle), each vertex factor includes a factor $T_3 - Qsin^2\theta W$, where T_3 is the third component of the weak isospin of the fermion, Q is the electric charge of the fermion (in units of the elementary charge), and θW is the weak mixing angle. Because the weak isospin is different for fermions of different chirality, either left-handed or right-handed, the coupling is different as well.

The **relative** strengths of each coupling can be estimated by considering that the decay rates include the square of these factors, and all possible diagrams (e.g. sum over quark families, and left and right contributions). This is just an estimate, as we are considering only tree-level diagrams in the Fermi theory.

Here, L and R denote the left- and right-handed chiralities of the fermions respectively. (The right-handed neutrinos do not exist in the standard model. However, in some extensions beyond the standard model they do.) The notation $x = sin^2\theta W$ is used.

20.6 See also

- Bose–Einstein statistics

- Boson

- List of particles

- Standard Model (mathematical formulation)

- W' and Z' bosons

- X and Y bosons: analogous pair of bosons predicted by the Grand Unified Theory

20.7 References

[1] J. Beringer; et al. (2012). "2012 Review of Particle Physics - Gauge and Higgs Bosons" (PDF). *Physical Review D* **86**: 1. Bibcode:2012PhRvD..86a0001B. doi:10.1103/PhysRevD.86.010001.

[2] (PDF) http://pdg.lbl.gov/2013/reviews/rpp2013-rev-w-mass.pdf. Missing or empty |title= (help)

[3] Steven Weinberg, A Model of Leptons, Phys. Rev. Lett. 19, 1264–1266 (1967) – the electroweak unification paper.

[4] Weinberg, Steven (1993). *Dreams of a Final Theory: the search for the fundamental laws of nature*. Vintage Press. p. 94. ISBN 0-09-922391-0.

[5] "Forty years of the first attempt at the electroweak unification and of the prediction of the weak neutral boson".

[6] "The Nobel Prize in Physics 1979". Nobel Foundation. Retrieved 2008-09-10.

[7] Nobel Prize in Physics for 1979 (see also Nobel Prize in Physics on Wikipedia)

[8] The UA2 Collaboration collection

[9] 1984 Nobel Prize in physics

[10] C. Amsler et al. (Particle Data Group), PL B667, 1 (2008) and 2009 partial update for the 2010 edition

20.8   External links

- The Review of Particle Physics, the ultimate source of information on particle properties.

- The W and Z particles: a personal recollection by Pierre Darriulat

- When CERN saw the end of the alphabet by Daniel Denegri

- W and Z particles at Hyperphysics

Chapter 21

Alternatives to the Standard Model Higgs

Although the Higgs boson, as included in the Standard Model, is arguably the simplest method of achieving the Higgs mechanism, it is not without problems. Consequently, particle physicists have searched for alternative models which solve one or more of these problems, including the Higgs hierarchy problem and Quantum triviality.

21.1   Overview

See also: Introduction to the Higgs field

In particle physics, elementary particles and forces give rise to the world around us. Physicists explain the behaviors of these particles and how they interact using the Standard Model—a widely accepted framework believed to explain most of the world we see around us.[1] Initially, when these models were being developed and tested, it seemed that the mathematics behind those models, which were satisfactory in areas already tested, would also forbid elementary particles from having any mass, which showed clearly that these initial models were incomplete. In 1964 three groups of physicists almost simultaneously released papers describing how masses could be given to these particles, using approaches known as symmetry breaking. This approach allowed the particles to obtain a mass, without breaking other parts of particle physics theory that were already believed reasonably correct. This idea became known as the Higgs mechanism (not the same as the Higgs boson), and later experiments confirmed that such a mechanism does exist—but they could not show exactly *how* it happens.

The simplest theory for how this effect takes place in nature, and the theory that became incorporated into the Standard Model, was that if one or more of a particular kind of "field" (known as a Higgs field) happened to permeate space, and if it could interact with elementary particles in a particular way, then this would give rise to a Higgs mechanism in nature. In the basic Standard Model there is one field and one related Higgs boson; in some extensions to the Standard Model there are multiple fields and multiple Higgs bosons.

In the years since the Higgs field and boson were proposed as a way to explain the origins of symmetry breaking, several alternatives have been proposed that suggest how a symmetry breaking mechanism could occur without requiring a Higgs field to exist. Models which do not include a Higgs field or a Higgs boson are known as Higgsless models. In these models, strongly interacting dynamics rather than an additional (Higgs) field produce the non-zero vacuum expectation value that breaks electroweak symmetry.

21.2   List of alternative models

A partial list of proposed alternatives to a Higgs field as a source for symmetry breaking includes:

- Technicolor models break electroweak symmetry through new gauge interactions, which were originally modeled on quantum chromodynamics.[2][3]

- Extra-dimensional Higgsless models use the fifth component of the gauge fields to play the role of the Higgs fields. It is possible to produce electroweak symmetry breaking by imposing certain boundary conditions on the extra dimensional fields, increasing the unitarity breakdown scale up to the energy scale of the extra dimension.[4][5] Through the AdS/QCD correspondence this model can be related to technicolor models and to "UnHiggs" models in which the Higgs field is of unparticle nature.[6]

- Models of composite W and Z vector bosons.[7][8]

- Top quark condensate.

- "Unitary Weyl gauge". If one adds a suitable gravitational term to the standard model action with gravitational coupling, the theory becomes locally scale invariant (i.e. Weyl invariant). Weyl transformations act multiplicatively on the Higgs field, so one can fix the Weyl gauge by requiring the Higgs scalar to be a constant.[9][10]

- Asymptotically safe weak interactions[11][12] based on some nonlinear sigma models.[13]

- Preon and models inspired by preons such as Ribbon model of Standard Model particles by Sundance Bilson-Thompson, based in braid theory and compatible with loop quantum gravity and similar theories.[14] This model not only explains mass but leads to an interpretation of electric charge as a topological quantity (twists carried on the individual ribbons) and colour charge as modes of twisting.

- Symmetry breaking driven by non-equilibrium dynamics of quantum fields above the electroweak scale.[15][16]

- Unparticle physics and the unhiggs.[17][18] These are models that posit that the Higgs sector and Higgs boson are scaling invariant, also known as unparticle physics.

- In theory of superfluid vacuum masses of elementary particles can arise as a result of interaction with the physical vacuum, similarly to the gap generation mechanism in superconductors.[19][20]

- UV-completion by classicalization, in which the unitarization of the WW scattering happens by creation of classical configurations.[21]

21.3   See also

- Higgs boson

21.4   References

[1] Heath, Nick, *The Cern tech that helped track down the God particle*, TechRepublic, July 4, 2012

[2] Steven Weinberg (1976), "Implications of dynamical symmetry breaking",*Physical Review***D13**(4): 974–996,Bibcode:, doi:10.1103/PhysRevD.13.974.
   S. Weinberg (1979), "Implications of dynamical symmetry breaking: An addendum", *Physical Review* **D19** (4): 1277–1280, Bibcode:1979PhRvD..19.1277W, doi:10.1103/PhysRevD.19.1277.

[3] Leonard Susskind (1979), "Dynamics of spontaneous symmetry breaking in the Weinberg-Salam theory", *Physical Review* **D20** (10): 2619–2625, Bibcode:1979PhRvD..20.2619S, doi:10.1103/PhysRevD.20.2619.

[4] Csaki, C.; Grojean, C.; Pilo, L.; Terning, J. (2004), "Towards a realistic model of Higgsless electroweak symmetry breaking", *Physical Review Letters* **92** (10): 101802, arXiv:hep-ph/0308038, Bibcode:2004PhRvL..92j1802C, doi:10.1103/PhysRevLe, PMID 15089195

[5] Csaki, C.; Grojean, C.; Pilo, L.; Terning, J.; Terning, John (2004), "Gauge theories on an interval: Unitarity without a Higgs", *Physical Review D* **69** (5): 055006, arXiv:hep-ph/0305237, Bibcode:2004PhRvD..69e5006C, doi:10.1103/PhysRevD.69.055006

[6] Calmet, X.; Deshpande, N. G.; He, X. G.; Hsu, S. D. H. (2008), "Invisible Higgs boson, continuous mass fields and unHiggs mechanism", *Physical Review D* **79** (5): 055021, arXiv:0810.2155, Bibcode:2009PhRvD..79e5021C, doi:10.1103/PhysRe

[7] Abbott, L. F.; Farhi, E. (1981), "Are the Weak Interactions Strong?", *Physics Letters B* **101** (1–2): 69, Bibcode:1981PhLB.., doi:10.1016/0370-2693(81)90492-5

[8] Speirs, Neil Alexander (1985), "Composite models of weak gauge bosons", *Doctoral thesis, Durham University*

[9] Montag, J. Lee (1992), "Spontaneously Broken Conformal Symmetry and the Standard Model"

[10] Pawlowski, M.; Raczka, R. (1994), "A Unified Conformal Model for Fundamental Interactions without Dynamical Higgs Field", *Foundations of Physics* **24** (9): 1305–1327, arXiv:hep-th/9407137, Bibcode:1994FoPh...24.1305P, doi:10.1007/BF02148570

[11] Calmet, X. (2011), "Asymptotically safe weak interactions", *Mod. Phys. Lett.* **A26** (21): 1571–1576, arXiv:1012.5529, Bibcode:2011MPLA...26.1571C, doi:10.1142/S0217732311035900

[12] Calmet, X. (2011), "An Alternative view on the electroweak interactions", *Int.J.Mod.Phys.* **A26** (17): 2855–2864, arXiv:100, Bibcode:2011IJMPA..26.2855C, doi:10.1142/S0217751X11053699

[13] Codello, A.; Percacci, R. (2008), "Fixed Points of Nonlinear Sigma Models in d>2", *Physics Letters B* **672** (3): 280–283, arXiv:0810.0715, Bibcode:2009PhLB..672..280C, doi:10.1016/j.physletb.2009.01.032

[14] Bilson-Thompson, Sundance O.; Markopoulou, Fotini; Smolin, Lee (2007), "Quantum gravity and the standard model", *Class. Quantum Grav.* **24** (16): 3975–3993, arXiv:hep-th/0603022, Bibcode:2007CQGra..24.3975B, doi:10.1088/0264-9381/24/1.

[15] Goldfain, E. (2008), "Bifurcations and pattern formation in particle physics: An introductory study", *EPL (Europhysics Letters)* **82**: 11001, Bibcode:2008EL.....8211001G, doi:10.1209/0295-5075/82/11001

[16] Goldfain (2010), "Non-equilibrium Dynamics as Source of Asymmetries in High Energy Physics" (PDF), *Electronic Journal of Theoretical Physics* **7** (24): 219

[17] Stancato, David; Terning, John (2008), "The Unhiggs", *Journal of High Energy Physics* **2009** (11): 101, arXiv:0807.3961, Bibcode:2009JHEP...11..101S, doi:10.1088/1126-6708/2009/11/101

[18] Falkowski, Adam; Perez-Victoria, Manuel (2009), "Electroweak Precision Observables and the Unhiggs", *Journal of High Energy Physics* **2009** (12): 061, arXiv:0901.3777, Bibcode:2009JHEP...12..061F, doi:10.1088/1126-6708/2009/12/061

[19] Zloshchastiev, Konstantin G. (2009), "Spontaneous symmetry breaking and mass generation as built-in phenomena in logarithmic nonlinear quantum theory", *Acta Physica Polonica B* **42** (2): 261–292, arXiv:0912.4139, Bibcode:2011AcPPB..42..261Z, doi:10.5506/APhysPolB.42.261

[20] Avdeenkov, Alexander V.; Zloshchastiev, Konstantin G. (2011), "Quantum Bose liquids with logarithmic nonlinearity: Self-sustainability and emergence of spatial extent", *Journal of Physics B: Atomic, Molecular and Optical Physics* **44** (19): 195303, arXiv:1108.0847, Bibcode:2011JPhB...44s5303A, doi:10.1088/0953-4075/44/19/195303

[21] Dvali, Gia; Giudice, Gian F.; Gomez, Cesar; Kehagias, Alex (2011), "UV-Completion by Classicalization", *Journal of High Energy Physics* **2011** (8): 108, arXiv:1010.1415, Bibcode:2011JHEP...08..108D, doi:10.1007/JHEP08(2011)108

21.5 External links

- Higgsless model on arxiv.org

Chapter 22

Fermion

In particle physics, a **fermion** (a name coined by Paul Dirac[1] from the surname of Enrico Fermi) is any particle characterized by Fermi–Dirac statistics. These particles obey the Pauli exclusion principle. Fermions include all quarks and leptons, as well as any composite particle made of an odd number of these, such as all baryons and many atoms and nuclei. Fermions differ from bosons, which obey Bose–Einstein statistics.

A fermion can be an elementary particle, such as the electron, or it can be a composite particle, such as the proton. According to the spin-statistics theorem in any reasonable relativistic quantum field theory, particles with integer spin are bosons, while particles with half-integer spin are fermions.

Besides this spin characteristic, fermions have another specific property: they possess conserved baryon or lepton quantum numbers. Therefore what is usually referred as the spin statistics relation is in fact a spin statistics-quantum number relation.[2]

As a consequence of the Pauli exclusion principle, only one fermion can occupy a particular quantum state at any given time. If multiple fermions have the same spatial probability distribution, then at least one property of each fermion, such as its spin, must be different. Fermions are usually associated with matter, whereas bosons are generally force carrier particles, although in the current state of particle physics the distinction between the two concepts is unclear. At low temperature fermions show superfluidity for uncharged particles and superconductivity for charged particles. Composite fermions, such as protons and neutrons, are the key building blocks of everyday matter. Weakly interacting fermions can also display bosonic behavior under extreme conditions, such as superconductivity.

22.1   Elementary fermions

The Standard Model recognizes two types of elementary fermions, quarks and leptons. In all, the model distinguishes 24 different fermions. There are six quarks (up, down, strange, charm, bottom and top quarks), and six leptons (electron, electron neutrino, muon, muon neutrino, tau particle and tau neutrino), along with the corresponding antiparticle of each of these.

Mathematically, fermions come in three types - Weyl fermions (massless), Dirac fermions (massive), and Majorana fermions (each its own antiparticle). Most Standard Model fermions are believed to be Dirac fermions, although it is unknown at this time whether the neutrinos are Dirac or Majorana fermions. Dirac fermions can be treated as a combination of two Weyl fermions.[3]:106 So far there is no known example of Weyl fermion in particle physics. In July 2015, Weyl fermions have been experimentally realized in Weyl semimetals.

Enrico Fermi

22.2   Composite fermions

See also: List of particles § Composite particles

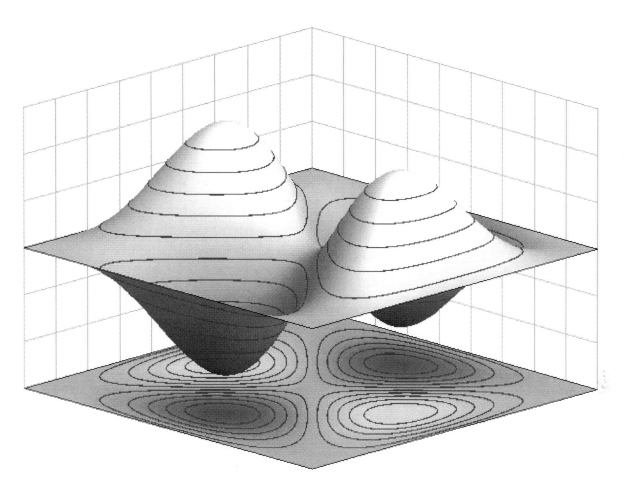

Antisymmetric wavefunction for a (fermionic) 2-particle state in an infinite square well potential.

Composite particles (such as hadrons, nuclei, and atoms) can be bosons or fermions depending on their constituents. More precisely, because of the relation between spin and statistics, a particle containing an odd number of fermions is itself a fermion. It will have half-integer spin.

Examples include the following:

- A baryon, such as the proton or neutron, contains three fermionic quarks and thus it is a fermion.

- The nucleus of a carbon-13 atom contains six protons and seven neutrons and is therefore a fermion.

- The atom helium-3 (^{3}He) is made of two protons, one neutron, and two electrons, and therefore it is a fermion.

The number of bosons within a composite particle made up of simple particles bound with a potential has no effect on whether it is a boson or a fermion.

Fermionic or bosonic behavior of a composite particle (or system) is only seen at large (compared to size of the system) distances. At proximity, where spatial structure begins to be important, a composite particle (or system) behaves according to its constituent makeup.

Fermions can exhibit bosonic behavior when they become loosely bound in pairs. This is the origin of superconductivity and the superfluidity of helium-3: in superconducting materials, electrons interact through the exchange of phonons, forming Cooper pairs, while in helium-3, Cooper pairs are formed via spin fluctuations.

The quasiparticles of the fractional quantum Hall effect are also known as composite fermions, which are electrons with an even number of quantized vortices attached to them.

22.2.1 Skyrmions

Main article: Skyrmion

In a quantum field theory, there can be field configurations of bosons which are topologically twisted. These are coherent states (or solitons) which behave like a particle, and they can be fermionic even if all the constituent particles are bosons. This was discovered by Tony Skyrme in the early 1960s, so fermions made of bosons are named *skyrmions* after him.

Skyrme's original example involved fields which take values on a three-dimensional sphere, the original nonlinear sigma model which describes the large distance behavior of pions. In Skyrme's model, reproduced in the large N or string approximation to quantum chromodynamics (QCD), the proton and neutron are fermionic topological solitons of the pion field.

Whereas Skyrme's example involved pion physics, there is a much more familiar example in quantum electrodynamics with a magnetic monopole. A bosonic monopole with the smallest possible magnetic charge and a bosonic version of the electron will form a fermionic dyon.

The analogy between the Skyrme field and the Higgs field of the electroweak sector has been used[4] to postulate that all fermions are skyrmions. This could explain why all known fermions have baryon or lepton quantum numbers and provide a physical mechanism for the Pauli exclusion principle.

22.3 See also

22.4 Notes

[1] Notes on Dirac's lecture *Developments in Atomic Theory* at Le Palais de la Découverte, 6 December 1945, UKNATARCHI Dirac Papers BW83/2/257889. See note 64 on page 331 in "The Strangest Man: The Hidden Life of Paul Dirac, Mystic of the Atom" by Graham Farmelo

[2] Physical Review D volume 87, page 0550003, year 2013, author Weiner, Richard M., title "Spin-statistics-quantum number connection and supersymmetry" arxiv:1302.0969

[3] T. Morii; C. S. Lim; S. N. Mukherjee (1 January 2004). *The Physics of the Standard Model and Beyond*. World Scientific. ISBN 978-981-279-560-1.

[4] Weiner, Richard M. (2010). "The Mysteries of Fermions". *International Journal of Theoretical Physics* **49** (5): 1174–1180. arXiv:0901.3816. Bibcode:2010IJTP...49.1174W. doi:10.1007/s10773-010-0292-7.

Chapter 23

Yukawa interaction

In particle physics, **Yukawa's interaction**, named after Hideki Yukawa, is an interaction between a scalar field φ and a Dirac field ψ of the type

$$V \approx g\bar{\psi}\phi\psi \text{ (scalar) or } g\bar{\psi}i\gamma^5\phi\psi \text{ (pseudoscalar).}$$

The Yukawa interaction can be used to describe the nuclear force between nucleons (which are fermions), mediated by pions (which are pseudoscalar mesons). The Yukawa interaction is also used in the Standard Model to describe the coupling between the Higgs field and massless quark and lepton fields (i.e., the fundamental fermion particles). Through spontaneous symmetry breaking, these fermions acquire a mass proportional to the vacuum expectation value of the Higgs field.

23.1 The action

The action for a meson field φ interacting with a Dirac baryon field ψ is

$$S[\phi, \psi] = \int d^d x \; [\mathcal{L}_{\text{meson}}(\phi) + \mathcal{L}_{\text{Dirac}}(\psi) + \mathcal{L}_{\text{Yukawa}}(\phi, \psi)]$$

where the integration is performed over d dimensions (typically 4 for four-dimensional spacetime). The meson Lagrangian is given by

$$\mathcal{L}_{\text{meson}}(\phi) = \frac{1}{2}\partial^\mu \phi \partial_\mu \phi - V(\phi).$$

Here, $V(\phi)$ is a self-interaction term. For a free-field massive meson, one would have $V(\phi) = \frac{1}{2}\mu^2\phi^2$ where μ is the mass for the meson. For a (renormalizable, polynomial) self-interacting field, one will have $V(\phi) = \frac{1}{2}\mu^2\phi^2 + \lambda\phi^4$ where λ is a coupling constant. This potential is explored in detail in the article on the quartic interaction.

The free-field Dirac Lagrangian is given by

$$\mathcal{L}_{\text{Dirac}}(\psi) = \bar{\psi}(i\partial\!\!\!/ - m)\psi$$

where m is the positive, real mass of the fermion.

The Yukawa interaction term is

$\mathcal{L}_{\text{Yukawa}}(\phi, \psi) = -g\bar{\psi}\phi\psi$

where g is the (real) coupling constant for scalar mesons and

$\mathcal{L}_{\text{Yukawa}}(\phi, \psi) = -g\bar{\psi}i\gamma^5\phi\psi$

for pseudoscalar mesons. Putting it all together one can write the above more explicitly as

$$S[\phi, \psi] = \int d^d x \left[\frac{1}{2}\partial^\mu\phi\partial_\mu\phi - V(\phi) + \bar{\psi}(i\partial\!\!\!/ - m)\psi - g\bar{\psi}\phi\psi \right].$$

23.2 Classical potential

If two fermions interact through a Yukawa interaction with Yukawa particle mass μ, the potential between the two particles, known as the Yukawa potential, will be:

$$V(r) = -\frac{g^2}{4\pi}\frac{1}{r}e^{-\mu r}$$

which is the same as a Coulomb potential except for the sign and the exponential factor. The sign will make the interaction attractive between all particles (the electromagnetic interaction is repulsive for identical particles). This is explained by the fact that the Yukawa particle has spin zero and even spin always results in an attractive potential. The exponential will give the interaction a finite range, so that particles at great distances will hardly interact any longer.

23.3 Spontaneous symmetry breaking

Now suppose that the potential $V(\phi)$ has a minimum not at $\phi = 0$ but at some non-zero value ϕ_0. This can happen if one writes (for example) $V(\phi) = \mu^2\phi^2 + \lambda\phi^4$ and then sets μ to an imaginary value. In this case, one says that the Lagrangian exhibits spontaneous symmetry breaking. The non-zero value of ϕ is called the vacuum expectation value of ϕ. In the Standard Model, this non-zero value is responsible for the fermion masses, as shown below.

To exhibit the mass term, one re-expresses the action in terms of the field $\tilde{\phi} = \phi - \phi_0$, where ϕ_0 is now understood to be a constant independent of position. We now see that the Yukawa term has a component

$g\phi_0\bar{\psi}\psi$

and since both g and ϕ_0 are constants, this term looks exactly like a mass term for a fermion with mass $g\phi_0$. This is the mechanism by which spontaneous symmetry breaking gives mass to fermions. The field $\tilde{\phi}$ is known as the Higgs field.

23.4 Majorana form

It is also possible to have a Yukawa interaction between a scalar and a Majorana field. In fact, the Yukawa interaction involving a scalar and a Dirac spinor can be thought of as a Yukawa interaction involving a scalar with two Majorana spinors of the same mass. Broken out in terms of the two chiral Majorana spinors, one has

$$S[\phi, \chi] = \int d^d x \left[\frac{1}{2}\partial^\mu\phi\partial_\mu\phi - V(\phi) + \chi^\dagger i\bar{\sigma}\cdot\partial\chi + \frac{i}{2}(m+g\phi)\chi^T\sigma^2\chi - \frac{i}{2}(m+g\phi)^*\chi^\dagger\sigma^2\chi^* \right]$$

where g is a complex coupling constant and m is a complex number.

23.5   Feynman rules

The article Yukawa potential provides a simple example of the Feynman rules and a calculation of a scattering amplitude from a Feynman diagram involving the Yukawa interaction.

23.6   References

- Itzykson, Claude; Zuber, Jean-Bernard (1980). *Quantum Field Theory*. New York: McGraw-Hill. ISBN 0-07-032071-3.

- Bjorken, James D.; Drell, Sidney D. (1964). *Relativistic Quantum Mechanics*. New York: McGraw-Hill. ISBN 0-07-232002-8.

- Peskin, Michael E.; Schroeder, Daniel V. (1995). *An Introduction to Quantum Field Theory*. Addison-Wesley. ISBN 0-201-50397-2.

Chapter 24

Boson

For other uses, see Boson (disambiguation).

In quantum mechanics, a **boson** (/ˈboʊsɒn/,[1] /ˈboʊzɒn/[2]) is a particle that follows Bose–Einstein statistics. Bosons make up one of the two classes of particles, the other being fermions.[3] The name boson was coined by Paul Dirac[4] to commemorate the contribution of the Indian physicist Satyendra Nath Bose[5][6] in developing, with Einstein, Bose–Einstein statistics—which theorizes the characteristics of elementary particles.[7] Examples of bosons include fundamental particles such as photons, gluons, and W and Z bosons (the four force-carrying gauge bosons of the Standard Model), the Higgs boson, and the still-theoretical graviton of quantum gravity; composite particles (e.g. mesons and stable nuclei of even mass number such as deuterium (with one proton and one neutron, mass number = 2), helium-4, or lead-208[Note 1]); and some quasiparticles (e.g. Cooper pairs, plasmons, and phonons).[8]:130

An important characteristic of bosons is that their statistics do not restrict the number of them that occupy the same quantum state. This property is exemplified by helium-4 when it is cooled to become a superfluid.[9] Unlike bosons, two identical fermions cannot occupy the same quantum space. Whereas the elementary particles that make up matter (i.e. leptons and quarks) are fermions, the elementary bosons are force carriers that function as the 'glue' holding matter together.[10] This property holds for all particles with integer spin (s = 0, 1, 2 etc.) as a consequence of the spin–statistics theorem. When a gas of Bose particles is cooled down to absolute zero then the kinetic energy of the particles decreases to a negligible amount and they condense into a lowest energy level state. This state is called Bose-Einstein condensation. It is believed that this phenomenon is the secret behind superfluidity of liquids.

24.1 Types

Bosons may be either elementary, like photons, or composite, like mesons.

While most bosons are composite particles, in the Standard Model there are five bosons which are elementary:

- the four gauge bosons ($\gamma \cdot g \cdot Z \cdot W\pm$)

- the only scalar boson (the Higgs boson (H0))

Additionally, the graviton (G) is a hypothetical elementary particle not incorporated in the Standard Model. If it exists, a graviton must be a boson, and could conceivably be a gauge boson.

Composite bosons are important in superfluidity and other applications of Bose–Einstein condensates. When a gas of Bose particles is cooled to absolute zero its kinetic energy decreases up to a negligible amount then the particles would condense into the lowest energy state. This phenomenon is known as Bose-Einstein condensation and it is believed that this phenomenon is the secret behind superfluidity of liquids.

Satyendra Nath Bose

24.2 Properties

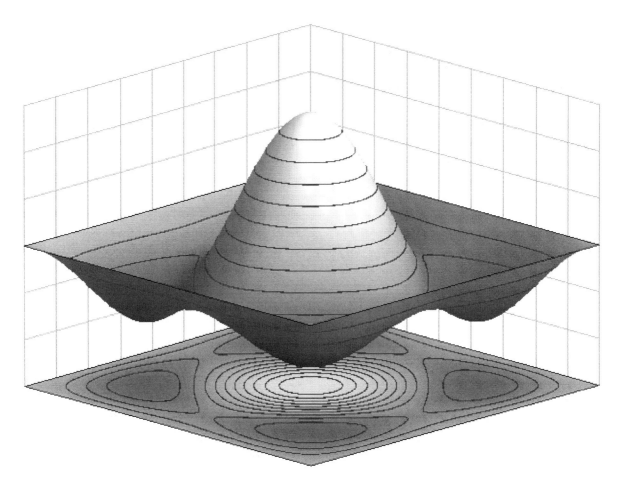

Symmetric wavefunction for a (bosonic) 2-particle state in an infinite square well potential.

Bosons differ from fermions, which obey Fermi–Dirac statistics. Two or more identical fermions cannot occupy the same quantum state (see Pauli exclusion principle).

Since bosons with the same energy can occupy the same place in space, bosons are often force carrier particles. Fermions are usually associated with matter (although in quantum physics the distinction between the two concepts is not clear cut).

Bosons are particles which obey Bose–Einstein statistics: when one swaps two bosons (of the same species), the wave function of the system is unchanged.[11] Fermions, on the other hand, obey Fermi–Dirac statistics and the Pauli exclusion principle: two fermions cannot occupy the same quantum state, resulting in a "rigidity" or "stiffness" of matter which includes fermions. Thus fermions are sometimes said to be the constituents of matter, while bosons are said to be the particles that transmit interactions (force carriers), or the constituents of radiation. The quantum fields of bosons are bosonic fields, obeying canonical commutation relations.

The properties of lasers and masers, superfluid helium-4 and Bose–Einstein condensates are all consequences of statistics of bosons. Another result is that the spectrum of a photon gas in thermal equilibrium is a Planck spectrum, one example of which is black-body radiation; another is the thermal radiation of the opaque early Universe seen today as microwave background radiation. Interactions between elementary particles are called fundamental interactions. The fundamental interactions of virtual bosons with real particles result in all forces we know.

All known elementary and composite particles are bosons or fermions, depending on their spin: particles with half-integer spin are fermions; particles with integer spin are bosons. In the framework of nonrelativistic quantum mechanics, this is a purely empirical observation. However, in relativistic quantum field theory, the spin–statistics theorem shows that half-integer spin particles cannot be bosons and integer spin particles cannot be fermions.[12]

In large systems, the difference between bosonic and fermionic statistics is only apparent at large densities—when their wave functions overlap. At low densities, both types of statistics are well approximated by Maxwell–Boltzmann statistics, which is described by classical mechanics.

24.3   Elementary bosons

See also: List of particles: Bosons

All observed elementary particles are either fermions or bosons. The observed elementary bosons are all gauge bosons: photons, W and Z bosons, gluons, and the Higgs boson.

- Photons are the force carriers of the electromagnetic field.

- W and Z bosons are the force carriers which mediate the weak force.

- Gluons are the fundamental force carriers underlying the strong force.

- Higgs Bosons give W and Z bosons mass via the Higgs mechanism. Their existence was confirmed by CERN on 14 March 2013.

Finally, many approaches to quantum gravity postulate a force carrier for gravity, the graviton, which is a boson of spin plus or minus two.

24.4   Composite bosons

See also: List of particles: Composite particles

Composite particles (such as hadrons, nuclei, and atoms) can be bosons or fermions depending on their constituents. More precisely, because of the relation between spin and statistics, a particle containing an even number of fermions is a boson, since it has integer spin.

Examples include the following:

- Any meson, since mesons contain one quark and one antiquark.

- The nucleus of a carbon-12 atom, which contains 6 protons and 6 neutrons.

- The helium-4 atom, consisting of 2 protons, 2 neutrons and 2 electrons.

The number of bosons within a composite particle made up of simple particles bound with a potential has no effect on whether it is a boson or a fermion.

24.5   To which states can bosons crowd?

Bose–Einstein statistics encourages identical bosons to crowd into one quantum state, but not any state is necessarily convenient for it. Aside of statistics, bosons can interact – for example, helium-4 atoms are repulsed by intermolecular force on a very close approach, and if one hypothesizes their condensation in a spatially-localized state, then gains from the statistics cannot overcome a prohibitive force potential. A spatially-delocalized state (i.e. with low $|\psi(x)|$) is preferable: if the number density of the condensate is about the same as in ordinary liquid or solid state, then the repulsive potential for the N-particle condensate in such state can be not higher than for a liquid or a crystalline lattice of the same N particles described without quantum statistics. Thus, Bose–Einstein statistics for a material particle is not a mechanism to bypass physical restrictions on the density of the corresponding substance, and superfluid liquid helium has the density comparable to the density of ordinary liquid matter. Spatially-delocalized states also permit for a low momentum according to uncertainty principle, hence for low kinetic energy; that's why superfluidity and superconductivity are usually observed in low temperatures.

Photons do not interact with themselves and hence do not experience this difference in states where to crowd (see squeezed coherent state).

24.6 See also

- Anyon
- Bose gas
- Identical particles
- Parastatistics
- Fermion

24.7 Notes

[1] Even-mass-number nuclides, which comprise $152/255 = \sim 60\%$ of all stable nuclides, are bosons, i.e. they have integer spin. Almost all (148 of the 152) are even-proton, even-neutron (EE) nuclides, which necessarily have spin 0 because of pairing. The remainder of the stable bosonic nuclides are 5 odd-proton, odd-neutron stable nuclides (see even and odd atomic nuclei#Odd proton, odd neutron); these odd–odd bosons are: 2
1H, 6
3Li,10
5B, 14
7N and 180m
73Ta). All have nonzero integer spin.

24.8 References

[1] Wells, John C. (1990). *Longman pronunciation dictionary*. Harlow, England: Longman. ISBN 0582053838. entry "Boson"

[2] "boson". *Collins Dictionary*.

[3] Carroll, Sean (2007) *Dark Matter, Dark Energy: The Dark Side of the Universe*, Guidebook Part 2 p. 43, The Teaching Company, ISBN 1598033506 "...boson: A force-carrying particle, as opposed to a matter particle (fermion). Bosons can be piled on top of each other without limit. Examples include photons, gluons, gravitons, weak bosons, and the Higgs boson. The spin of a boson is always an integer, such as 0, 1, 2, and so on..."

[4] Notes on Dirac's lecture *Developments in Atomic Theory* at Le Palais de la Découverte, 6 December 1945, UKNATARCHI Dirac Papers BW83/2/257889. See note 64 to p. 331 in "The Strangest Man" by Graham Farmelo

[5] Daigle, Katy (10 July 2012). "India: Enough about Higgs, let's discuss the boson". *AP News*. Retrieved 10 July 2012.

[6] Bal, Hartosh Singh (19 September 2012). "The Bose in the Boson". *New York Times blog*. Retrieved 21 September 2012.

[7] "Higgs boson: The poetry of subatomic particles". *BBC News*. 4 July 2012. Retrieved 6 July 2012.

[8] Charles P. Poole, Jr. (11 March 2004). *Encyclopedic Dictionary of Condensed Matter Physics*. Academic Press. ISBN 978-0-08-054523-3.

[9] "boson". *Merriam-Webster Online Dictionary*. Retrieved 21 March 2010.

[10] Carroll, Sean. "Explain it in 60 seconds: Bosons". *Symmetry Magazine*. Fermilab/SLAC. Retrieved 15 February 2013.

[11] Srednicki, Mark (2007). *Quantum Field Theory*, Cambridge University Press, pp. 28–29, ISBN 978-0-521-86449-7.

[12] Sakurai, J.J. (1994). *Modern Quantum Mechanics* (Revised Edition), p. 362. Addison-Wesley, ISBN 0-201-53929-2.

Chapter 25

Gluon

Gluons /'gluːɒnz/ are elementary particles that act as the exchange particles (or gauge bosons) for the strong force between quarks, analogous to the exchange of photons in the electromagnetic force between two charged particles.[6]

In technical terms, gluons are vector gauge bosons that mediate strong interactions of quarks in quantum chromodynamics (QCD). Gluons themselves carry the color charge of the strong interaction. This is unlike the photon, which mediates the electromagnetic interaction but lacks an electric charge. Gluons therefore participate in the strong interaction in addition to mediating it, making QCD significantly harder to analyze than QED (quantum electrodynamics).

25.1   Properties

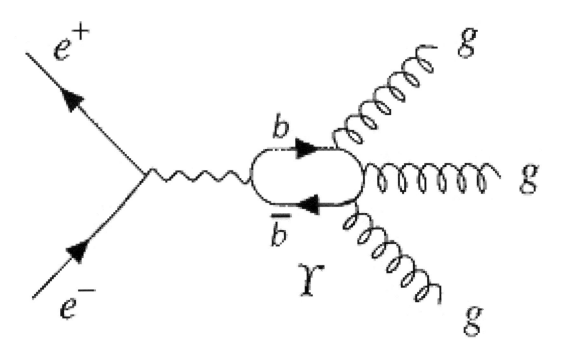

Diagram 2: $e^+e^- \to \Upsilon(9.46) \to 3g$

The gluon is a vector boson; like the photon, it has a spin of 1. While massive spin-1 particles have three polarization states, massless gauge bosons like the gluon have only two polarization states because gauge invariance requires the polarization to be transverse. In quantum field theory, unbroken gauge invariance requires that gauge bosons have zero mass (experiment limits the gluon's rest mass to less than a few meV/c^2). The gluon has negative intrinsic parity.

25.2 Numerology of gluons

Unlike the single photon of QED or the three W and Z bosons of the weak interaction, there are eight independent types of gluon in QCD.

This may be difficult to understand intuitively. Quarks carry three types of color charge; antiquarks carry three types of anticolor. Gluons may be thought of as carrying both color and anticolor, but to correctly understand how they are combined, it is necessary to consider the mathematics of color charge in more detail.

25.2.1 Color charge and superposition

In quantum mechanics, the states of particles may be added according to the principle of superposition; that is, they may be in a "combined state" with a *probability*, if some particular quantity is measured, of giving several different outcomes. A relevant illustration in the case at hand would be a gluon with a color state described by:

$$(r\bar{b} + b\bar{r})/\sqrt{2}.$$

This is read as "red–antiblue plus blue–antired". (The factor of the square root of two is required for normalization, a detail that is not crucial to understand in this discussion.) If one were somehow able to make a direct measurement of the color of a gluon in this state, there would be a 50% chance of it having red-antiblue color charge and a 50% chance of blue-antired color charge.

25.2.2 Color singlet states

It is often said that the stable strongly interacting particles (such as the proton and the neutron, i.e. hadrons) observed in nature are "colorless", but more precisely they are in a "color singlet" state, which is mathematically analogous to a *spin* singlet state.[7] Such states allow interaction with other color singlets, but not with other color states; because long-range gluon interactions do not exist, this illustrates that gluons in the singlet state do not exist either.[7]

The color singlet state is:[7]

$$(r\bar{r} + b\bar{b} + g\bar{g})/\sqrt{3}.$$

In words, if one could measure the color of the state, there would be equal probabilities of it being red-antired, blue-antiblue, or green-antigreen.

25.2.3 Eight gluon colors

There are eight remaining independent color states, which correspond to the "eight types" or "eight colors" of gluons. Because states can be mixed together as discussed above, there are many ways of presenting these states, which are known as the "color octet". One commonly used list is:[7]

These are equivalent to the Gell-Mann matrices; the translation between the two is that red-antired is the upper-left matrix entry, red-antiblue is the upper middle entry, blue-antigreen is the middle right entry, and so on. The critical feature of these particular eight states is that they are linearly independent, and also independent of the singlet state; there is no way to add any combination of states to produce any other. (It is also impossible to add them to make rr, gg, or bb[8] otherwise the forbidden singlet state could also be made.) There are many other possible choices, but all are mathematically equivalent, at least equally complex, and give the same physical results.

25.2.4 Group theory details

Technically, QCD is a gauge theory with SU(3) gauge symmetry. Quarks are introduced as spinor fields in N_f flavors, each in the fundamental representation (triplet, denoted **3**) of the color gauge group, SU(3). The gluons are vector fields in the adjoint representation (octets, denoted **8**) of color SU(3). For a general gauge group, the number of force-carriers (like photons or gluons) is always equal to the dimension of the adjoint representation. For the simple case of SU(N), the dimension of this representation is $N^2 - 1$.

In terms of group theory, the assertion that there are no color singlet gluons is simply the statement that quantum chromodynamics has an SU(3) rather than a U(3) symmetry. There is no known *a priori* reason for one group to be preferred over the other, but as discussed above, the experimental evidence supports SU(3).[7] The U(1) group for electromagnetic field combines with a slightly more complicated group known as SU(2),S stands for "special", which means the corresponding matrices have determinant 1.

25.3 Confinement

Main article: Color confinement

Since gluons themselves carry color charge, they participate in strong interactions. These gluon-gluon interactions constrain color fields to string-like objects called "flux tubes", which exert constant force when stretched. Due to this force, quarks are confined within composite particles called hadrons. This effectively limits the range of the strong interaction to 1×10^{-15} meters, roughly the size of an atomic nucleus. Beyond a certain distance, the energy of the flux tube binding two quarks increases linearly. At a large enough distance, it becomes energetically more favorable to pull a quark-antiquark pair out of the vacuum rather than increase the length of the flux tube.

Gluons also share this property of being confined within hadrons. One consequence is that gluons are not directly involved in the nuclear forces between hadrons. The force mediators for these are other hadrons called mesons.

Although in the normal phase of QCD single gluons may not travel freely, it is predicted that there exist hadrons that are formed entirely of gluons — called glueballs. There are also conjectures about other exotic hadrons in which real gluons (as opposed to virtual ones found in ordinary hadrons) would be primary constituents. Beyond the normal phase of QCD (at extreme temperatures and pressures), quark–gluon plasma forms. In such a plasma there are no hadrons; quarks and gluons become free particles.

25.4 Experimental observations

Quarks and gluons (colored) manifest themselves by fragmenting into more quarks and gluons, which in turn hadronize into normal (colorless) particles, correlated in jets. As shown in 1978 summer conferences[2] the PLUTO detector at the electron-positron collider DORIS (DESY) produced the first evidence that the hadronic decays of the very narrow resonance Υ(9.46) could be interpreted as three-jet event topologies produced by three gluons. Later published analyses by the same experiment confirmed this interpretation and also the spin 1 nature of the gluon[9][10] (see also the recollection[2] and PLUTO experiments).

In summer 1979 at higher energies at the electron-positron collider PETRA (DESY) again three-jet topologies were observed, now interpreted as qq gluon bremsstrahlung, now clearly visible, by TASSO,[11] MARK-J[12] and PLUTO experiments[13] (later in 1980 also by JADE[14]). The spin 1 of the gluon was confirmed in 1980 by TASSO[15] and PLUTO experiments[16] (see also the review[3]). In 1991 a subsequent experiment at the LEP storage ring at CERN again confirmed this result.[17]

The gluons play an important role in the elementary strong interactions between quarks and gluons, described by QCD and studied particularly at the electron-proton collider HERA at DESY. The number and momentum distribution of the gluons in the proton (gluon density) have been measured by two experiments, H1 and ZEUS,[18] in the years 1996 till today (2012). The gluon contribution to the proton spin has been studied by the HERMES experiment at HERA.[19] The gluon density in the proton (when behaving hadronically) also has been measured.[20]

Color confinement is verified by the failure of free quark searches (searches of fractional charges). Quarks are normally produced in pairs (quark + antiquark) to compensate the quantum color and flavor numbers; however at Fermilab single production of top quarks has been shown (technically this still involves a pair production, but quark and antiquark are of different flavor).[21] No glueball has been demonstrated.

Deconfinement was claimed in 2000 at CERN SPS[22] in heavy-ion collisions, and it implies a new state of matter: quark–gluon plasma, less interacting than in the nucleus, almost as in a liquid. It was found at the Relativistic Heavy Ion Collider (RHIC) at Brookhaven in the years 2004–2010 by four contemporaneous experiments.[23] A quark–gluon plasma state has been confirmed at the CERN Large Hadron Collider (LHC) by the three experiments ALICE, ATLAS and CMS in 2010.[24]

25.5   See also

- Quark

- Hadron

- Meson

- Gauge boson

- Quark model

- Quantum chromodynamics

- Quark–gluon plasma

- Color confinement

- Glueball

- Gluon field

- Gluon field strength tensor

- Exotic hadrons

- Standard Model

- Three-jet events

- Deep inelastic scattering

25.6   References

[1] M. Gell-Mann (1962). "Symmetries of Baryons and Mesons". *Physical Review* **125**(3): 1067–1084. Bibcode:1962PhRv..125. doi:10.1103/PhysRev.125.1067.

[2] B.R. Stella and H.-J. Meyer (2011). "ϒ(9.46 GeV) and the gluon discovery (a critical recollection of PLUTO results)". *European Physical Journal H* **36** (2): 203–243. arXiv:1008.1869v3. Bibcode:2011EPJH...36..203S. doi:10.1140/epjh/e2011-10029-3.

[3] P. Söding (2010). "On the discovery of the gluon". *European Physical Journal H* **35** (1): 3–28. Bibcode:2010EPJH...35....3S. doi:10.1140/epjh/e2010-00002-5.

[4] W.-M. Yao; et al. (2006). "Review of Particle Physics" (PDF). *Journal of Physics G* **33**: 1. arXiv:astro-ph/0601168.BibY. doi:10.1088/0954-3899/33/1/001.

[5] F. Yndurain (1995). "Limits on the mass of the gluon". *Physics Letters B* **345** (4): 524. Bibcode:1995PhLB..345..524Y. doi:10.1016/0370-2693(94)01677-5.

[6] C.R. Nave. "The Color Force". *HyperPhysics*. Georgia State University, Department of Physics. Retrieved 2012-04-02.

[7] David Griffiths (1987). *Introduction to Elementary Particles*. John Wiley & Sons. pp. 280–281. ISBN 0-471-60386-4.

[8] J. Baez. "Why are there eight gluons and not nine?". Retrieved 2009-09-13.

[9] Ch. Berger *et al.* (PLUTO Collaboration) (1979). "Jet analysis of the Υ(9.46) decay into charged hadrons". **82** (3–4): 449. Bibcode:1979PhLB...82..449B. doi:10.1016/0370-2693(79)90265-X.

[10] Ch. Berger *et al.* (PLUTO Collaboration) (1981). "Topology of the Υ decay". *Zeitschrift für Physik C* **8** (2): 101. Bibcode: doi:10.1007/BF01547873.

[11] R. Brandelik *et al.* (TASSO collaboration) (1979). "Evidence for Planar Events in e⁺e⁻ Annihilation at High Energies". *Physics Letters B* **86** (2): 243–249. Bibcode:1979PhLB...86..243B. doi:10.1016/0370-2693(79)90830-X.

[12] D.P. Barber *et al.* (MARK-J collaboration) (1979). "Discovery of Three-Jet Events and a Test of Quantum Chromodynamics at PETRA". *Physical Review Letters* **43** (12): 830. Bibcode:1979PhRvL..43..830B. doi:10.1103/PhysRevLett.43.830.

[13] Ch. Berger *et al.* (PLUTO Collaboration) (1979). "Evidence for Gluon Bremsstrahlung in e⁺e⁻ Annihilations at High Energies". *Physics Letters B* **86** (3–4): 418. Bibcode:1979PhLB...86..418B. doi:10.1016/0370-2693(79)90869-4.

[14] W. Bartel *et al.* (JADE Collaboration) (1980). "Observation of planar three-jet events in e⁺e⁻ annihilation and evidence for gluon bremsstrahlung". *Physics Letters B* **91**: 142. Bibcode:1980PhLB...91..142B. doi:10.1016/0370-2693(80)90680-2.

[15] R. Brandelik *et al.* (TASSO Collaboration) (1980). "Evidence for a spin-1 gluon in three-jet events". *Physics Letters B* **97** (3–4): 453. Bibcode:1980PhLB...97..453B. doi:10.1016/0370-2693(80)90639-5.

[16] Ch. Berger *et al.* (PLUTO Collaboration) (1980). "A study of multi-jet events in e⁺e⁻ annihilation". *Physics Letters B* **97** (3–4): 459. Bibcode:1980PhLB...97..459B. doi:10.1016/0370-2693(80)90640-1.

[17] G. Alexander *et al.* (OPAL Collaboration) (1991). "Measurement of Three-Jet Distributions Sensitive to the Gluon Spin in e⁺e⁻ Annihilations at √s = 91 GeV". *Zeitschrift für Physik C* **52** (4): 543. Bibcode:1991ZPhyC..52..543A. doi:10.1007/BF01562326.

[18] L. Lindeman (H1 and ZEUS collaborations) (1997). "Proton structure functions and gluon density at HERA". *Nuclear Physics B Proceedings Supplements* **64**: 179–183. Bibcode:1998NuPhS..64..179L. doi:10.1016/S0920-5632(97)01057-8.

[19] http://www-hermes.desy.de

[20] C. Adloff *et al.* (H1 collaboration) (1999). "Charged particle cross sections in the photoproduction and extraction of the gluon density in the photon". *European Physical Journal C* **10**: 363–372. arXiv:hep-ex/9810020. Bibcode:1999EPJC...10..363H. doi:10.1007/s100520050761.

[21] M. Chalmers (6 March 2009). "Top result for Tevatron". *Physics World*. Retrieved 2012-04-02.

[22] M.C. Abreu; et al. (2000). "Evidence for deconfinement of quark and antiquark from the J/Ψ suppression pattern measured in Pb-Pb collisions at the CERN SpS". *Physics Letters B* **477**: 28–36. Bibcode:2000PhLB..477...28A. doi:10.1016/S0370-2693(00)00237-9.

[23] D. Overbye (15 February 2010). "In Brookhaven Collider, Scientists Briefly Break a Law of Nature". *New York Times*. Retrieved 2012-04-02.

[24] "LHC experiments bring new insight into primordial universe" (Press release). CERN. 26 November 2010. Retrieved 2012-04-02.

25.7 Further reading

- A. Ali and G. Kramer (2011). "JETS and QCD: A historical review of the discovery of the quark and gluon jets and its impact on QCD". *European Physical Journal H* **36**(2): 245–326. arXiv:1012.2288. Bibcode:2011EPJH...36...doi: 10.1140/epjh/e2011-10047-1.

Chapter 26

Tau (particle)

Not to be confused with the τ^+ of the τ–θ puzzle, which is now identified as a kaon.

The **tau** (τ), also called the **tau lepton**, **tau particle**, or **tauon**, is an elementary particle similar to the electron, with negative electric charge and a spin of $^1/_2$. Together with the electron, the muon, and the three neutrinos, it is a lepton. Like all elementary particles with half-integral spin, the tau has a corresponding antiparticle of opposite charge but equal mass and spin, which in the tau's case is the **antitau** (also called the *positive tau*). Tau particles are denoted by τ– and the antitau by τ+.

Tau leptons have a lifetime of $2.9{\times}10^{-13}$ s and a mass of 1776.82 MeV/c^2 (compared to 105.7 MeV/c^2 for muons and 0.511 MeV/c^2 for electrons). Since their interactions are very similar to those of the electron, a tau can be thought of as a much heavier version of the electron. Because of their greater mass, tau particles do not emit as much bremsstrahlung radiation as electrons; consequently they are potentially highly penetrating, much more so than electrons. However, because of their short lifetime, the range of the tau is mainly set by their decay length, which is too small for bremsstrahlung to be noticeable: their penetrating power appears only at ultra high energy (above PeV energies).[4]

As with the case of the other charged leptons, the tau has an associated tau neutrino, denoted by ν_τ.

26.1   History

The tau was detected in a series of experiments between 1974 and 1977 by Martin Lewis Perl with his colleagues at the SLAC-LBL group.[2] Their equipment consisted of SLAC's then-new e+–e– colliding ring, called SPEAR, and the LBL magnetic detector. They could detect and distinguish between leptons, hadrons and photons. They did not detect the tau directly, but rather discovered anomalous events:

"We have discovered 64 events of the form

e+ + e– → e± + μ∓ + at least two undetected particles

for which we have no conventional explanation."

The need for at least two undetected particles was shown by the inability to conserve energy and momentum with only one. However, no other muons, electrons, photons, or hadrons were detected. It was proposed that this event was the production and subsequent decay of a new particle pair:

e+ + e– → τ+ + τ– → e± + μ∓ + 4ν

This was difficult to verify, because the energy to produce the τ+τ– pair is similar to the threshold for D meson production.

Work done at DESY-Hamburg, and with the Direct Electron Counter (DELCO) at SPEAR, subsequently established the mass and spin of the tau.

The symbol τ was derived from the Greek τρίτον (*triton*, meaning "third" in English), since it was the third charged lepton discovered.[5]

Martin Perl shared the 1995 Nobel Prize in Physics with Frederick Reines. The latter was awarded his share of the prize for experimental discovery of the neutrino.

26.2 Tau decay

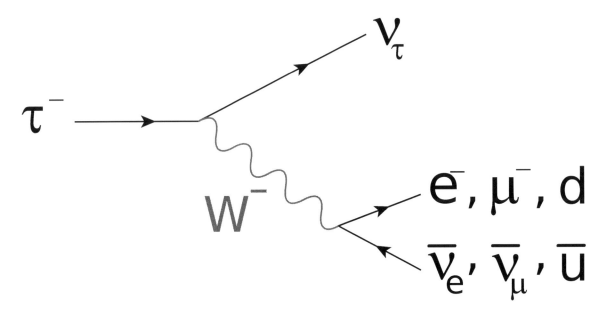

Feynman diagram of the common decays of the tau by emission of a W boson.

The tau is the only lepton that can decay into hadrons – the other leptons do not have the necessary mass. Like the other decay modes of the tau, the hadronic decay is through the weak interaction.[6]

The branching ratio of the dominant hadronic tau decays are:[3]

- 25.52% for decay into a charged pion, a neutral pion, and a tau neutrino;

- 10.83% for decay into a charged pion and a tau neutrino;

- 9.30% for decay into a charged pion, two neutral pions, and a tau neutrino;

- 8.99% for decay into three charged pions (of which two have the same electrical charge) and a tau neutrino;

- 2.70% for decay into three charged pions (of which two have the same electrical charge), a neutral pion, and a tau neutrino;

- 1.05% for decay into three neutral pions, a charged pion, and a tau neutrino.

In total, the tau lepton will decay hadronically approximately 64.79% of the time.

Since the tauonic lepton number is conserved in weak decays, a tau neutrino is always created when a tau decays.[6]

The branching ratio of the common purely leptonic tau decays are:[3]

- 17.82% for decay into a tau neutrino, electron and electron antineutrino;

- 17.39% for decay into a tau neutrino, muon and muon antineutrino.

The similarity of values of the two branching ratios is a consequence of lepton universality.

26.3 Exotic atoms

The tau lepton is predicted to form exotic atoms like other charged subatomic particles. One of such, called **tauonium** by the analogy to muonium, consists in antitauon and an electron: τ+e−.[7]

Another one is an onium atom τ+τ− called *true tauonium* and is difficult to detect due to tau's extremely short lifetime at low (non-relativistic) energies needed to form this atom. Its detection is important for quantum electrodynamics.[7]

26.4 See also

- Koide formula

26.5 References

[1] L. B. Okun (1980). *Leptons and Quarks.* V.I. Kisin (trans.). North-Holland Publishing. p. 103. ISBN 978-0444869241.

[2] Perl, M. L.; Abrams, G.; Boyarski, A.; Breidenbach, M.; Briggs, D.; Bulos, F.; Chinowsky, W.; Dakin, J.; et al. (1975). "Evidence for Anomalous Lepton Production in e+e− Annihilation".*Physical Review Letters***35**(22): 1489.Bibcode:197.doi: 10.1103/PhysRevLett.35.1489.

[3] J. Beringer *et al.* (Particle Data Group) (2012). "Review of Particle Physics". *Journal of Physics G* **86** (1): 581–651. Bibcode:2012PhRvD..86a0001B. doi:10.1103/PhysRevD.86.010001. |chapter= ignored (help)

[4] D. Fargion, P.G. De Sanctis Lucentini, M. De Santis, M. Grossi (2004). "Tau Air Showers from Earth". *The Astrophysical Journal* **613** (2): 1285. arXiv:hep-ph/0305128. Bibcode:2004ApJ...613.1285F. doi:10.1086/423124.

[5] M.L. Perl (1977). "Evidence for, and properties of, the new charged heavy lepton" (PDF). In T. Thanh Van (ed.). *Proceedings of the XII Rencontre de Moriond.* SLAC-PUB-1923.

[6] Riazuddin (2009). "Non-standard interactions" (PDF). *NCP 5th Particle Physics Sypnoisis* (Islamabad,: Riazuddin, Head of High-Energy Theory Group at National Center for Physics) **1** (1): 1–25.

[7] Brodsky, Stanley J.; Lebed, Richard F. (2009). "Production of the Smallest QED Atom: True Muonium ($\mu^+\mu^-$)". *Physical Review Letters* **102** (21): 213401. arXiv:0904.2225. Bibcode:2009PhRvL.102u3401B. doi:10.1103/PhysRevLett.102.213401.

26.6 External links

- Nobel Prize in Physics 1995

- Perl's logbook showing tau discovery

- A Tale of Three Papers gives the covers of the three original papers announcing the discovery.

Chapter 27

Photon

This article is about the elementary particle of light. For other uses, see Photon (disambiguation).

A **photon** is an elementary particle, the quantum of light and all other forms of electromagnetic radiation. It is the force carrier for the electromagnetic force, even when static via virtual photons. The effects of this force are easily observable at the microscopic and at the macroscopic level, because the photon has zero rest mass; this allows long distance interactions. Like all elementary particles, photons are currently best explained by quantum mechanics and exhibit wave–particle duality, exhibiting properties of waves and of particles. For example, a single photon may be refracted by a lens or exhibit wave interference with itself, but also act as a particle giving a definite result when its position is measured. Waves and quanta, being two observable aspects of a single phenomenon cannot have their true nature described in terms of any mechanical model. [2] A representation of this dual property of light, which assumes certain points on the wave front to be the seat of the energy is also impossible. Thus, the quanta in a light wave cannot be spatially localized. Some defined physical parameters of a photon are listed.

The modern photon concept was developed gradually by Albert Einstein in the first years of the 20th century to explain experimental observations that did not fit the classical wave model of light. In particular, the photon model accounted for the frequency dependence of light's energy, and explained the ability of matter and radiation to be in thermal equilibrium. It also accounted for anomalous observations, including the properties of black-body radiation, that other physicists, most notably Max Planck, had sought to explain using *semiclassical models*, in which light is still described by Maxwell's equations, but the material objects that emit and absorb light do so in amounts of energy that are *quantized* (i.e., they change energy only by certain particular discrete amounts and cannot change energy in any arbitrary way). Although these semiclassical models contributed to the development of quantum mechanics, many further experiments[3][4] starting with Compton scattering of single photons by electrons, first observed in 1923, validated Einstein's hypothesis that *light itself* is quantized. In 1926 the optical physicist Frithiof Wolfers and the chemist Gilbert N. Lewis coined the name *photon* for these particles, and after 1927, when Arthur H. Compton won the Nobel Prize for his scattering studies, most scientists accepted the validity that quanta of light have an independent existence, and the term *photon* for light quanta was accepted.

In the Standard Model of particle physics, photons and other elementary particles are described as a necessary consequence of physical laws having a certain symmetry at every point in spacetime. The intrinsic properties of particles, such as charge, mass and spin, are determined by the properties of this gauge symmetry. The photon concept has led to momentous advances in experimental and theoretical physics, such as lasers, Bose–Einstein condensation, quantum field theory, and the probabilistic interpretation of quantum mechanics. It has been applied to photochemistry, high-resolution microscopy, and measurements of molecular distances. Recently, photons have been studied as elements of quantum computers and for applications in optical imaging and optical communication such as quantum cryptography.

27.1 Nomenclature

In 1900, the German physicist Max Planck was working on black-body radiation and suggested that the energy in electromagnetic waves could only be released in "packets" of energy. In his 1901 article [5] in Annalen der Physik he called these packets "energy elements". The word *quanta* (singular *quantum*) was used even before 1900 to mean particles or amounts of different quantities, including electricity. Later, in 1905, Albert Einstein went further by suggesting that electromagnetic waves could only exist in these discrete wave-packets.[6] He called such a wave-packet *the light quantum* (German: *das Lichtquant*).[Note 1] The name *photon* derives from the Greek word for light, φῶς (transliterated *phôs*). Arthur Compton used *photon* in 1928, referring to Gilbert N. Lewis.[7] The same name was used earlier, by the American physicist and psychologist Leonard T. Troland, who coined the word in 1916, in 1921 by the Irish physicist John Joly, in 1924 by the French physiologist René Wurmser (1890-1993) and in 1926 by the French physicist Frithiof Wolfers (1891-1971).[8] The name was suggested initially as a unit related to the illumination of the eye and the resulting sensation of light and was used later on in a physiological context. Although Wolfers's and Lewis's theories were never accepted, as they were contradicted by many experiments, the new name was adopted very soon by most physicists after Compton used it.[8][Note 2]

In physics, a photon is usually denoted by the symbol γ (the Greek letter gamma). This symbol for the photon probably derives from gamma rays, which were discovered in 1900 by Paul Villard,[9][10] named by Ernest Rutherford in 1903, and shown to be a form of electromagnetic radiation in 1914 by Rutherford and Edward Andrade.[11] In chemistry and optical engineering, photons are usually symbolized by $h\nu$, the energy of a photon, where h is Planck's constant and the Greek letter ν (nu) is the photon's frequency. Much less commonly, the photon can be symbolized by hf, where its frequency is denoted by f.

27.2 Physical properties

See also: Special relativity and Photonic molecule

A photon is massless,[Note 3] has no electric charge,[12] and is stable. A photon has two possible polarization states. In the momentum representation, which is preferred in quantum field theory, a photon is described by its wave vector, which determines its wavelength λ and its direction of propagation. A photon's wave vector may not be zero and can be represented either as a spatial 3-vector or as a (relativistic) four-vector; in the latter case it belongs to the light cone (pictured). Different signs of the four-vector denote different circular polarizations, but in the 3-vector representation one should account for the polarization state separately; it actually is a spin quantum number. In both cases the space of possible wave vectors is three-dimensional.

The photon is the gauge boson for electromagnetism,[13]:29-30 and therefore all other quantum numbers of the photon (such as lepton number, baryon number, and flavour quantum numbers) are zero.[14] Also, the photon does not obey the Pauli exclusion principle.[15]:1221

Photons are emitted in many natural processes. For example, when a charge is accelerated it emits synchrotron radiation. During a molecular, atomic or nuclear transition to a lower energy level, photons of various energy will be emitted, from radio waves to gamma rays. A photon can also be emitted when a particle and its corresponding antiparticle are annihilated (for example, electron–positron annihilation).[15]:572, 1114, 1172

In empty space, the photon moves at c (the speed of light) and its energy and momentum are related by $E = pc$, where p is the magnitude of the momentum vector $\mathbf{p}$. This derives from the following relativistic relation, with $m = 0$:[16]

$$E^2 = p^2c^2 + m^2c^4.$$

The energy and momentum of a photon depend only on its frequency (ν) or inversely, its wavelength (λ):

$$E = \hbar\omega = h\nu = \frac{hc}{\lambda}$$

$$\boldsymbol{p} = \hbar\boldsymbol{k},$$

where $\boldsymbol{k}$ is the wave vector (where the wave number $k = |\boldsymbol{k}| = 2\pi/\lambda$), $\omega = 2\pi\nu$ is the angular frequency, and $\hbar = h/2\pi$ is the reduced Planck constant.[17]

Since $\boldsymbol{p}$ points in the direction of the photon's propagation, the magnitude of the momentum is

$$p = \hbar k = \frac{h\nu}{c} = \frac{h}{\lambda}.$$

The photon also carries spin angular momentum that does not depend on its frequency.[18] The magnitude of its spin is $\sqrt{2}\hbar$ and the component measured along its direction of motion, its helicity, must be $\pm\hbar$. These two possible helicities, called right-handed and left-handed, correspond to the two possible circular polarization states of the photon.[19]

To illustrate the significance of these formulae, the annihilation of a particle with its antiparticle in free space must result in the creation of at least *two* photons for the following reason. In the center of momentum frame, the colliding antiparticles have no net momentum, whereas a single photon always has momentum (since, as we have seen, it is determined by the photon's frequency or wavelength, which cannot be zero). Hence, conservation of momentum (or equivalently, translational invariance) requires that at least two photons are created, with zero net momentum. (However, it is possible if the system interacts with another particle or field for annihilation to produce one photon, as when a positron annihilates with a bound atomic electron, it is possible for only one photon to be emitted, as the nuclear Coulomb field breaks translational symmetry.)[20]:64-65 The energy of the two photons, or, equivalently, their frequency, may be determined from conservation of four-momentum. Seen another way, the photon can be considered as its own antiparticle. The reverse process, pair production, is the dominant mechanism by which high-energy photons such as gamma rays lose energy while passing through matter.[21] That process is the reverse of "annihilation to one photon" allowed in the electric field of an atomic nucleus.

The classical formulae for the energy and momentum of electromagnetic radiation can be re-expressed in terms of photon events. For example, the pressure of electromagnetic radiation on an object derives from the transfer of photon momentum per unit time and unit area to that object, since pressure is force per unit area and force is the change in momentum per unit time.[22]

27.2.1 Experimental checks on photon mass

Current commonly accepted physical theories imply or assume the photon to be strictly massless. If the photon is not a strictly massless particle, it would not move at the exact speed of light in vacuum, c. Its speed would be lower and depend on its frequency. Relativity would be unaffected by this; the so-called speed of light, c, would then not be the actual speed at which light moves, but a constant of nature which is the maximum speed that any object could theoretically attain in space-time.[23] Thus, it would still be the speed of space-time ripples (gravitational waves and gravitons), but it would not be the speed of photons.

If a photon did have non-zero mass, there would be other effects as well. Coulomb's law would be modified and the electromagnetic field would have an extra physical degree of freedom. These effects yield more sensitive experimental probes of the photon mass than the frequency dependence of the speed of light. If Coulomb's law is not exactly valid, then that would cause the presence of an electric field inside a hollow conductor when it is subjected to an external electric field. This thus allows one to test Coulomb's law to very high precision.[24] A null result of such an experiment has set a limit of $m \lesssim 10^{-14}$ eV/c^2.[25]

Sharper upper limits have been obtained in experiments designed to detect effects caused by the galactic vector potential. Although the galactic vector potential is very large because the galactic magnetic field exists on very long length scales, only the magnetic field is observable if the photon is massless. In case of a massive photon, the mass term $\frac{1}{2}m^2 A_\mu A^\mu$ would affect the galactic plasma. The fact that no such effects are seen implies an upper bound on the photon mass of $m < 3\times10^{-27}$ eV/c^2.[26] The galactic vector potential can also be probed directly by measuring the torque exerted on a magnetized ring.[27] Such methods were used to obtain the sharper upper limit of 10^{-18}eV/c^2 (the equivalent of 1.07×10^{-27} atomic mass units) given by the Particle Data Group.[28]

These sharp limits from the non-observation of the effects caused by the galactic vector potential have been shown to be model dependent.[29] If the photon mass is generated via the Higgs mechanism then the upper limit of $m \lesssim 10^{-14}$ eV/c^2 from the test of Coulomb's law is valid.

Photons inside superconductors do develop a nonzero effective rest mass; as a result, electromagnetic forces become short-range inside superconductors.[30]

See also: Supernova/Acceleration Probe

27.3   Historical development

Main article: Light
 In most theories up to the eighteenth century, light was pictured as being made up of particles. Since particle models cannot easily account for the refraction, diffraction and birefringence of light, wave theories of light were proposed by René Descartes (1637),[31] Robert Hooke (1665),[32] and Christiaan Huygens (1678);[33] however, particle models remained dominant, chiefly due to the influence of Isaac Newton.[34] In the early nineteenth century, Thomas Young and August Fresnel clearly demonstrated the interference and diffraction of light and by 1850 wave models were generally accepted.[35] In 1865, James Clerk Maxwell's prediction[36] that light was an electromagnetic wave—which was confirmed experimentally in 1888 by Heinrich Hertz's detection of radio waves[37]—seemed to be the final blow to particle models of light.

The Maxwell wave theory, however, does not account for *all* properties of light. The Maxwell theory predicts that the energy of a light wave depends only on its intensity, not on its frequency; nevertheless, several independent types of experiments show that the energy imparted by light to atoms depends only on the light's frequency, not on its intensity. For example, some chemical reactions are provoked only by light of frequency higher than a certain threshold; light of frequency lower than the threshold, no matter how intense, does not initiate the reaction. Similarly, electrons can be ejected from a metal plate by shining light of sufficiently high frequency on it (the photoelectric effect); the energy of the ejected electron is related only to the light's frequency, not to its intensity.[38][Note 4]

At the same time, investigations of blackbody radiation carried out over four decades (1860–1900) by various researchers[culminated in Max Planck's hypothesis[5][40]that the energy of *any* system that absorbs or emits electromagnetic radiation of frequency ν is an integer multiple of an energy quantum $E=h\nu$. As shown by Albert Einstein, some form of energy quantization *must* be assumed to account for the thermal equilibrium observed between matter and electromagnetic radiation; for this explanation of the photoelectric effect, Einstein received the 1921 Nobel Prize in physics.[42]
Since the Maxwell theory of light allows for all possible energies of electromagnetic radiation, most physicists assumed initially that the energy quantization resulted from some unknown constraint on the matter that absorbs or emits the radiation. In 1905, Einstein was the first to propose that energy quantization was a property of electromagnetic radiation itself.[6] Although he accepted the validity of Maxwell's theory, Einstein pointed out that many anomalous experiments could be explained if the *energy* of a Maxwellian light wave were localized into point-like quanta that move independently of one another, even if the wave itself is spread continuously over space.[6] In 1909[41] and 1916,[43] Einstein showed that, if Planck's law of black-body radiation is accepted, the energy quanta must also carry momentum $p = h/\lambda$, making them full-fledged particles. This photon momentum was observed experimentally[44] by Arthur Compton, for which he received the Nobel Prize in 1927. The pivotal question was then: how to unify Maxwell's wave theory of light with its experimentally observed particle nature? The answer to this question occupied Albert Einstein for the rest of his life,[45] and was solved in quantum electrodynamics and its successor, the Standard Model (see Second quantization and The photon as a gauge boson, below).

27.4   Einstein's light quantum

Unlike Planck, Einstein entertained the possibility that there might be actual physical quanta of light—what we now call photons. He noticed that a light quantum with energy proportional to its frequency would explain a number of troubling puzzles and paradoxes, including an unpublished law by Stokes, the ultraviolet catastrophe, and of course the photoelectric effect. Stokes's law sais imply that the frequency of fluorescent light cannot be greater than the frequency of the light (usually ultraviolet) inducing it. Einstein eliminated the ultraviolet catastrophe by imagining a gas of photons behaving like a gas of electrons that he had previously considered. He was advised by a colleague to be careful how he wrote up this paper, in order to not challenge Planck too directly, as he was a powerful figure, and indeed the warning was justified, as Planck never forgave him for writing it.[46]

27.5 Early objections

Einstein's 1905 predictions were verified experimentally in several ways in the first two decades of the 20th century, as recounted in Robert Millikan's Nobel lecture.[47] However, before Compton's experiment[44] showing that photons carried momentum proportional to their wave number (or frequency) (1922), most physicists were reluctant to believe that electromagnetic radiation itself might be particulate. (See, for example, the Nobel lectures of Wien,[39] Planck[40] and Millikan.[47]) Instead, there was a widespread belief that energy quantization resulted from some unknown constraint on the matter that absorbs or emits radiation. Attitudes changed over time. In part, the change can be traced to experiments such as Compton scattering, where it was much more difficult not to ascribe quantization to light itself to explain the observed results.[48]

Even after Compton's experiment, Niels Bohr, Hendrik Kramers and John Slater made one last attempt to preserve the Maxwellian continuous electromagnetic field model of light, the so-called BKS model.[49] To account for the data then available, two drastic hypotheses had to be made:

1. **Energy and momentum are conserved only on the average in interactions between matter and radiation, not in elementary processes such as absorption and emission.** This allows one to reconcile the discontinuously changing energy of the atom (jump between energy states) with the continuous release of energy into radiation.

2. **Causality is abandoned**. For example, spontaneous emissions are merely emissions induced by a "virtual" electromagnetic field.

However, refined Compton experiments showed that energy–momentum is conserved extraordinarily well in elementary processes; and also that the jolting of the electron and the generation of a new photon in Compton scattering obey causality to within 10 ps. Accordingly, Bohr and his co-workers gave their model "as honorable a funeral as possible".[45] Nevertheless, the failures of the BKS model inspired Werner Heisenberg in his development of matrix mechanics.[50]

A few physicists persisted[51] in developing semiclassical models in which electromagnetic radiation is not quantized, but matter appears to obey the laws of quantum mechanics. Although the evidence for photons from chemical and physical experiments was overwhelming by the 1970s, this evidence could not be considered as *absolutely* definitive; since it relied on the interaction of light with matter, a sufficiently complicated theory of matter could in principle account for the evidence. Nevertheless, *all* semiclassical theories were refuted definitively in the 1970s and 1980s by photon-correlation experiments.[Note 5] Hence, Einstein's hypothesis that quantization is a property of light itself is considered to be proven.

27.6 Wave–particle duality and uncertainty principles

See also: Wave–particle duality, Squeezed coherent state, Uncertainty principle and De Broglie–Bohm theory
Photons, like all quantum objects, exhibit wave-like and particle-like properties. Their dual wave–particle nature can be difficult to visualize. The photon displays clearly wave-like phenomena such as diffraction and interference on the length scale of its wavelength. For example, a single photon passing through a double-slit experiment lands on the screen exhibiting interference phenomena but only if no measure was made on the actual slit being run across. To account for the particle interpretation that phenomenon is called probability distribution but behaves according to Maxwell's equations.[52] However, experiments confirm that the photon is *not* a short pulse of electromagnetic radiation; it does not spread out as it propagates, nor does it divide when it encounters a beam splitter.[53] Rather, the photon seems to be a point-like particle since it is absorbed or emitted *as a whole* by arbitrarily small systems, systems much smaller than its wavelength, such as an atomic nucleus ($\approx 10^{-15}$ m across) or even the point-like electron. Nevertheless, the photon is *not* a point-like particle whose trajectory is shaped probabilistically by the electromagnetic field, as conceived by Einstein and others; that hypothesis was also refuted by the photon-correlation experiments cited above. According to our present understanding, the electromagnetic field itself is produced by photons, which in turn result from a local gauge symmetry and the laws of quantum field theory (see the Second quantization and Gauge boson sections below).

A key element of quantum mechanics is Heisenberg's uncertainty principle, which forbids the simultaneous measurement of the position and momentum of a particle along the same direction. Remarkably, the uncertainty principle for charged, material particles *requires* the quantization of light into photons, and even the frequency dependence of the photon's energy and momentum. An elegant illustration is Heisenberg's thought experiment for locating an electron with an ideal microscope.[54] The position of the electron can be determined to within the resolving power of the microscope, which is given by a formula from classical optics

$$\Delta x \sim \frac{\lambda}{\sin \theta}$$

where θ is the aperture angle of the microscope. Thus, the position uncertainty Δx can be made arbitrarily small by reducing the wavelength λ. The momentum of the electron is uncertain, since it received a "kick" Δp from the light scattering from it into the microscope. If light were *not* quantized into photons, the uncertainty Δp could be made arbitrarily small by reducing the light's intensity. In that case, since the wavelength and intensity of light can be varied independently, one could simultaneously determine the position and momentum to arbitrarily high accuracy, violating the uncertainty principle. By contrast, Einstein's formula for photon momentum preserves the uncertainty principle; since the photon is scattered anywhere within the aperture, the uncertainty of momentum transferred equals

$$\Delta p \sim p_{\text{photon}} \sin \theta = \frac{h}{\lambda} \sin \theta$$

giving the product $\Delta x \Delta p \sim h$, which is Heisenberg's uncertainty principle. Thus, the entire world is quantized; both matter and fields must obey a consistent set of quantum laws, if either one is to be quantized.[55]

The analogous uncertainty principle for photons forbids the simultaneous measurement of the number n of photons (see Fock state and the Second quantization section below) in an electromagnetic wave and the phase ϕ of that wave

$$\Delta n \Delta \phi > 1$$

See coherent state and squeezed coherent state for more details.

Both (photons and material) particles such as electrons create analogous interference patterns when passing through a double-slit experiment. For photons, this corresponds to the interference of a Maxwell light wave whereas, for material particles, this corresponds to the interference of the Schrödinger wave equation. Although this similarity might suggest that Maxwell's equations are simply Schrödinger's equation for photons, most physicists do not agree.[56][57] For one thing, they are mathematically different; most obviously, Schrödinger's one equation solves for a complex field, whereas Maxwell's four equations solve for real fields. More generally, the normal concept of a Schrödinger probability wave function cannot be applied to photons.[58] Being massless, they cannot be localized without being destroyed; technically, photons cannot have a position eigenstate $|\mathbf{r}\rangle$, and, thus, the normal Heisenberg uncertainty principle $\Delta x \Delta p > h/2$ does not pertain to photons. A few substitute wave functions have been suggested for the photon,[59][60][61][62] but they have not come into general use. Instead, physicists generally accept the second-quantized theory of photons described below, quantum electrodynamics, in which photons are quantized excitations of electromagnetic modes.

Another interpretation, that avoids duality, is the De Broglie–Bohm theory: known also as the *pilot-wave model*, the photon in this theory is both, wave and particle.[63] *"This idea seems to me so natural and simple, to resolve the wave-particle dilemma in such a clear and ordinary way, that it is a great mystery to me that it was so generally ignored"*,[64] J.S.Bell.

27.7   Bose–Einstein model of a photon gas

Main articles: Bose gas, Bose–Einstein statistics, Spin-statistics theorem and Gas in a box

In 1924, Satyendra Nath Bose derived Planck's law of black-body radiation without using any electromagnetism, but rather a modification of coarse-grained counting of phase space.[65] Einstein showed that this modification is equivalent to assuming that photons are rigorously identical and that it implied a "mysterious non-local interaction",[66][67] now understood as the requirement for a symmetric quantum mechanical state. This work led to the concept of coherent states and the development of the laser. In the same papers, Einstein extended Bose's formalism to material particles (bosons) and predicted that they would condense into their lowest quantum state at low enough temperatures; this Bose–Einstein condensation was observed experimentally in 1995.[68] It was later used by Lene Hau to slow, and then completely stop, light in 1999[69] and 2001.[70]

The modern view on this is that photons are, by virtue of their integer spin, bosons (as opposed to fermions with half-integer spin). By the spin-statistics theorem, all bosons obey Bose–Einstein statistics (whereas all fermions obey Fermi–Dirac statistics).[71]

27.8 Stimulated and spontaneous emission

Main articles: Stimulated emission and Laser

 In 1916, Einstein showed that Planck's radiation law could be derived from a semi-classical, statistical treatment of photons and atoms, which implies a relation between the rates at which atoms emit and absorb photons. The condition follows from the assumption that light is emitted and absorbed by atoms independently, and that the thermal equilibrium is preserved by interaction with atoms. Consider a cavity in thermal equilibrium and filled with electromagnetic radiation and atoms that can emit and absorb that radiation. Thermal equilibrium requires that the energy density $\rho(\nu)$ of photons with frequency ν (which is proportional to their number density) is, on average, constant in time; hence, the rate at which photons of any particular frequency are *emitted* must equal the rate of *absorbing* them.[72]

Einstein began by postulating simple proportionality relations for the different reaction rates involved. In his model, the rate R_{ji} for a system to *absorb* a photon of frequency ν and transition from a lower energy E_j to a higher energy E_i is proportional to the number N_j of atoms with energy E_j and to the energy density $\rho(\nu)$ of ambient photons with that frequency,

$$R_{ji} = N_j B_{ji} \rho(\nu)$$

where B_{ji} is the rate constant for absorption. For the reverse process, there are two possibilities: spontaneous emission of a photon, and a return to the lower-energy state that is initiated by the interaction with a passing photon. Following Einstein's approach, the corresponding rate R_{ij} for the emission of photons of frequency ν and transition from a higher energy E_i to a lower energy E_j is

$$R_{ij} = N_i A_{ij} + N_i B_{ij} \rho(\nu)$$

where A_{ij} is the rate constant for emitting a photon spontaneously, and B_{ij} is the rate constant for emitting it in response to ambient photons (induced or stimulated emission). In thermodynamic equilibrium, the number of atoms in state i and that of atoms in state j must, on average, be constant; hence, the rates R_{ji} and R_{ij} must be equal. Also, by arguments analogous to the derivation of Boltzmann statistics, the ratio of N_i and N_j is $g_i/g_j \exp\left(E_j - E_i\right)/kT$, where $g_{i,j}$ are the degeneracy of the state i and that of j, respectively, $E_{i,j}$ their energies, k the Boltzmann constant and T the system's temperature. From this, it is readily derived that $g_i B_{ij} = g_j B_{ji}$ and

$$A_{ij} = \frac{8\pi h\nu^3}{c^3} B_{ij}.$$

The A and Bs are collectively known as the *Einstein coefficients*.[73]

Einstein could not fully justify his rate equations, but claimed that it should be possible to calculate the coefficients A_{ij}, B_{ji} and B_{ij} once physicists had obtained "mechanics and electrodynamics modified to accommodate the quantum hypothesis".[74] In fact, in 1926, Paul Dirac derived the B_{ij} rate constants in using a semiclassical approach,[75] and, in 1927, succeeded in deriving *all* the rate constants from first principles within the framework of quantum theory.[76][77] Dirac's work was the foundation of quantum electrodynamics, i.e., the quantization of the electromagnetic field itself. Dirac's approach is also called *second quantization* or quantum field theory;[78][79][80] earlier quantum mechanical treatments only treat material particles as quantum mechanical, not the electromagnetic field.

Einstein was troubled by the fact that his theory seemed incomplete, since it did not determine the *direction* of a spontaneously emitted photon. A probabilistic nature of light-particle motion was first considered by Newton in his treatment of birefringence and, more generally, of the splitting of light beams at interfaces into a transmitted beam and a reflected beam. Newton hypothesized that hidden variables in the light particle determined which path it would follow.[34] Similarly, Einstein hoped for a more complete theory that would leave nothing to chance, beginning his separation[45] from quantum mechanics. Ironically, Max Born's probabilistic interpretation of the wave function[81][82] was inspired by Einstein's later work searching for a more complete theory.[83]

27.9   Second quantization and high energy photon interactions

Main article: Quantum field theory

In 1910, Peter Debye derived Planck's law of black-body radiation from a relatively simple assumption.[84] He correctly decomposed the electromagnetic field in a cavity into its Fourier modes, and assumed that the energy in any mode was an integer multiple of $h\nu$, where ν is the frequency of the electromagnetic mode. Planck's law of black-body radiation follows immediately as a geometric sum. However, Debye's approach failed to give the correct formula for the energy fluctuations of blackbody radiation, which were derived by Einstein in 1909.[41]

In 1925, Born, Heisenberg and Jordan reinterpreted Debye's concept in a key way.[85] As may be shown classically, the Fourier modes of the electromagnetic field—a complete set of electromagnetic plane waves indexed by their wave vector **k** and polarization state—are equivalent to a set of uncoupled simple harmonic oscillators. Treated quantum mechanically, the energy levels of such oscillators are known to be $E = n h\nu$, where ν is the oscillator frequency. The key new step was to identify an electromagnetic mode with energy $E = n h\nu$ as a state with n photons, each of energy $h\nu$. This approach gives the correct energy fluctuation formula.

Dirac took this one step further.[76][77] He treated the interaction between a charge and an electromagnetic field as a small perturbation that induces transitions in the photon states, changing the numbers of photons in the modes, while conserving energy and momentum overall. Dirac was able to derive Einstein's A_{ij} and B_{ij} coefficients from first principles, and showed that the Bose–Einstein statistics of photons is a natural consequence of quantizing the electromagnetic field correctly (Bose's reasoning went in the opposite direction; he derived Planck's law of black-body radiation by *assuming* B–E statistics). In Dirac's time, it was not yet known that all bosons, including photons, must obey Bose–Einstein statistics.

Dirac's second-order perturbation theory can involve virtual photons, transient intermediate states of the electromagnetic field; the static electric and magnetic interactions are mediated by such virtual photons. In such quantum field theories, the probability amplitude of observable events is calculated by summing over *all* possible intermediate steps, even ones that are unphysical; hence, virtual photons are not constrained to satisfy $E = pc$, and may have extra polarization states; depending on the gauge used, virtual photons may have three or four polarization states, instead of the two states of real photons. Although these transient virtual photons can never be observed, they contribute measurably to the probabilities of observable events. Indeed, such second-order and higher-order perturbation calculations can give apparently infinite contributions to the sum. Such unphysical results are corrected for using the technique of renormalization.

Other virtual particles may contribute to the summation as well; for example, two photons may interact indirectly through virtual electron–positron pairs.[86] In fact, such photon-photon scattering (see two-photon physics), as well as electron-photon scattering, is meant to be one of the modes of operations of the planned particle accelerator, the International Linear Collider.[87]

In modern physics notation, the quantum state of the electromagnetic field is written as a Fock state, a tensor product of the states for each electromagnetic mode

$$|n_{k_0}\rangle \otimes |n_{k_1}\rangle \otimes \cdots \otimes |n_{k_n}\rangle \ldots$$

where $|n_{k_i}\rangle$ represents the state in which n_{k_i} photons are in the mode k_i. In this notation, the creation of a new photon in mode k_i (e.g., emitted from an atomic transition) is written as $|n_{k_i}\rangle \rightarrow |n_{k_i} + 1\rangle$. This notation merely expresses the concept of Born, Heisenberg and Jordan described above, and does not add any physics.

27.10 The hadronic properties of the photon

Measurements of the interaction between energetic photons and hadrons show that the interaction is much more intense than expected by the interaction of merely photons with the hadron's electric charge. Furthermore, the interaction of energetic photons with protons is similar to the interaction of photons with neutrons[88] in spite of the fact that the electric charge structures of protons and neutrons are substantially different.

A theory called Vector Meson Dominance (VMD) was developed to explain this effect. According to VMD, the photon is a superposition of the pure electromagnetic photon (which interacts only with electric charges) and vector meson.[89]

However, if experimentally probed at very short distances, the intrinsic structure of the photon is recognized as a flux of quark and gluon components, quasi-free according to asymptotic freedom in QCD and described by the photon structure function.[90][91] A comprehensive comparison of data with theoretical predictions is presented in a recent review.[92]

27.11 The photon as a gauge boson

Main article: Gauge theory

The electromagnetic field can be understood as a gauge field, i.e., as a field that results from requiring that a gauge symmetry holds independently at every position in spacetime.[93] For the electromagnetic field, this gauge symmetry is the Abelian U(1) symmetry of complex numbers of absolute value 1, which reflects the ability to vary the phase of a complex number without affecting observables or real valued functions made from it, such as the energy or the Lagrangian.

The quanta of an Abelian gauge field must be massless, uncharged bosons, as long as the symmetry is not broken; hence, the photon is predicted to be massless, and to have zero electric charge and integer spin. The particular form of the electromagnetic interaction specifies that the photon must have spin ± 1; thus, its helicity must be $\pm \hbar$. These two spin components correspond to the classical concepts of right-handed and left-handed circularly polarized light. However, the transient virtual photons of quantum electrodynamics may also adopt unphysical polarization states.[93]

In the prevailing Standard Model of physics, the photon is one of four gauge bosons in the electroweak interaction; the other three are denoted W^+, W^- and Z^0 and are responsible for the weak interaction. Unlike the photon, these gauge bosons have mass, owing to a mechanism that breaks their SU(2) gauge symmetry. The unification of the photon with W and Z gauge bosons in the electroweak interaction was accomplished by Sheldon Glashow, Abdus Salam and Steven Weinberg, for which they were awarded the 1979 Nobel Prize in physics.[94][95][96] Physicists continue to hypothesize grand unified theories that connect these four gauge bosons with the eight gluon gauge bosons of quantum chromodynamics; however, key predictions of these theories, such as proton decay, have not been observed experimentally.[97]

27.12 Contributions to the mass of a system

See also: Mass in special relativity and General relativity

The energy of a system that emits a photon is *decreased* by the energy E of the photon as measured in the rest frame of the emitting system, which may result in a reduction in mass in the amount E/c^2. Similarly, the mass of a system that absorbs a photon is *increased* by a corresponding amount. As an application, the energy balance of nuclear reactions involving photons is commonly written in terms of the masses of the nuclei involved, and terms of the form E/c^2 for the gamma photons (and for other relevant energies, such as the recoil energy of nuclei).[98]

This concept is applied in key predictions of quantum electrodynamics (QED, see above). In that theory, the mass of electrons (or, more generally, leptons) is modified by including the mass contributions of virtual photons, in a technique known as renormalization. Such "radiative corrections" contribute to a number of predictions of QED, such as the magnetic dipole moment of leptons, the Lamb shift, and the hyperfine structure of bound lepton pairs, such as muonium and positronium.[99]

Since photons contribute to the stress–energy tensor, they exert a gravitational attraction on other objects, according to the theory of general relativity. Conversely, photons are themselves affected by gravity; their normally straight trajectories may be bent by warped spacetime, as in gravitational lensing, and their frequencies may be lowered by moving to a higher gravitational potential, as in the Pound–Rebka experiment. However, these effects are not specific to photons; exactly the same effects would be predicted for classical electromagnetic waves.[100]

27.13 Photons in matter

See also: Group velocity and Photochemistry

Any 'explanation' of how photons travel through matter has to explain why different arrangements of matter are transparent or opaque at different wavelengths (light through carbon as diamond or not, as graphite) and why individual photons behave in the same way as large groups. Explanations that invoke 'absorption' and 're-emission' have to provide an explanation for the directionality of the photons (diffraction, reflection) and further explain how entangled photon pairs can travel through matter without their quantum state collapsing.

The simplest explanation is that light that travels through transparent matter does so at a lower speed than c, the speed of light in a vacuum. In addition, light can also undergo scattering and absorption. There are circumstances in which heat transfer through a material is mostly radiative, involving emission and absorption of photons within it. An example would be in the core of the Sun. Energy can take about a million years to reach the surface.[101] However, this phenomenon is distinct from scattered radiation passing diffusely through matter, as it involves local equilibrium between the radiation and the temperature. Thus, the time is how long it takes the *energy* to be transferred, not the *photons* themselves. Once in open space, a photon from the Sun takes only 8.3 minutes to reach Earth. The factor by which the speed of light is decreased in a material is called the refractive index of the material. In a classical wave picture, the slowing can be explained by the light inducing electric polarization in the matter, the polarized matter radiating new light, and the new light interfering with the original light wave to form a delayed wave. In a particle picture, the slowing can instead be described as a blending of the photon with quantum excitation of the matter (quasi-particles such as phonons and excitons) to form a polariton; this polariton has a nonzero effective mass, which means that it cannot travel at c.

Alternatively, photons may be viewed as *always* traveling at c, even in matter, but they have their phase shifted (delayed or advanced) upon interaction with atomic scatters: this modifies their wavelength and momentum, but not speed.[102] A light wave made up of these photons does travel slower than the speed of light. In this view the photons are "bare", and are scattered and phase shifted, while in the view of the preceding paragraph the photons are "dressed" by their interaction with matter, and move without scattering or phase shifting, but at a lower speed.

Light of different frequencies may travel through matter at different speeds; this is called dispersion. In some cases, it can result in extremely slow speeds of light in matter. The effects of photon interactions with other quasi-particles may be observed directly in Raman scattering and Brillouin scattering.[103]

Photons can also be absorbed by nuclei, atoms or molecules, provoking transitions between their energy levels. A classic example is the molecular transition of retinal $C_{20}H_{28}O$, which is responsible for vision, as discovered in 1958 by Nobel laureate biochemist George Wald and co-workers. The absorption provokes a cis-trans isomerization that, in combination with other such transitions, is transduced into nerve impulses. The absorption of photons can even break chemical bonds, as in the photodissociation of chlorine; this is the subject of photochemistry.[104][105] Analogously, gamma rays can in some circumstances dissociate atomic nuclei in a process called photodisintegration.

27.14 Technological applications

Photons have many applications in technology. These examples are chosen to illustrate applications of photons *per se*, rather than general optical devices such as lenses, etc. that could operate under a classical theory of light. The laser is an extremely important application and is discussed above under stimulated emission.

Individual photons can be detected by several methods. The classic photomultiplier tube exploits the photoelectric effect: a photon landing on a metal plate ejects an electron, initiating an ever-amplifying avalanche of electrons. Charge-coupled device chips use a similar effect in semiconductors: an incident photon generates a charge on a microscopic capacitor that can be detected. Other detectors such as Geiger counters use the ability of photons to ionize gas molecules, causing a detectable change in conductivity.[106]

Planck's energy formula $E = h\nu$ is often used by engineers and chemists in design, both to compute the change in energy resulting from a photon absorption and to predict the frequency of the light emitted for a given energy transition. For example, the emission spectrum of a fluorescent light bulb can be designed using gas molecules with different electronic energy levels and adjusting the typical energy with which an electron hits the gas molecules within the bulb.[Note 6]

Under some conditions, an energy transition can be excited by "two" photons that individually would be insufficient. This allows for higher resolution microscopy, because the sample absorbs energy only in the region where two beams of different colors overlap significantly, which can be made much smaller than the excitation volume of a single beam (see two-photon excitation microscopy). Moreover, these photons cause less damage to the sample, since they are of lower energy.[107]

In some cases, two energy transitions can be coupled so that, as one system absorbs a photon, another nearby system "steals" its energy and re-emits a photon of a different frequency. This is the basis of fluorescence resonance energy transfer, a technique that is used in molecular biology to study the interaction of suitable proteins.[108]

Several different kinds of hardware random number generator involve the detection of single photons. In one example, for each bit in the random sequence that is to be produced, a photon is sent to a beam-splitter. In such a situation, there are two possible outcomes of equal probability. The actual outcome is used to determine whether the next bit in the sequence is "0" or "1".[109][110]

27.15 Recent research

See also: Quantum optics

Much research has been devoted to applications of photons in the field of quantum optics. Photons seem well-suited to be elements of an extremely fast quantum computer, and the quantum entanglement of photons is a focus of research. Nonlinear optical processes are another active research area, with topics such as two-photon absorption, self-phase modulation, modulational instability and optical parametric oscillators. However, such processes generally do not require the assumption of photons *per se*; they may often be modeled by treating atoms as nonlinear oscillators. The nonlinear process of spontaneous parametric down conversion is often used to produce single-photon states. Finally, photons are essential in some aspects of optical communication, especially for quantum cryptography.[Note 7]

27.16 See also

- Advanced Photon Source at Argonne National Laboratory

- Ballistic photon

- Doppler shift

- Electromagnetic radiation

- HEXITEC

- Laser
- Light
- Luminiferous aether
- Medipix
- Phonons
- Photon counting
- Photon energy
- Photon polarization
- Photonic molecule
- Photography
- Photonics
- Quantum optics
- Single photon sources
- Static forces and virtual-particle exchange
- Two-photon physics
- EPR paradox
- Dirac equation

27.17   Notes

[1] Although the 1967 Elsevier translation of Planck's Nobel Lecture interprets Planck's *Lichtquant* as "photon", the more literal 1922 translation by Hans Thacher Clarke and Ludwik Silberstein *The origin and development of the quantum theory*, The Clarendon Press, 1922 (here) uses "light-quantum". No evidence is known that Planck himself used the term "photon" by 1926 (see also this note).

[2] Isaac Asimov credits Arthur Compton with defining quanta of energy as photons in 1923. Asimov, I. (1966). *The Neutrino, Ghost Particle of the Atom*. Garden City (NY): Doubleday. ISBN 0-380-00483-6. LCCN 66017073. and Asimov, I. (1966). *The Universe From Flat Earth To Quasar*. New York (NY): Walker. ISBN 0-8027-0316-X. LCCN 66022515.

[3] The mass of the photon is believed to be exactly zero, based on experiment and theoretical considerations described in the article. Some sources also refer to the *relativistic mass* concept, which is just the energy scaled to units of mass. For a photon with wavelength λ or energy E, this is $h/\lambda c$ or E/c^2. This usage for the term "mass" is no longer common in scientific literature. Further info: What is the mass of a photon? http://math.ucr.edu/home/baez/physics/ParticleAndNuclear/photon_mass.html

[4] The phrase "no matter how intense" refers to intensities below approximately 10^{13} W/cm^2 at which point perturbation theory begins to break down. In contrast, in the intense regime, which for visible light is above approximately 10^{14} W/cm^2, the classical wave description correctly predicts the energy acquired by electrons, called ponderomotive energy. (See also: Boreham *et al.* (1996). "Photon density and the correspondence principle of electromagnetic interaction".) By comparison, sunlight is only about 0.1 W/cm^2.

[5] These experiments produce results that cannot be explained by any classical theory of light, since they involve anticorrelations that result from the quantum measurement process. In 1974, the first such experiment was carried out by Clauser, who reported a violation of a classical Cauchy–Schwarz inequality. In 1977, Kimble *et al.* demonstrated an analogous anti-bunching effect of photons interacting with a beam splitter; this approach was simplified and sources of error eliminated in the photon-anticorrelation experiment of Grangier *et al.* (1986). This work is reviewed and simplified further in Thorn *et al.* (2004). (These references are listed below under #Additional references.)

[6] An example is US Patent Nr. 5212709.

[7] Introductory-level material on the various sub-fields of quantum optics can be found in Fox, M. (2006). *Quantum Optics: An Introduction*. Oxford University Press. ISBN 0-19-856673-5.

27.18 References

[1] Amsler, C. (Particle Data Group); Amsler; Doser; Antonelli; Asner; Babu; Baer; Band; Barnett; Bergren; Beringer; Bernardi; Bertl; Bichsel; Biebel; Bloch; Blucher; Blusk; Cahn; Carena; Caso; Ceccucci; Chakraborty; Chen; Chivukula; Cowan; Dahl; d'Ambrosio; Damour; et al. (2008). "Review of Particle Physics: Gauge and Higgs bosons" (PDF). *Physics Letters B* **667**: 1. Bibcode:2008PhLB..667....1P. doi:10.1016/j.physletb.2008.07.018.

[2] Joos, George (1951). *Theoretical Physics*. London and Glasgow: Blackie and Son Limited. p. 679.

[3] Kimble, H.J.; Dagenais, M.; Mandel, L.; Dagenais; Mandel (1977). "Photon Anti-bunching in Resonance Fluorescence". *Physical Review Letters* **39** (11): 691–695. Bibcode:1977PhRvL..39..691K. doi:10.1103/PhysRevLett.39.691.

[4] Grangier, P.; Roger, G.; Aspect, A.; Roger; Aspect (1986). "Experimental Evidence for a Photon Anticorrelation Effect on a Beam Splitter: A New Light on Single-Photon Interferences". *Europhysics Letters* **1** (4): 173–179. Bibcode:1986EL......1..173G. doi:10.1209/0295-5075/1/4/004.

[5] Planck, M. (1901). "On the Law of Distribution of Energy in the Normal Spectrum". *Annalen der Physik* **4** (3): 553–563. Bibcode:1901AnP...309..553P. doi:10.1002/andp.19013090310. Archived from the original on 2008-04-18.

[6] Einstein, A. (1905). "Über einen die Erzeugung und Verwandlung des Lichtes betreffenden heuristischen Gesichtspunkt" (PDF). *Annalen der Physik* (in German) **17** (6): 132–148. Bibcode:1905AnP...322..132E. doi:10.1002/andp.19053220607.. An English translation is available from Wikisource.

[7] "Discordances entre l'expérience et la théorie électromagnétique du rayonnement." In Électrons et Photons. Rapports et Discussions de Cinquième Conseil de Physique, edited by Institut International de Physique Solvay. Paris: Gauthier-Villars, pp. 55-85.

[8] Helge Kragh: *Photon: New light on an old name*. Arxiv, 2014-2-28

[9] Villard, P. (1900). "Sur la réflexion et la réfraction des rayons cathodiques et des rayons déviables du radium". *Comptes Rendus des Séances de l'Académie des Sciences* (in French) **130**: 1010–1012.

[10] Villard, P. (1900). "Sur le rayonnement du radium". *Comptes Rendus des Séances de l'Académie des Sciences* (in French) **130**: 1178–1179.

[11] Rutherford, E.; Andrade, E.N.C. (1914). "The Wavelength of the Soft Gamma Rays from Radium B". *Philosophical Magazine* **27** (161): 854–868. doi:10.1080/14786440508635156.

[12] Kobychev, V.V.; Popov, S.B. (2005). "Constraints on the photon charge from observations of extragalactic sources". *Astronomy Letters* **31** (3): 147–151. arXiv:hep-ph/0411398. Bibcode:2005AstL...31..147K. doi:10.1134/1.1883345.

[13] Role as gauge boson and polarization section 5.1 inAitchison, I.J.R.; Hey, A.J.G. (1993). *Gauge Theories in Particle Physics*. IOP Publishing. ISBN 0-85274-328-9.

[14] See p.31 inAmsler, C.; et al. (2008). "Review of Particle Physics". *Physics Letters B* **667**: 1–1340. Bibcode:2008PhLB..667....1P. doi:10.1016/j.physletb.2008.07.018.

[15] Halliday, David; Resnick, Robert; Walker, Jerl (2005), *Fundamental of Physics* (7th ed.), USA: John Wiley and Sons, Inc., ISBN 0-471-23231-9

[16] See section 1.6 in Alonso, M.; Finn, E.J. (1968). *Fundamental University Physics Volume III: Quantum and Statistical Physics*. Addison-Wesley. ISBN 0-201-00262-0.

[17] Davison E. Soper, Electromagnetic radiation is made of photons, Institute of Theoretical Science, University of Oregon

[18] This property was experimentally verified by Raman and Bhagavantam in 1931: Raman, C.V.; Bhagavantam, S. (1931). "Experimental proof of the spin of the photon" (PDF). *Indian Journal of Physics* **6**: 353.

[19] Burgess, C.; Moore, G. (2007). "1.3.3.2". *The Standard Model. A Primer*. Cambridge University Press. ISBN 0-521-86036-9.

[20] Griffiths, David J. (2008), *Introduction to Elementary Particles* (2nd revised ed.), WILEY-VCH, ISBN 978-3-527-40601-2

[21] E.g., section 9.3 in Alonso, M.; Finn, E.J. (1968). *Fundamental University Physics Volume III: Quantum and Statistical Physics*. Addison-Wesley.

[22] E.g., Appendix XXXII in Born, M. (1962). *Atomic Physics*. Blackie & Son. ISBN 0-486-65984-4.

[23] Mermin, David (February 1984). "Relativity without light". *American Journal of Physics* **52**(2): 119–124. Bibcode:1984AmJ. doi:10.1119/1.13917.

[24] Plimpton, S.; Lawton, W. (1936). "A Very Accurate Test of Coulomb's Law of Force Between Charges". *Physical Review* **50** (11): 1066. Bibcode:1936PhRv...50.1066P. doi:10.1103/PhysRev.50.1066.

[25] Williams, E.; Faller, J.; Hill, H. (1971). "New Experimental Test of Coulomb's Law: A Laboratory Upper Limit on the Photon Rest Mass". *Physical Review Letters* **26** (12): 721. Bibcode:1971PhRvL..26..721W. doi:10.1103/PhysRevLett.26.721.

[26] Chibisov, G V (1976). "Astrophysical upper limits on the photon rest mass". *Soviet Physics Uspekhi* **19**(7): 624. Bibcode:19. doi:10.1070/PU1976v019n07ABEH005277.

[27] Lakes, Roderic (1998). "Experimental Limits on the Photon Mass and Cosmic Magnetic Vector Potential". *Physical Review Letters* **80** (9): 1826. Bibcode:1998PhRvL..80.1826L. doi:10.1103/PhysRevLett.80.1826.

[28] Amsler, C; Doser, M; Antonelli, M; Asner, D; Babu, K; Baer, H; Band, H; Barnett, R; et al. (2008). "Review of Particle Physics*". *Physics Letters B* **667**: 1. Bibcode:2008PhLB..667....1P. doi:10.1016/j.physletb.2008.07.018. Summary Table

[29] Adelberger, Eric; Dvali, Gia; Gruzinov, Andrei (2007). "Photon-Mass Bound Destroyed by Vortices". *Physical Review Letters* **98** (1): 010402. arXiv:hep-ph/0306245. Bibcode:2007PhRvL..98a0402A. doi:10.1103/PhysRevLett.98.010402. PMID 17358459. preprint

[30] Wilczek, Frank (2010). *The Lightness of Being: Mass, Ether, and the Unification of Forces*. Basic Books. p. 212. ISBN 978-0-465-01895-6.

[31] Descartes, R. (1637). *Discours de la méthode (Discourse on Method)* (in French). Imprimerie de Ian Maire. ISBN 0-268-00870-1.

[32] Hooke, R. (1667). *Micrographia: or some physiological descriptions of minute bodies made by magnifying glasses with observations and inquiries thereupon ...* London (UK): Royal Society of London. ISBN 0-486-49564-7.

[33] Huygens, C. (1678). *Traité de la lumière* (in French).. An English translation is available from Project Gutenberg

[34] Newton, I. (1952) [1730]. *Opticks* (4th ed.). Dover (NY): Dover Publications. Book II, Part III, Propositions XII–XX; Queries 25–29. ISBN 0-486-60205-2.

[35] Buchwald, J.Z. (1989). *The Rise of the Wave Theory of Light: Optical Theory and Experiment in the Early Nineteenth Century*. University of Chicago Press. ISBN 0-226-07886-8. OCLC 18069573.

[36] Maxwell, J.C. (1865). "A Dynamical Theory of the Electromagnetic Field". *Philosophical Transactions of the Royal Society* **155**: 459–512. Bibcode:1865RSPT..155..459C. doi:10.1098/rstl.1865.0008. This article followed a presentation by Maxwell on 8 December 1864 to the Royal Society.

[37] Hertz, H. (1888). "Über Strahlen elektrischer Kraft". *Sitzungsberichte der Preussischen Akademie der Wissenschaften (Berlin)* (in German) **1888**: 1297–1307.

[38] Frequency-dependence of luminiscence p. 276f., photoelectric effect section 1.4 in Alonso, M.; Finn, E.J. (1968). *Fundamental University Physics Volume III: Quantum and Statistical Physics*. Addison-Wesley. ISBN 0-201-00262-0.

[39] Wien, W. (1911). "Wilhelm Wien Nobel Lecture".

[40] Planck, M. (1920). "Max Planck's Nobel Lecture".

[41] Einstein, A. (1909). "Über die Entwicklung unserer Anschauungen über das Wesen und die Konstitution der Strahlung" (PDF). *Physikalische Zeitschrift* (in German) **10**: 817–825.. An English translation is available from Wikisource.

[42] Presentation speech by Svante Arrhenius for the 1921 Nobel Prize in Physics, December 10, 1922. Online text from [nobel-prize.org], The Nobel Foundation 2008. Access date 2008-12-05.

[43] Einstein, A. (1916). "Zur Quantentheorie der Strahlung". *Mitteilungen der Physikalischen Gesellschaft zu Zürich* **16**: 47. Also *Physikalische Zeitschrift*, **18**, 121–128 (1917). (German)

[44] Compton, A. (1923). "A Quantum Theory of the Scattering of X-rays by Light Elements". *Physical Review* **21** (5): 483–502. Bibcode:1923PhRv...21..483C. doi:10.1103/PhysRev.21.483.

[45] Pais, A. (1982). *Subtle is the Lord: The Science and the Life of Albert Einstein*. Oxford University Press. ISBN 0-19-853907-X.

[46] *Einstein and the Quantum: The Quest of the Valiant Swabian*, A. Douglas Stone, Princeton University Press, 2013.

[47] Millikan, R.A (1924). "Robert A. Millikan's Nobel Lecture".

[48] Hendry, J. (1980). "The development of attitudes to the wave-particle duality of light and quantum theory, 1900–1920". *Annals of Science* **37** (1): 59–79. doi:10.1080/00033798000200121.

[49] Bohr, N.; Kramers, H.A.; Slater, J.C. (1924). "The Quantum Theory of Radiation". *Philosophical Magazine* **47**: 785–802. doi:10.1080/14786442408565262. Also *Zeitschrift für Physik*, **24**, 69 (1924).

[50] Heisenberg, W. (1933). "Heisenberg Nobel lecture".

[51] Mandel, L. (1976). E. Wolf, ed. "The case for and against semiclassical radiation theory". *Progress in Optics*. Progress in Optics (North-Holland) **13**: 27–69. doi:10.1016/S0079-6638(08)70018-0. ISBN 978-0-444-10806-7.

[52] Taylor, G.I. (1909). *Interference fringes with feeble light*. Proceedings of the Cambridge Philosophical Society **15**: 114–115.

[53] Saleh, B. E. A. and Teich, M. C. (2007). *Fundamentals of Photonics*. Wiley. ISBN 0-471-35832-0.

[54] Heisenberg, W. (1927). "Über den anschaulichen Inhalt der quantentheoretischen Kinematik und Mechanik". *Zeitschrift für Physik* (in German) **43** (3–4): 172–198. Bibcode:1927ZPhy...43..172H. doi:10.1007/BF01397280.

[55] E.g., p. 10f. in Schiff, L.I. (1968). *Quantum Mechanics* (3rd ed.). McGraw-Hill. ASIN B001B3MINM. ISBN 0-07-055287-8.

[56] Kramers, H.A. (1958). *Quantum Mechanics*. Amsterdam: North-Holland. ASIN B0006AUW5C. ISBN 0-486-49533-7.

[57] Bohm, D. (1989) [1954]. *Quantum Theory*. Dover Publications. ISBN 0-486-65969-0.

[58] Newton, T.D.; Wigner, E.P. (1949). "Localized states for elementary particles". *Reviews of Modern Physics* **21** (3): 400–406. Bibcode:1949RvMP...21..400N. doi:10.1103/RevModPhys.21.400.

[59] Bialynicki-Birula, I. (1994). "On the wave function of the photon" (PDF). *Acta Physica Polonica A* **86**: 97–116.

[60] Sipe, J.E. (1995). "Photon wave functions". *Physical Review A* **52** (3): 1875–1883. Bibcode:1995PhRvA..52.1875S. doi:

[61] Bialynicki-Birula, I. (1996). "Photon wave function". *Progress in Optics*. Progress in Optics **36**: 245–294. doi:10.1016/S0079-6638(08)70316-0. ISBN 978-0-444-82530-8.

[62] Scully, M.O.; Zubairy, M.S. (1997). *Quantum Optics*. Cambridge (UK): Cambridge University Press. ISBN 0-521-43595-1.

[63] The best illustration is the Couder experiment, demonstrating the behaviour of a mechanical analog, see https://www.youtube.com/watch?v=W9yWv5dqSKk

[64] Bell, J. S., "Speakable and Unspeakable in Quantum Mechanics", Cambridge: Cambridge University Press, 1987.

[65] Bose, S.N. (1924). "Plancks Gesetz und Lichtquantenhypothese". *Zeitschrift für Physik* (in German) **26**: 178–181. Bibcode: doi:10.1007/BF01327326.

[66] Einstein, A. (1924). "Quantentheorie des einatomigen idealen Gases". *Sitzungsberichte der Preussischen Akademie der Wissenschaften (Berlin), Physikalisch-mathematische Klasse* (in German) **1924**: 261–267.

[67] Einstein, A. (1925). "Quantentheorie des einatomigen idealen Gases, Zweite Abhandlung". *Sitzungsberichte der Preussischen Akademie der Wissenschaften (Berlin), Physikalisch-mathematische Klasse* (in German) **1925**: 3–14. doi:10.10 .ISBN978-3-527-60895-9.

[68] Anderson, M.H.; Ensher, J.R.; Matthews, M.R.; Wieman, C.E.; Cornell, E.A. (1995). "Observation of Bose–Einstein Conden-sation in a Dilute Atomic Vapor". *Science* **269** (5221): 198–201. Bibcode:1995Sci...269..198A. doi:10.1126/science.269. .JSTOR2888436.PMID17789847.

[69] "Physicists Slow Speed of Light". News.harvard.edu (1999-02-18). Retrieved on 2015-05-11.

[70] "Light Changed to Matter, Then Stopped and Moved". photonics.com (February 2007). Retrieved on 2015-05-11.

[71] Streater, R.F.; Wightman, A.S. (1989). *PCT, Spin and Statistics, and All That*. Addison-Wesley. ISBN 0-201-09410-X.

[72] Einstein, A. (1916). "Strahlungs-emission und -absorption nach der Quantentheorie". *Verhandlungen der Deutschen Physikalis-chen Gesellschaft* (in German) **18**: 318–323. Bibcode:1916DPhyG..18..318E.

[73] Section 1.4 in Wilson, J.; Hawkes, F.J.B. (1987). *Lasers: Principles and Applications*. New York: Prentice Hall. ISBN 0-13-523705-X.

[74] P. 322 in Einstein, A. (1916). "Strahlungs-emission und -absorption nach der Quantentheorie". *Verhandlungen der Deutschen Physikalischen Gesellschaft* (in German) **18**: 318–323. Bibcode:1916DPhyG..18..318E.:

> Die Konstanten A_m^n and B_m^n würden sich direkt berechnen lassen, wenn wir im Besitz einer im Sinne der Quantenhypothese modifizierten Elektrodynamik und Mechanik wären."

[75] Dirac, P.A.M. (1926). "On the Theory of Quantum Mechanics". *Proceedings of the Royal Society A* **112** (762): 661–677. Bibcode:1926RSPSA.112..661D. doi:10.1098/rspa.1926.0133.

[76] Dirac, P.A.M. (1927). "The Quantum Theory of the Emission and Absorption of Radiation" (PDF). *Proceedings of the Royal Society A* **114** (767): 243–265. Bibcode:1927RSPSA.114..243D. doi:10.1098/rspa.1927.0039.

[77] Dirac, P.A.M. (1927b). *The Quantum Theory of Dispersion*. *Proceedings of the Royal Society A* **114**: 710–728. doi:

[78] Heisenberg, W.; Pauli, W. (1929). "Zur Quantentheorie der Wellenfelder". *Zeitschrift für Physik* (in German) **56**: 1. Bibcode: doi:10.1007/BF01340129.

[79] Heisenberg, W.; Pauli, W. (1930). "Zur Quantentheorie der Wellenfelder". *Zeitschrift für Physik* (in German) **59** (3–4): 139. Bibcode:1930ZPhy...59..168H. doi:10.1007/BF01341423.

[80] Fermi, E. (1932). "Quantum Theory of Radiation" (PDF). *Reviews of Modern Physics* **4**: 87. Bibcode:1932RvMP....4...87F. doi:10.1103/RevModPhys.4.87.

[81] Born, M. (1926). "Zur Quantenmechanik der Stossvorgänge" (PDF). *Zeitschrift für Physik* (in German) **37** (12): 863–867. Bibcode:1926ZPhy...37..863B. doi:10.1007/BF01397477.

[82] Born, M. (1926). "Quantenmechanik der Stossvorgänge". *Zeitschrift für Physik* (in German) **38** (11–12): 803. Bibcode: doi:10.1007/BF01397184.

[83] Pais, A. (1986). *Inward Bound: Of Matter and Forces in the Physical World*. Oxford University Press. p. 260. ISBN 0-19-851997-4. Specifically, Born claimed to have been inspired by Einstein's never-published attempts to develop a "ghost-field" theory, in which point-like photons are guided probabilistically by ghost fields that follow Maxwell's equations.

[84] Debye, P. (1910). "Der Wahrscheinlichkeitsbegriff in der Theorie der Strahlung". *Annalen der Physik* (in German) **33** (16): 1427–1434. Bibcode:1910AnP...338.1427D. doi:10.1002/andp.19103381617.

[85] Born, M.; Heisenberg, W.; Jordan, P. (1925). "Quantenmechanik II". *Zeitschrift für Physik* (in German) **35** (8–9): 557–615. Bibcode:1926ZPhy...35..557B. doi:10.1007/BF01379806.

[86] Photon-photon-scattering section 7-3-1, renormalization chapter 8-2 in Itzykson, C.; Zuber, J.-B. (1980). *Quantum Field Theory*. McGraw-Hill. ISBN 0-07-032071-3.

[87] Weiglein, G. (2008). "Electroweak Physics at the ILC". *Journal of Physics: Conference Series* **110** (4): 042033. arXiv:0711.3003. Bibcode:2008JPhCS.110d2033W. doi:10.1088/1742-6596/110/4/042033.

[88] Bauer, T. H.; Spital, R. D.; Yennie, D. R.; Pipkin, F. M. (1978). "The hadronic properties of the photon in high-energy interactions". *Reviews of Modern Physics* **50** (2): 261. Bibcode:1978RvMP...50..261B. doi:10.1103/RevModPhys.50.261.

[89] Sakurai, J. J. (1960). "Theory of strong interactions". *Annals of Physics* **11**: 1. Bibcode:1960AnPhy..11....1S. doi:10.1016/0003-4916(60)90126-3.

[90] Walsh, T. F.; Zerwas, P. (1973). "Two-photon processes in the parton model". *Physics Letters B* **44**(2): 195. Bibcode:. doi:10.1016/0370-2693(73)90520-0.

[91] Witten, E. (1977). "Anomalous cross section for photon-photon scattering in gauge theories". *Nuclear Physics B* **120** (2): 189. Bibcode:1977NuPhB.120..189W. doi:10.1016/0550-3213(77)90038-4.

[92] Nisius, R. (2000). "The photon structure from deep inelastic electron–photon scattering". *Physics Reports* **332** (4–6): 165. Bibcode:2000PhR...332..165N. doi:10.1016/S0370-1573(99)00115-5.

[93] Ryder, L.H. (1996). *Quantum field theory* (2nd ed.). Cambridge University Press. ISBN 0-521-47814-6.

[94] Sheldon Glashow Nobel lecture, delivered 8 December 1979.

[95] Abdus Salam Nobel lecture, delivered 8 December 1979.

[96] Steven Weinberg Nobel lecture, delivered 8 December 1979.

[97] E.g., chapter 14 in Hughes, I. S. (1985). *Elementary particles* (2nd ed.). Cambridge University Press. ISBN 0-521-26092-2.

[98] E.g., section 10.1 in Dunlap, R.A. (2004). *An Introduction to the Physics of Nuclei and Particles*. Brooks/Cole. ISBN 0-534-39294-6.

[99] Radiative correction to electron mass section 7-1-2, anomalous magnetic moments section 7-2-1, Lamb shift section 7-3-2 and hyperfine splitting in positronium section 10-3 in Itzykson, C.; Zuber, J.-B. (1980). *Quantum Field Theory*. McGraw-Hill. ISBN 0-07-032071-3.

[100] E. g. sections 9.1 (gravitational contribution of photons) and 10.5 (influence of gravity on light) in Stephani, H.; Stewart, J. (1990). *General Relativity: An Introduction to the Theory of Gravitational Field*. Cambridge University Press. pp. 86 ff, 108 ff. ISBN 0-521-37941-5.

[101] Naeye, R. (1998). *Through the Eyes of Hubble: Birth, Life and Violent Death of Stars*. CRC Press. ISBN 0-7503-0484-7. OCLC 40180195.

[102] Ch 4 in Hecht, Eugene (2001). *Optics*. Addison Wesley. ISBN 978-0-8053-8566-3.

[103] Polaritons section 10.10.1, Raman and Brillouin scattering section 10.11.3 in Patterson, J.D.; Bailey, B.C. (2007). *Solid-State Physics: Introduction to the Theory*. Springer. pp. 569 ff, 580 ff. ISBN 3-540-24115-9.

[104] E.g., section 11-5 C in Pine, S.H.; Hendrickson, J.B.; Cram, D.J.; Hammond, G.S. (1980). *Organic Chemistry* (4th ed.). McGraw-Hill. ISBN 0-07-050115-7.

[105] Nobel lecture given by G. Wald on December 12, 1967, online at nobelprize.org: The Molecular Basis of Visual Excitation.

[106] Photomultiplier section 1.1.10, CCDs section 1.1.8, Geiger counters section 1.3.2.1 in Kitchin, C.R. (2008). *Astrophysical Techniques*. Boca Raton (FL): CRC Press. ISBN 1-4200-8243-4.

[107] Denk, W.; Svoboda, K. (1997). "Photon upmanship: Why multiphoton imaging is more than a gimmick". *Neuron* **18** (3): 351–357. doi:10.1016/S0896-6273(00)81237-4. PMID 9115730.

[108] Lakowicz, J.R. (2006). *Principles of Fluorescence Spectroscopy*. Springer. pp. 529 ff. ISBN 0-387-31278-1.

[109] Jennewein, T.; Achleitner, U.; Weihs, G.; Weinfurter, H.; Zeilinger, A. (2000). "A fast and compact quantum random number generator". *Review of Scientific Instruments* **71** (4): 1675–1680. arXiv:quant-ph/9912118. Bibcode:2000RScI...71.1675J. doi:10.1063/1.1150518.

[110] Stefanov, A.; Gisin, N.; Guinnard, O.; Guinnard, L.; Zbiden, H. (2000). "Optical quantum random number generator". *Journal of Modern Optics* **47** (4): 595–598. doi:10.1080/095003400147908.

27.19   Additional references

By date of publication:

- Clauser, J.F. (1974). "Experimental distinction between the quantum and classical field-theoretic predictions for the photoelectric effect". *Physical Review D* **9** (4): 853–860. Bibcode:1974PhRvD...9..853C. doi:

- Kimble, H.J.; Dagenais, M.; Mandel, L. (1977). "Photon Anti-bunching in Resonance Fluorescence". *Physical Review Letters* **39** (11): 691–695. Bibcode:1977PhRvL..39..691K. doi:10.1103/PhysRevLett.39.691.

- Pais, A. (1982). *Subtle is the Lord: The Science and the Life of Albert Einstein*. Oxford University Press.

- Feynman, Richard (1985). *QED: The Strange Theory of Light and Matter*. Princeton University Press. ISBN 978-0-691-12575-6.

- Grangier, P.; Roger, G.; Aspect, A. (1986). "Experimental Evidence for a Photon Anticorrelation Effect on a Beam Splitter: A New Light on Single-Photon Interferences". *Europhysics Letters* **1** (4): 173–179. Bibcode:1986E .doi:10.1209/0295-5075/1/4/004.

- Lamb, W.E. (1995). "Anti-photon". *Applied Physics B* **60** (2–3): 77–84. Bibcode:1995ApPhB..60...77L.

- Special supplemental issue of *Optics and Photonics News* (vol. 14, October 2003) article web link

  - Roychoudhuri, C.; Rajarshi, R. (2003). "The nature of light: what is a photon?". *Optics and Photonics News* **14**: S1 (Supplement).

  - Zajonc, A. "Light reconsidered". *Optics and Photonics News* **14**: S2–S5 (Supplement).

  - Loudon, R. "What is a photon?". *Optics and Photonics News* **14**: S6–S11 (Supplement).

  - Finkelstein, D. "What is a photon?". *Optics and Photonics News* **14**: S12–S17 (Supplement).

  - Muthukrishnan, A.; Scully, M.O.; Zubairy, M.S. "The concept of the photon—revisited". *Optics and Photonics News* **14**: S18–S27 (Supplement).

  - Mack, H.; Schleich, W.P.. "A photon viewed from Wigner phase space". *Optics and Photonics News* **14**: S28–S35 (Supplement).

- Glauber, R. (2005). "One Hundred Years of Light Quanta" (PDF). *2005 Physics Nobel Prize Lecture*.

- Hentschel, K. (2007). "Light quanta: The maturing of a concept by the stepwise accretion of meaning". *Physics and Philosophy* **1** (2): 1–20.

Education with single photons:

- Thorn, J.J.; Neel, M.S.; Donato, V.W.; Bergreen, G.S.; Davies, R.E.; Beck, M. (2004). "Observing the quantum behavior of light in an undergraduate laboratory" (PDF). *American Journal of Physics* **72** (9): 1210–1219. Bibcode:2004AmJPh..72.1210T. doi:10.1119/1.1737397.

- Bronner, P.; Strunz, Andreas; Silberhorn, Christine; Meyn, Jan-Peter (2009). "Interactive screen experiments with single photons". *European Journal of Physics* **30** (2): 345–353. Bibcode:2009EJPh...30..345B. doi:10.1088/0143-0807/30/2/014.

27.20   External links

- The dictionary definition of photon at Wiktionary

- Media related to Photon at Wikimedia Commons

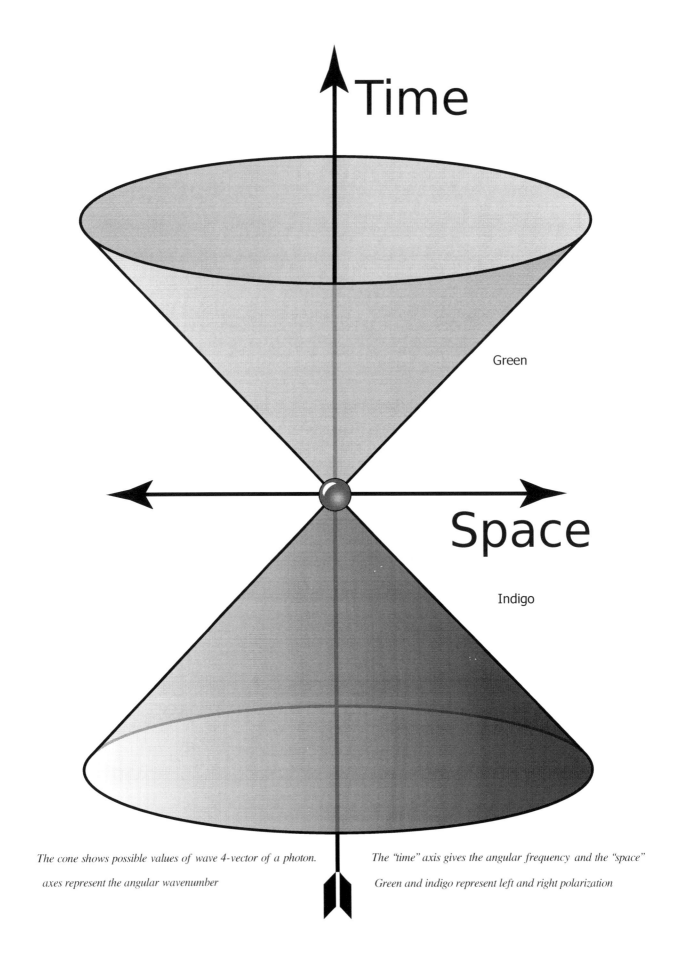

The cone shows possible values of wave 4-vector of a photon. The "time" axis gives the angular frequency and the "space" axes represent the angular wavenumber Green and indigo represent left and right polarization

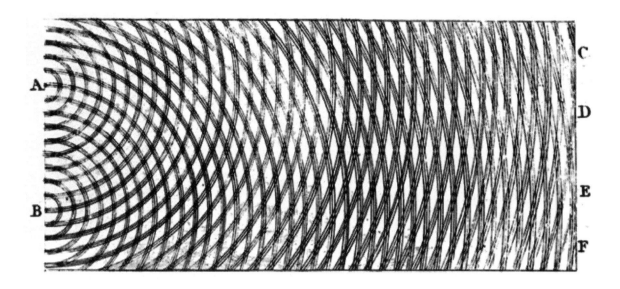

Thomas Young's double-slit experiment in 1801 showed that light can act as a wave, helping to invalidate early particle theories of light.[151:964]

Light wave

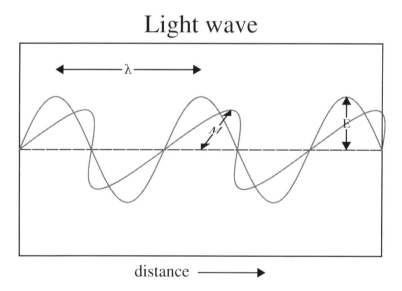

λ = wave length

E = amplitude of electric field

M = amplitude of magnetic field

distance ⟶

In 1900, Maxwell's theoretical model of light as oscillating electric and magnetic fields seemed complete. However, several observations could not be explained by any wave model of electromagnetic radiation, leading to the idea that light-energy was packaged into quanta *described by E=hν. Later experiments showed that these light-quanta also carry momentum and, thus, can be considered particles: the* photon *concept was born, leading to a deeper understanding of the electric and magnetic fields themselves.*

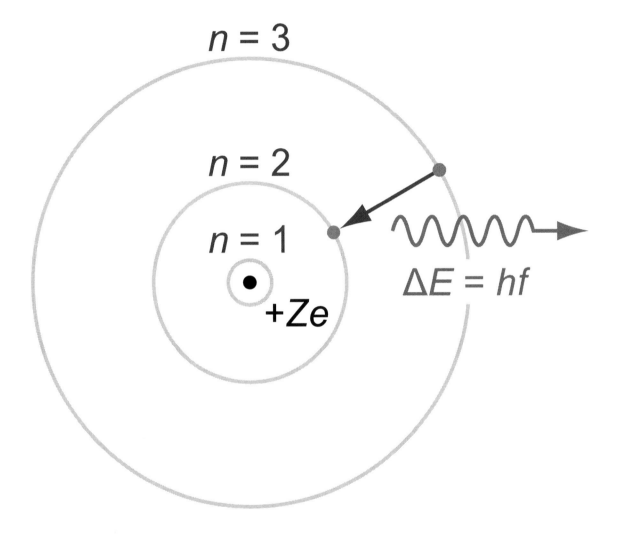

Up to 1923, most physicists were reluctant to accept that light itself was quantized. Instead, they tried to explain photon behavior by quantizing only matter, as in the Bohr model of the hydrogen atom (shown here). Even though these semiclassical models were only a first approximation, they were accurate for simple systems and they led to quantum mechanics.

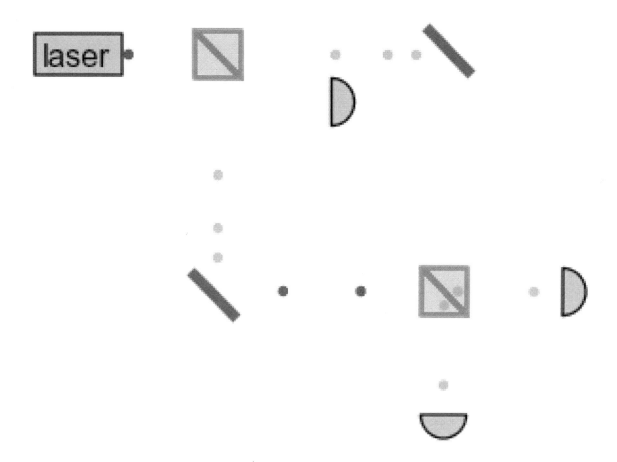

Photons in a Mach–Zehnder interferometer exhibit wave-like interference and particle-like detection at single-photon detectors.

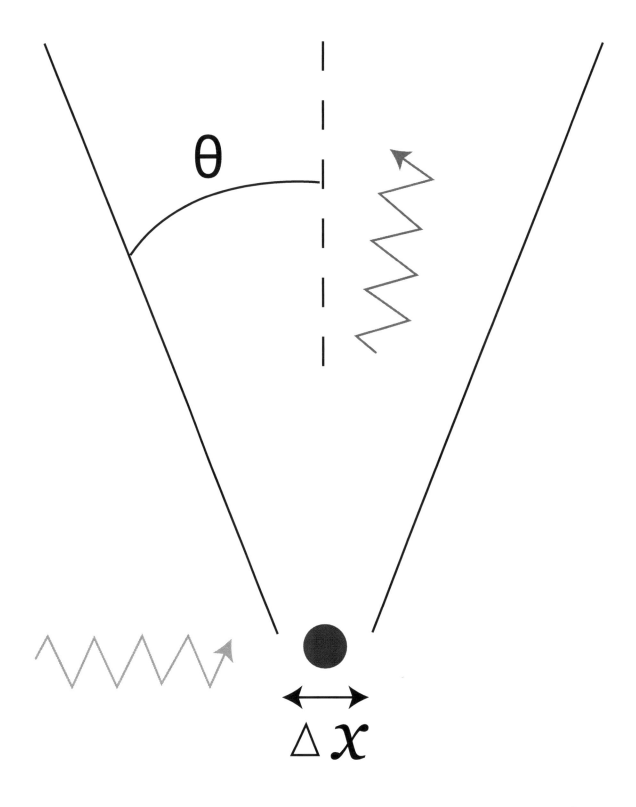

Heisenberg's thought experiment for locating an electron (shown in blue) with a high-resolution gamma-ray microscope. The incoming gamma ray (shown in green) is scattered by the electron up into the microscope's aperture angle θ. The scattered gamma ray is shown in red. Classical optics shows that the electron position can be resolved only up to an uncertainty Δx that depends on θ and the wavelength λ of the incoming light.

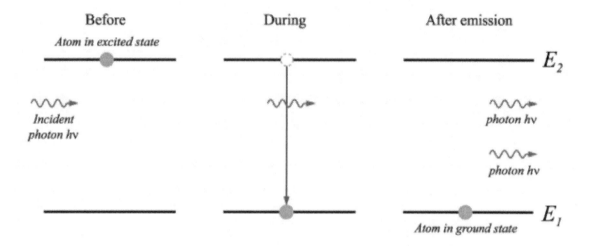

Stimulated emission (in which photons "clone" themselves) was predicted by Einstein in his kinetic analysis, and led to the development of the laser. Einstein's derivation inspired further developments in the quantum treatment of light, which led to the statistical interpretation of quantum mechanics.

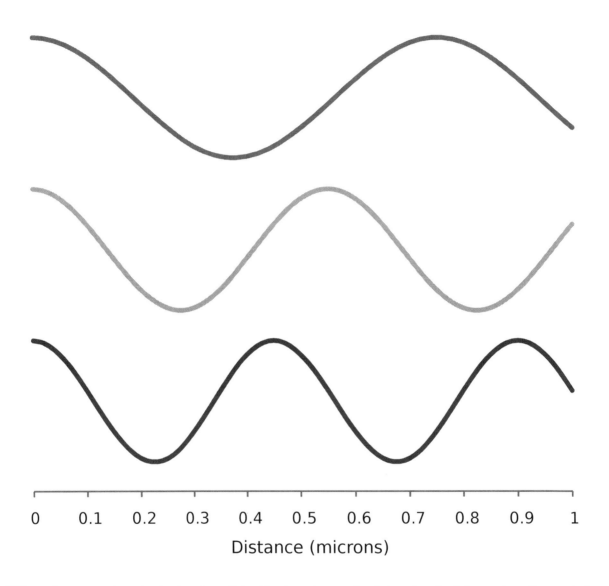

Different electromagnetic modes (such as those depicted here) can be treated as independent simple harmonic oscillators. A photon corresponds to a unit of energy E=hν in its electromagnetic mode.

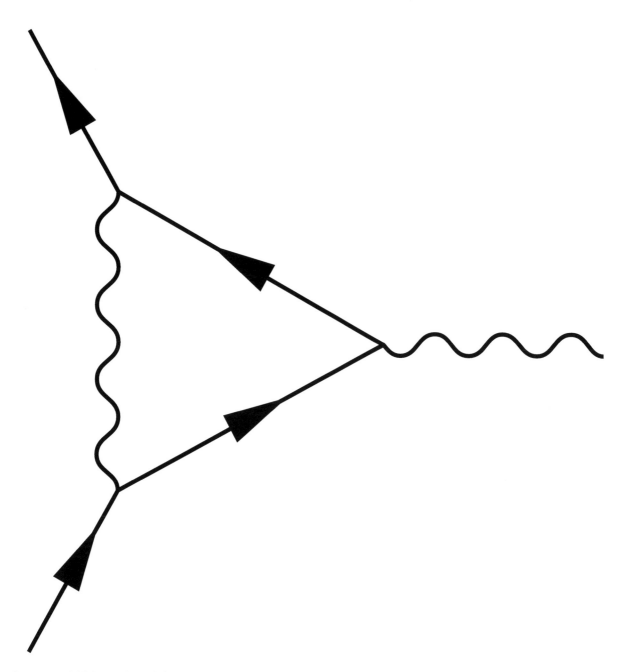

In quantum field theory, the probability of an event is computed by summing the probability amplitude (a complex number) for all possible ways in which the event can occur, as in the Feynman diagram shown here; the probability equals the square of the modulus of the total amplitude.

Chapter 28

Force carrier

In particle physics, **force carriers** are particles that give rise to forces between other particles. These particles are bundles of energy (quanta) of a particular kind of field. There is one kind of field for every species of elementary particle. For instance, there is an electron field whose quanta are electrons, and an electromagnetic field whose quanta are photons.[1] The force carrier particles that mediate the electromagnetic, weak, and strong interactions are called gauge bosons.

28.1   Particle and field viewpoints

Main article: Wave–particle duality

In particle physics, quantum field theories such as the Standard Model describe nature in terms of fields. Each field has a complementary description as the set of particles of a particular type. A force between two particles can be described either as the action of a force field generated by one particle on the other, or in terms of the exchange of virtual force carrier particles between them.

The energy of a wave in a field (for example, electromagnetic waves in the electromagnetic field) is quantized, and the quantum excitations of the field can be interpreted as particles. The Standard Model contains the following particles, each of which is an excitation of a particular field:

- Gluons, excitations of the strong gauge field.

- Photons, W bosons, and Z bosons, excitations of the electroweak gauge fields.

- Higgs bosons, excitations of one component of the Higgs field, which gives mass to fundamental particles.

- Several types of fermions, described as excitations of fermionic fields.

In addition, composite particles such as mesons can be described as excitations of an effective field.

Gravity is not a part of the Standard Model, but it is thought that there may be particles called gravitons which are the excitations of gravitational waves. The status of this particle is still tentative, because the theory is incomplete and because the interactions of *single* gravitons may be too weak to be detected.[2]

28.2   Forces from the particle viewpoint

Main article: Static forces and virtual-particle exchange
 When one particle scatters off another, altering its trajectory, there are two ways to think about the process. In the field picture, we imagine that the field generated by one particle caused a force on the other. Alternatively, we can imagine

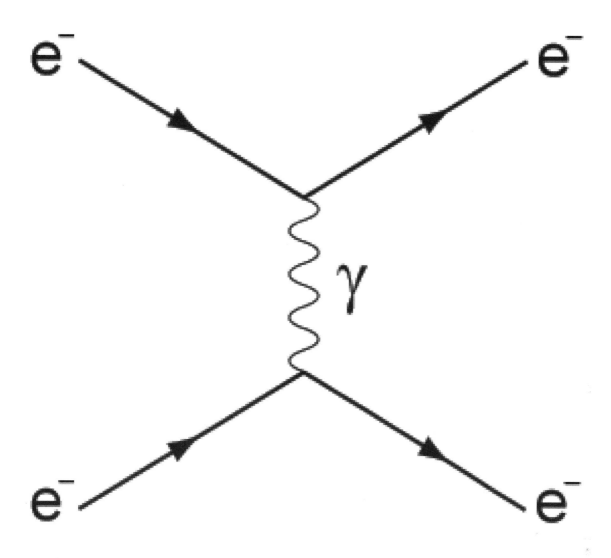

A Feynman diagram of scattering between two electrons by emission of a virtual photon.

one particle emitting a virtual particle which is absorbed by the other. The virtual particle transfers momentum from one particle to the other. This particle viewpoint is especially helpful when there are a large number of complicated quantum corrections to the calculation since these corrections can be visualized as Feynman diagrams containing additional virtual particles.

The description of forces in terms of virtual particles is limited by the applicability of the perturbation theory from which it is derived. In certain situations, such as low-energy QCD and the description of bound states, perturbation theory breaks down.

28.3 Examples

- The electromagnetic force can be described by the exchange of virtual photons.

- The nuclear force binding protons and neutrons can be described by an effective field of which mesons are the excitations.

- At sufficiently large energies, the strong interaction between quarks can be described by the exchange of virtual gluons.

- Beta decay is an example of an interaction due to the exchange of a W boson, but not an example of a force.

- Gravitation may be due to the exchange of virtual gravitons.

28.4 History

The concept of messenger particles dates back to the 18th century when the French physicist Charles Coulomb showed that the electrostatic force between electrically charged objects follows a law similar to Newton's Law of Gravitation. In time, this relationship became known as Coulomb's law. By 1862, Hermann von Helmholtz had described a ray of light as the "quickest of all the messengers". In 1905, Albert Einstein proposed the existence of a light-particle in answer to the question: "what are light quanta?"

In 1923, at the Washington University in St. Louis, Arthur Holly Compton demonstrated an effect now known as Compton scattering. This effect is only explainable if light can behave as a stream of particles and it convinced the physics community of the existence of Einstein's light-particle. Lastly, in 1926, one year before the theory of quantum mechanics was published, Gilbert N. Lewis introduced the term "photon", which soon became the name for Einstein's light particle. From there, the concept of messenger particles developed further.

28.5 See also

- Virtual particle

- Fundamental interaction

- Particle physics

28.6 References

[1] Steven Weinberg, Dreams of a Final Theory, Hutchinson, 1993

[2] Rothman, Tony; Stephen Boughn (November 2006). "Can Gravitons be Detected?". *Foundations of Physics* **36** (12): 1801–1825. arXiv:gr-qc/0601043. Bibcode:2006FoPh...36.1801R. doi:10.1007/s10701-006-9081-9.

28.7 External links

- Messenger Particles - Cern Interactive Slide Show

Chapter 29

Spontaneous symmetry breaking

Spontaneous symmetry breaking[1][2][3] is a mode of realization of symmetry breaking in a physical system, where the underlying laws are invariant under a symmetry transformation, but the system as a whole changes under such transformations, in contrast to explicit symmetry breaking. It is a spontaneous process by which a system in a symmetrical state ends up in an asymmetrical state. It thus describes systems where the equations of motion or the Lagrangian obey certain symmetries, but the lowest-energy solutions do not exhibit that symmetry.

Consider a symmetrical upward dome with a trough circling the bottom. If a ball is put at the very peak of the dome, the system is symmetrical with respect to a rotation around the center axis. But the ball may *spontaneously break* this symmetry by rolling down the dome into the trough, a point of lowest energy. Afterward, the ball has come to a rest at some fixed point on the perimeter. The dome and the ball retain their individual symmetry, but the system does not.[4]

Most simple phases of matter and phase transitions, like crystals, magnets, and conventional superconductors can be simply understood from the viewpoint of spontaneous symmetry breaking. Notable exceptions include topological phases of matter like the fractional quantum Hall effect.

29.1   Spontaneous symmetry breaking in physics

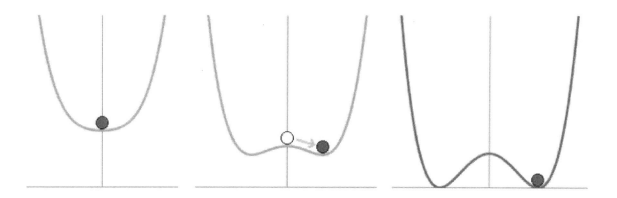

Spontaneous symmetry breaking simplified: *– At high energy levels* (left) *the ball settles in the center, and the result is symmetrical. At lower energy levels* (right), *the overall "rules" remain symmetrical, but the "Mexican hat" potential comes into effect: "local" symmetry is inevitably broken since eventually the ball must roll one way (at random) and not another.*

29.1.1 Particle physics

In particle physics the force carrier particles are normally specified by field equations with gauge symmetry; their equations predict that certain measurements will be the same at any point in the field. For instance, field equations might predict that the mass of two quarks is constant. Solving the equations to find the mass of each quark might give two solutions. In one solution, quark A is heavier than quark B. In the second solution, quark B is heavier than quark A *by the same amount*. The symmetry of the equations is not reflected by the individual solutions, but it is reflected by the range of solutions. An actual measurement reflects only one solution, representing a breakdown in the symmetry of the underlying theory. "Hidden" is perhaps a better term than "broken" because the symmetry is always there in these equations. This phenomenon is called *spontaneous* symmetry breaking because *nothing* (that we know) breaks the symmetry in the equations.[5]:194–195

Chiral symmetry

Main article: Chiral symmetry breaking

Chiral symmetry breaking is an example of spontaneous symmetry breaking affecting the chiral symmetry of the strong interactions in particle physics. It is a property of quantum chromodynamics, the quantum field theory describing these interactions, and is responsible for the bulk of the mass (over 99%) of the nucleons, and thus of all common matter, as it converts very light bound quarks into 100 times heavier constituents of baryons. The approximate Nambu–Goldstone bosons in this spontaneous symmetry breaking process are the pions, whose mass is an order of magnitude lighter than the mass of the nucleons. It served as the prototype and significant ingredient of the Higgs mechanism underlying the electroweak symmetry breaking.

Higgs mechanism

Main articles: Brout–Englert–Higgs mechanism and Yukawa interaction

The strong, weak, and electromagnetic forces can all be understood as arising from gauge symmetries. The Higgs mechanism, the spontaneous symmetry breaking of gauge symmetries, is an important component in understanding the superconductivity of metals and the origin of particle masses in the standard model of particle physics. One important consequence of the distinction between true symmetries and *gauge symmetries*, is that the spontaneous breaking of a gauge symmetry does not give rise to characteristic massless Nambu–Goldstone modes, but only massive modes, like the plasma mode in a superconductor, or the Higgs mode observed in particle physics.

In the standard model of particle physics, spontaneous symmetry breaking of the $SU(2) \times U(1)$ gauge symmetry associated with the electro-weak force generates masses for several particles, and separates the electromagnetic and weak forces. The W and Z bosons are the elementary particles that mediate the weak interaction, while the photon mediates the electromagnetic interaction. At energies much greater than 100 GeV all these particles behave in a similar manner. The Weinberg–Salam theory predicts that, at lower energies, this symmetry is broken so that the photon and the massive W and Z bosons emerge.[6] In addition, fermions develop mass consistently.

Without spontaneous symmetry breaking, the Standard Model of elementary particle interactions requires the existence of a number of particles. However, some particles (the W and Z bosons) would then be predicted to be massless, when, in reality, they are observed to have mass. To overcome this, spontaneous symmetry breaking is augmented by the Higgs mechanism to give these particles mass. It also suggests the presence of a new particle, the Higgs boson, reported as possibly identifiable with a boson detected in 2012. (If the Higgs boson were not confirmed to have been found, it would mean that the simplest implementation of the Higgs mechanism and spontaneous symmetry breaking *as they are currently formulated* require modification.)

Superconductivity of metals is a condensed-matter analog of the Higgs phenomena, in which a condensate of Cooper pairs of electrons spontaneously breaks the U(1) gauge "symmetry" associated with light and electromagnetism.

29.1.2 Condensed matter physics

Most phases of matter can be understood through the lens of spontaneous symmetry breaking. For example, crystals are periodic arrays of atoms that are not invariant under all translations (only under a small subset of translations by a lattice vector). Magnets have north and south poles that are oriented in a specific direction, breaking rotational symmetry. In addition to these examples, there are a whole host of other symmetry-breaking phases of matter including nematic phases of liquid crystals, charge- and spin-density waves, superfluids and many others.

There are several known examples of matter that cannot be described by spontaneous symmetry breaking, including: topologically ordered phases of matter like fractional quantum Hall liquids, and spin-liquids. These states do not break any symmetry, but are distinct phases of matter. Unlike the case of spontaneous symmetry breaking, there is not a general framework for describing such states.

Continuous symmetry

The ferromagnet is the canonical system which spontaneously breaks the continuous symmetry of the spins below the Curie temperature and at $h = 0$, where h is the external magnetic field. Below the Curie temperature the energy of the system is invariant under inversion of the magnetization $m(\mathbf{x})$ such that $m(\mathbf{x}) = -m(-\mathbf{x})$. The symmetry is spontaneously broken as $h \to 0$ when the Hamiltonian becomes invariant under the inversion transformation, but the expectation value is not invariant.

Spontaneously, symmetry broken phases of matter are characterized by an order parameter that describes the quantity which breaks the symmetry under consideration. For example, in a magnet, the order parameter is the local magnetization.

Spontaneously breaking of a continuous symmetry is inevitably accompanied by gapless (meaning that these modes do not cost any energy to excite) Nambu–Goldstone modes associated with slow long-wavelength fluctuations of the order parameter. For example, vibrational modes in a crystal, known as phonons, are associated with slow density fluctuations of the crystal's atoms. The associated Goldstone mode for magnets are oscillating waves of spin known as spin-waves. For symmetry-breaking states, whose order parameter is not a conserved quantity, Nambu–Goldstone modes are typically massless and propagate at a constant velocity.

An important theorem, due to Mermin and Wagner, states that, at finite temperature, thermally activated fluctuations of Nambu–Goldstone modes destroy the long-range order, and prevent spontaneous symmetry breaking in one- and two-dimensional systems. Similarly, quantum fluctuations of the order parameter prevent most types of continuous symmetry breaking in one-dimensional systems even at zero temperature (an important exception is ferromagnets, whose order parameter, magnetization, is an exactly conserved quantity and does not have any quantum fluctuations).

Other long-range interacting systems such as cylindrical curved surfaces interacting via the Coulomb potential or Yukawa potential has been shown to break translational and rotational symmetries.[7] It was shown, in the presence of a symmetric Hamiltonian, and in the limit of infinite volume, the system spontaneously adopts a chiral configuration, i.e. breaks mirror plane symmetry.

29.1.3 Dynamical symmetry breaking

Dynamical symmetry breaking (DSB) is a special form of spontaneous symmetry breaking where the ground state of the system has reduced symmetry properties compared to its theoretical description (Lagrangian).

Dynamical breaking of a global symmetry is a spontaneous symmetry breaking, that happens not at the (classical) tree level (i.e. at the level of the bare action), but due to quantum corrections (i.e. at the level of the effective action).

Dynamical breaking of a gauge symmetry is subtler. In the conventional spontaneous gauge symmetry breaking, there exists an unstable Higgs particle in the theory, which drives the vacuum to a symmetry-broken phase (see e.g. Electroweak interaction). In dynamical gauge symmetry breaking, however, no unstable Higgs particle operates in the theory, but the bound states of the system itself provide the unstable fields that render the phase transition. For example, Bardeen, Hill, and Lindner published a paper which attempts to replace the conventional Higgs mechanism in the standard model, by a DSB that is driven by a bound state of top-antitop quarks (such models, where a composite particle plays the role of the Higgs boson, are often referred to as "Composite Higgs models").[8] Dynamical breaking of gauge symmetries is often due to creation of a fermionic condensate; for example the quark condensate, which is connected to the dynamical breaking of chiral symmetry in quantum chromodynamics. Conventional superconductivity is the paradigmatic example from the condensed matter side, where phonon-mediated attractions lead electrons to become bound in pairs and then condense, thereby breaking the electromagnetic gauge symmetry.

29.2 Generalisation and technical usage

For spontaneous symmetry breaking to occur, there must be a system in which there are several equally likely outcomes. The system as a whole is therefore symmetric with respect to these outcomes. (If we consider any two outcomes, the probability is the same. This contrasts sharply to explicit symmetry breaking.) However, if the system is sampled (i.e. if the system is actually used or interacted with in any way), a specific outcome must occur. Though the system as a whole is symmetric, it is never encountered with this symmetry, but only in one specific asymmetric state. Hence, the symmetry is said to be spontaneously broken in that theory. Nevertheless, the fact that each outcome is equally likely is a reflection of the underlying symmetry, which is thus often dubbed "hidden symmetry", and has crucial formal consequences. (See the article on the Goldstone boson).

When a theory is symmetric with respect to a symmetry group, but requires that one element of the group be distinct, then spontaneous symmetry breaking has occurred. The theory must not dictate *which* member is distinct, only that *one is*. From this point on, the theory can be treated as if this element actually is distinct, with the proviso that any results found in this way must be resymmetrized, by taking the average of each of the elements of the group being the distinct one.

The crucial concept in physics theories is the order parameter. If there is a field (often a background field) which acquires an expectation value (not necessarily a *vacuum* expectation value) which is not invariant under the symmetry in question, we say that the system is in the ordered phase, and the symmetry is spontaneously broken. This is because other subsystems interact with the order parameter, which specifies a "frame of reference" to be measured against. In that case, the vacuum state does not obey the initial symmetry (which would keep it invariant, in the linearly realized **Wigner mode** in which it would be a singlet), and, instead changes under the (hidden) symmetry, now implemented in the (nonlinear) **Nambu–Goldstone mode**. Normally, in the absence of the Higgs mechanism, massless Goldstone bosons arise.

The symmetry group can be discrete, such as the space group of a crystal, or continuous (e.g., a Lie group), such as the rotational symmetry of space. However, if the system contains only a single spatial dimension, then only discrete symmetries may be broken in a vacuum state of the full quantum theory, although a classical solution may break a continuous symmetry.

29.3 A pedagogical example: the Mexican hat potential

In the simplest idealized relativistic model, the spontaneously broken symmetry is summarized through an illustrative scalar field theory. The relevant Lagrangian, which essentially dictates how a system behaves, can be split up into kinetic and potential terms,

It is in this potential term $V(\Phi)$ that the symmetry breaking is triggered. An example of a potential, due to Jeffrey Goldstone[9] is illustrated in the graph at the right.

This potential has an infinite number of possible minima (vacuum states) given by

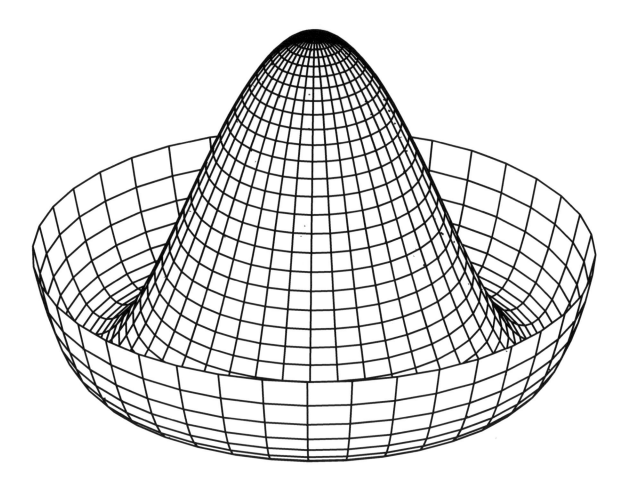

Graph of Goldstone's "Mexican hat" potential function V *versus* φ.

for any real θ between 0 and 2π. The system also has an unstable vacuum state corresponding to $\Phi = 0$. This state has a U(1) symmetry. However, once the system falls into a specific stable vacuum state (amounting to a choice of θ), this symmetry will appear to be lost, or "spontaneously broken".

In fact, any other choice of θ would have exactly the same energy, implying the existence of a massless Nambu–Goldstone boson, the mode running around the circle at the minimum of this potential, and indicating there is some memory of the original symmetry in the Lagrangian.

29.4   Other examples

- For ferromagnetic materials, the underlying laws are invariant under spatial rotations. Here, the order parameter is the magnetization, which measures the magnetic dipole density. Above the Curie temperature, the order parameter is zero, which is spatially invariant, and there is no symmetry breaking. Below the Curie temperature, however, the magnetization acquires a constant nonvanishing value, which points in a certain direction (in the idealized situation where we have full equilibrium; otherwise, translational symmetry gets broken as well). The residual rotational symmetries which leave the orientation of this vector invariant remain unbroken, unlike the other rotations which do not and are thus spontaneously broken.

- The laws describing a solid are invariant under the full Euclidean group, but the solid itself spontaneously breaks this group down to a space group. The displacement and the orientation are the order parameters.

- General relativity has a Lorentz symmetry, but in FRW cosmological models, the mean 4-velocity field defined by averaging over the velocities of the galaxies (the galaxies act like gas particles at cosmological scales) acts as an order parameter breaking this symmetry. Similar comments can be made about the cosmic microwave background.

- For the electroweak model, as explained earlier, a component of the Higgs field provides the order parameter breaking the electroweak gauge symmetry to the electromagnetic gauge symmetry. Like the ferromagnetic example, there is a phase transition at the electroweak temperature. The same comment about us not tending to notice broken symmetries suggests why it took so long for us to discover electroweak unification.

- In superconductors, there is a condensed-matter collective field ψ, which acts as the order parameter breaking the electromagnetic gauge symmetry.

- Take a thin cylindrical plastic rod and push both ends together. Before buckling, the system is symmetric under rotation, and so visibly cylindrically symmetric. But after buckling, it looks different, and asymmetric. Nevertheless, features of the cylindrical symmetry are still there: ignoring friction, it would take no force to freely spin the rod around, displacing the ground state in time, and amounting to an oscillation of vanishing frequency, unlike the radial oscillations in the direction of the buckle. This spinning mode is effectively the requisite Nambu–Goldstone boson.

- Consider a uniform layer of fluid over an infinite horizontal plane. This system has all the symmetries of the Euclidean plane. But now heat the bottom surface uniformly so that it becomes much hotter than the upper surface. When the temperature gradient becomes large enough, convection cells will form, breaking the Euclidean symmetry.

- Consider a bead on a circular hoop that is rotated about a vertical diameter. As the rotational velocity is increased gradually from rest, the bead will initially stay at its initial equilibrium point at the bottom of the hoop (intuitively stable, lowest gravitational potential). At a certain critical rotational velocity, this point will become unstable and the bead will jump to one of two other newly created equilibria, equidistant from the center. Initially, the system is symmetric with respect to the diameter, yet after passing the critical velocity, the bead ends up in one of the two new equilibrium points, thus breaking the symmetry.

29.5   Nobel Prize

On October 7, 2008, the Royal Swedish Academy of Sciences awarded the 2008 Nobel Prize in Physics to three scientists for their work in subatomic physics symmetry breaking. Yoichiro Nambu, of the University of Chicago, won half of the prize for the discovery of the mechanism of spontaneous broken symmetry in the context of the strong interactions, specifically chiral symmetry breaking. Physicists Makoto Kobayashi and Toshihide Maskawa shared the other half of the prize for discovering the origin of the explicit breaking of CP symmetry in the weak interactions.[10] This origin is ultimately reliant on the Higgs mechanism, but, so far understood as a "just so" feature of Higgs couplings, not a spontaneously broken symmetry phenomenon.

29.6   See also

- Autocatalytic reactions and order creation

- Catastrophe theory

- Chiral symmetry breaking

- CP-violation

- Explicit symmetry breaking

- Gauge gravitation theory

- Goldstone boson

- Grand unified theory

- Higgs mechanism

- Higgs boson

- Higgs field (classical)

- Irreversibility

- Magnetic catalysis of chiral symmetry breaking

- Mermin–Wagner theorem

- Quantum fluctuation

- Sakurai Prize for Theoretical Particle Physics

- Second-order phase transition

- Symmetry breaking

- Tachyon condensation

- Tachyonic field

- Wheeler–Feynman absorber theory

- 1964 PRL symmetry breaking papers

29.7   Notes

- ^ Note that (as in fundamental Higgs driven spontaneous gauge symmetry breaking) the term "symmetry breaking" is a misnomer when applied to gauge symmetries.

29.8   References

[1] *Dynamical Symmetry Breaking in Quantum Field Theories*. By Vladimir A. Miranskij. Pg 15.

[2] Patterns of Symmetry Breaking. Edited by Henryk Arodz, Jacek Dziarmaga, Wojciech Hubert Zurek. Pg 141.

[3] Bubbles, Voids and Bumps in Time: The New Cosmology. Edited by James Cornell. Pg 125.

[4] Gerald M. Edelman, Bright Air, Brilliant Fire: On the Matter of the Mind (New York: BasicBooks, 1992) 203.

[5] Steven Weinberg (20 April 2011). *Dreams of a Final Theory: The Scientist's Search for the Ultimate Laws of Nature*. Knopf Doubleday Publishing Group. ISBN 978-0-307-78786-6.

[6] A Brief History of Time, Stephen Hawking, Bantam; 10th anniversary edition (September 1, 1998). pp. 73–74.

[7] Kohlstedt, K.L.; Vernizzi, G.; Solis, F.J.; Olvera de la Cruz, M. (2007). "Spontaneous Chirality via Long-range Electrostatic Forces". *Physical Review Letters* **99**: 030602. arXiv:0704.3435. Bibcode:2007PhRvL..99c0602K. doi:10.1103/PhysRevLet.

[8] William A. Bardeen; Christopher T. Hill; Manfred Lindner (1990). "Minimal dynamical symmetry breaking of the standard model". *Physical Review D* **41** (5): 1647–1660. Bibcode:1990PhRvD..41.1647B. doi:10.1103/PhysRevD.41.1647.

[9] Goldstone, J. (1961). "Field theories with " Superconductor " solutions". *Il Nuovo Cimento* **19**: 154–164. doi:10.1007/BF.

[10] The Nobel Foundation. "The Nobel Prize in Physics 2008". *nobelprize.org*. Retrieved January 15, 2008.

29.9 External links

- Spontaneous symmetry breaking

- Physical Review Letters – 50th Anniversary Milestone Papers

- In CERN Courier, Steven Weinberg reflects on spontaneous symmetry breaking

- Englert–Brout–Higgs–Guralnik–Hagen–Kibble Mechanism on Scholarpedia

- History of Englert–Brout–Higgs–Guralnik–Hagen–Kibble Mechanism on Scholarpedia

- The History of the Guralnik, Hagen and Kibble development of the Theory of Spontaneous Symmetry Breaking and Gauge Particles

- International Journal of Modern Physics A: The History of the Guralnik, Hagen and Kibble development of the Theory of Spontaneous Symmetry Breaking and Gauge Particles

- Guralnik, G S; Hagen, C R and Kibble, T W B (1967). Broken Symmetries and the Goldstone Theorem. Advances in Physics, vol. 2 Interscience Publishers, New York. pp. 567–708 ISBN 0-470-17057-3

- Spontaneous Symmetry Breaking in Gauge Theories: a Historical Survey

Chapter 30

1964 PRL symmetry breaking papers

In 1964, three teams wrote scientific papers which proposed related but different approaches to explain how mass could arise in local gauge theories. These three now famous papers were written by Robert Brout and François Englert,[1][2] Peter Higgs,[3] and Gerald Guralnik, C. Richard Hagen, and Tom Kibble (GHK),[4][5] and are credited with the theory of the Higgs mechanism and the prediction of the Higgs field and Higgs boson. Together, these provide a theoretical means by which Goldstone's theorem (a problematic limitation affecting early modern particle physics theories) can be avoided. They show how gauge bosons can acquire non-zero masses as a result of spontaneous symmetry breaking within gauge invariant models of the universe.[6]

As such, these form the key element of the electroweak theory that forms part of the Standard Model of particle physics, and of many models, such as the Grand Unified Theory, that go beyond it. The papers that introduce this mechanism were published in *Physical Review Letters* (*PRL*) and were each recognized as milestone papers by *PRL* 's 50th anniversary celebration.[7] All of the six physicists were awarded the 2010 J. J. Sakurai Prize for Theoretical Particle Physics for this work,[8] and in 2013 Englert and Higgs received the Nobel Prize in Physics.[9]

On 4 July 2012, the two main experiments at the LHC (ATLAS and CMS) both reported independently the confirmed existence of a previously unknown particle with a mass of about 125 GeV/c^2 (about 133 proton masses, on the order of 10^{-25} kg), which is "consistent with the Higgs boson" and widely believed to be the Higgs boson.[10]

30.1 Introduction

A gauge theory of elementary particles is a very attractive potential framework for constructing the ultimate theory. Such a theory has the very desirable property of being potentially renormalizable—shorthand for saying that all calculational infinities encountered can be consistently absorbed into a few parameters of the theory. However, as soon as one gives mass to the gauge fields, renormalizability is lost, and the theory rendered useless. Spontaneous symmetry breaking is a promising mechanism, which could be used to give mass to the vector gauge particles. A significant difficulty which one encounters, however, is Goldstone's theorem, which states that in any quantum field theory which has a spontaneously broken symmetry there must occur a zero-mass particle. So the problem arises—how can one break a symmetry and at the same time not introduce unwanted zero-mass particles. The resolution of this dilemma lies in the observation that in the case of gauge theories, the Goldstone theorem can be avoided by working in the so-called radiation gauge. This is because the proof of Goldstone's theorem requires manifest Lorentz covariance, a property not possessed by the radiation gauge.

30.2 History

Particle physicists study matter made from fundamental particles whose interactions are mediated by exchange particles known as force carriers. At the beginning of the 1960s a number of these particles had been discovered or proposed, along with theories suggesting how they relate to each other, some of which had already been reformulated as field theories in which the objects of study are not particles and forces, but quantum fields and their symmetries. However, attempts to unify known fundamental forces such as the electromagnetic force and the weak nuclear force were known to be incomplete. One known omission was that gauge invariant approaches, including non-abelian models such as Yang–Mills theory (1954), which held great promise for unified theories, also seemed to predict known massive particles as massless.[11] Goldstone's theorem, relating to continuous symmetries within some theories, also appeared to rule out many obvious solutions,[12] since it appeared to show that zero-mass particles would have to also exist that were "simply not seen".[13] According to Guralnik, physicists had "no understanding" how these problems could be overcome in 1964.[13] In 2014, Guralnik and Hagen wrote a paper that contended that even after 50 years there is still widespread misunderstanding, by physicists and the Nobel Committee, of the Goldstone boson role.[14] This paper, published in *Modern Physics Letters A*, turned out to be Guralnik's last published work.[15]

Particle physicist and mathematician Peter Woit summarised the state of research at the time:

> "Yang and Mills work on non-abelian gauge theory had one huge problem: in perturbation theory it has massless particles which don't correspond to anything we see. One way of getting rid of this problem is now fairly well-understood, the phenomenon of confinement realized in QCD, where the strong interactions get rid of the massless "gluon" states at long distances. By the very early sixties, people had begun to understand another source of massless particles: spontaneous symmetry breaking of a continuous symmetry. What Philip Anderson realized and worked out in the summer of 1962 was that, when you have *both* gauge symmetry *and* spontaneous symmetry breaking, the Nambu–Goldstone massless mode can combine with the massless gauge field modes to produce a physical massive vector field. This is what happens in superconductivity, a subject about which Anderson was (and is) one of the leading experts." *[text condensed]* [11]

The Higgs mechanism is a process by which vector bosons can get rest mass *without* explicitly breaking gauge invariance, as a byproduct of spontaneous symmetry breaking.[6][16] The mathematical theory behind spontaneous symmetry breaking was initially conceived and published within particle physics by Yoichiro Nambu in 1960,[17] the concept that such a mechanism could offer a possible solution for the "mass problem" was originally suggested in 1962 by Philip Anderson,[18]:4-5[19] and Abraham Klein and Benjamin Lee showed in March 1964 that Goldstone's theorem could be avoided this way in at least some non-relativistic cases and speculated it might be possible in truly relativistic cases.[20]

These approaches were quickly developed into a full relativistic model, independently and almost simultaneously, by three groups of physicists: by François Englert and Robert Brout in August 1964;[1] by Peter Higgs in October 1964;[3] and by Gerald Guralnik, Carl Hagen, and Tom Kibble (GHK) in November 1964.[4] Higgs also wrote a short but important[6] response published in September 1964 to an objection by Gilbert,[21] which showed that if calculating within the radiation gauge, Goldstone's theorem and Gilbert's objection would become inapplicable.[Note 1] (Higgs later described Gilbert's objection as prompting his own paper.[22]) Properties of the model were further considered by Guralnik in 1965,[23] by Higgs in 1966,[24] by Kibble in 1967,[25] and further by GHK in 1967.[26] The original three 1964 papers showed that when a gauge theory is combined with an additional field that spontaneously breaks the symmetry, the gauge bosons can consistently acquire a finite mass.[6][16][27] In 1967, Steven Weinberg[28] and Abdus Salam[29] independently showed how a Higgs mechanism could be used to break the electroweak symmetry of Sheldon Glashow's unified model for the weak and electromagnetic interactions[30] (itself an extension of work by Schwinger), forming what became the Standard Model of particle physics. Weinberg was the first to observe that this would also provide mass terms for the fermions.[31] [Note 2]

However, the seminal papers on spontaneous breaking of gauge symmetries were at first largely ignored, because it was widely believed that the (non-Abelian gauge) theories in question were a dead-end, and in particular that they could not be renormalised. In 1971–72, Martinus Veltman and Gerard 't Hooft proved renormalisation of Yang–Mills was possible in two papers covering massless, and then massive, fields.[31] Their contribution, and others' work on the renormalization group, was eventually "enormously profound and influential",[32] but even with all key elements of the eventual theory published there was still almost no wider interest. For example, Coleman found in a study that "essentially no-one paid any attention" to Weinberg's paper prior to 1971[33] – now the most cited in particle physics[34] – and even in 1970 according to Politzer, Glashow's teaching of the weak interaction contained no mention of Weinberg's, Salem's, or Glashow's own work.[32] In practice, Politzer states, almost everyone learned of the theory due to physicist Benjamin Lee, who combined the work of Veltman and 't Hooft with insights by others, and popularised the completed theory.[32] In this way, from 1971, interest and acceptance "exploded" [32] and the ideas were quickly absorbed in the mainstream.[31][32]

30.2.1   The significance of requiring manifest covariance

Most students who have taken a course in electromagnetism have encountered the Coulomb potential. It basically states that two charged particles attract or repel each other by a force which varies according to the inverse square of their separation. This is fairly unambiguous for particles at rest, but if one or the other is following an arbitrary trajectory the question arises whether one should compute the force using the instantaneous positions of the particles or the so-called retarded positions. The latter recognizes that information cannot propagate instantaneously, rather it propagates at the speed of light. However, the radiation gauge says that one uses the instantaneous positions of the particles, but doesn't violate causality because there are compensating terms in the force equation. In contrast, the Lorenz gauge imposes manifest covariance (and thus causality) at all stages of a calculation. Predictions of observable quantities are identical in the two gauges, but the radiation gauge formulation of quantum field theory avoids Goldstone's theorem.[35]

30.2.2   Summary and impact of the *PRL* papers

The three papers written in 1964 were each recognised as milestone papers during *Physical Review Letters* 's 50th anniversary celebration.[27] Their six authors were also awarded the 2010 J. J. Sakurai Prize for Theoretical Particle Physics for this work.[36] (A controversy also arose the same year, because in the event of a Nobel Prize only up to three scientists could be recognised, with six being credited for the papers.[37]) Two of the three *PRL* papers (by Higgs and by GHK) contained equations for the hypothetical field that eventually would become known as the Higgs field and its hypothetical quantum, the Higgs boson.[3][4] Higgs's subsequent 1966 paper showed the decay mechanism of the boson; only a massive boson can decay and the decays can prove the mechanism.

Each of these papers is unique and demonstrates different approaches to showing how mass arise in gauge particles. Over the years, the differences between these papers are no longer widely understood, due to the passage of time and acceptance of end-results by the particle physics community. A study of citation indices is interesting—more than 40 years after the 1964 publication in *Physical Review Letters* there is little noticeable pattern of preference among them, with the vast majority of researchers in the field mentioning all three milestone papers.

In the paper by Higgs the boson is massive, and in a closing sentence Higgs writes that "an essential feature" of the theory "is the prediction of incomplete multiplets of scalar and vector bosons".[3] (Frank Close comments that 1960s gauge theorists were focused on the problem of massless *vector* bosons, and the implied existence of a massive *scalar* boson was not seen as important; only Higgs directly addressed it.[38]:154, 166, 175) In the paper by GHK the boson is massless and decoupled from the massive states.[4] In reviews dated 2009 and 2011, Guralnik states that in the GHK model the boson is massless only in a lowest-order approximation, but it is not subject to any constraint and acquires mass at higher orders, and adds that the GHK paper was the only one to show that there are no massless Goldstone bosons in the model and to give a complete analysis of the general Higgs mechanism.[13]<ref name="[14][5] All three reached similar conclusions, despite their very different approaches: Higgs' paper essentially used classical techniques, Englert and Brout's involved calculating vacuum polarization in perturbation theory around an assumed symmetry-breaking vacuum state, and GHK used operator formalism and conservation laws to explore in depth the ways in which Goldstone's theorem explicitly fails.[6]

In addition to explaining how mass is acquired by vector bosons, the Higgs mechanism also predicts the ratio between the W boson and Z boson masses as well as their couplings with each other and with the Standard Model quarks and leptons. Subsequently, many of these predictions have been verified by precise measurements performed at the LEP and the SLC colliders, thus overwhelmingly confirming that some kind of Higgs mechanism does take place in nature,[39] but the exact manner by which it happens has not yet been discovered. The results of searching for the Higgs boson are expected to provide evidence about how this is realized in nature.

30.2.3   Consequences of the papers

The resulting electroweak theory and Standard Model have correctly predicted (among other discoveries) weak neutral currents, three bosons, the top and charm quarks, and with great precision, the mass and other properties of some of these.[Note 3] Many of those involved eventually won Nobel Prizes or other renowned awards. A 1974 paper in *Reviews of Modern Physics* commented that "while no one doubted the [mathematical] correctness of these arguments, no one quite believed that nature was diabolically clever enough to take advantage of them".[40] By 1986 and again in the 1990s it became possible to write that understanding and proving the Higgs sector of the Standard Model was "the central problem today in particle physics." [41][42]

30.3 See also

- Higgs mechanism

- Higgs boson

- Standard Model

- Symmetry breaking

- Large Hadron Collider

- Fermilab

- Tevatron

- J. J. Sakurai Prize for Theoretical Particle Physics

- *The God Particle*, a popular science book on the Higgs boson, written by Leon M. Lederman

30.4 Notes

[1] Goldstone's theorem only applies to gauges having manifest Lorentz covariance, a condition that took time to become questioned. But the process of quantisation requires a gauge to be fixed and at this point it becomes possible to choose a gauge such as the 'radiation' gauge which is not invariant over time, so that these problems can be avoided.

[2] A field with the "Mexican hat" potential $V(\phi) = \mu^2\phi^2 + \lambda\phi^4$ and $\mu^2 < 0$ has a minimum not at zero but at some non-zero value ϕ_0 . By expressing the action in terms of the field $\tilde{\phi} = \phi - \phi_0$ (where ϕ_0 is a constant independent of position), we find the Yukawa term has a component $g\phi_0\bar{\psi}\psi$. Since both g and ϕ_0 are constants, this looks exactly like the mass term for a fermion of mass $g\phi_0$. The field $\tilde{\phi}$ is then the Higgs field.

[3] The success of the Higgs based electroweak theory and Standard Model is illustrated by their predictions of the mass of two particles later detected: the W boson (predicted mass: 80.390 ± 0.018 GeV, experimental measurement: 80.387 ± 0.019 GeV), and the Z boson (predicted mass: 91.1874 ± 0.0021, experimental measurement: 91.1876 ± 0.0021 GeV). The existence of the Z boson was itself another prediction. Other correct predictions included the weak neutral current, the gluon, and the top and charm quarks, all later proven to exist as the theory said.

30.5 References

[1] Englert, François; Brout, Robert (1964). "Broken Symmetry and the Mass of Gauge Vector Mesons". *Physical Review Letters* **13** (9): 321–23. Bibcode:1964PhRvL..13..321E. doi:10.1103/PhysRevLett.13.321.

[2] Brout, R.; Englert, F. (1998). "Spontaneous Symmetry Breaking in Gauge Theories: A Historical Survey". arXiv:hep-th/9802142 [hep-th].

[3] Higgs, Peter (1964). "Broken Symmetries and the Masses of Gauge Bosons". *Physical Review Letters* **13** (16): 508–509. Bibcode:1964PhRvL..13..508H. doi:10.1103/PhysRevLett.13.508.

[4] Guralnik, Gerald; Hagen, C. R.; Kibble, T. W. B. (1964). "Global Conservation Laws and Massless Particles". *Physical Review Letters* **13** (20): 585–587. Bibcode:1964PhRvL..13..585G. doi:10.1103/PhysRevLett.13.585.

[5] G.S. Guralnik (2009). "The History of the Guralnik, Hagen and Kibble development of the Theory of Spontaneous Symmetry Breaking and Gauge Particles". *International Journal of Modern Physics A* **24** (14): 2601–2627. arXiv:0907.3466. Bibcode:2009IJMPA..24.2601G. doi:10.1142/S0217751X09045431.

[6] Kibble, T. (2009). "Englert-Brout-Higgs-Guralnik-Hagen-Kibble mechanism". *Scholarpedia* **4**: 6441–6410. Bibcode:2. doi:10.4249/scholarpedia.6441.

[7] Blume, M.; Brown, S.; Millev, Y. (2008). "Letters from the past, a PRL retrospective (1964)". Physical Review Letters. Archived from the original on 10 January 2010. Retrieved 2010-01-30.

[8] "J. J. Sakurai Prize Winners". American Physical Society. 2010. Archived from the original on 12 February 2010. Retrieved 2010-01-30.

[9] http://www.nobelprize.org/nobel_prizes/physics/laureates/2013/

[10] "CERN experiments observe particle consistent with long-sought Higgs boson" (Press release). CERN. 4 July 2012. Retrieved 2015-06-02.

[11] Woit, P. (13 November 2010). "The Anderson–Higgs Mechanism". *Not Even Wrong*. Columbia University. Retrieved 2012-11-12.

[12] Goldstone, J.; Salam, A.; Weinberg, S. (1962). "Broken Symmetries".*Physical Review***127**(3): 965–970.Bibcode:1962PhG. doi:10.1103/PhysRev.127.965.

[13] Guralnik, G. S. (2011). "The Beginnings of Spontaneous Symmetry Breaking in Particle Physics — Derived From My on the Spot "Intellectual Battlefield Impressions"". arXiv:1110.2253 [physics.hist-ph].

[14] Guralnik, G.; Hagen, C. R. (2014). "Where have all the Goldstone bosons gone?". *Modern Physics Letters A* **29**: 1450046. arXiv:1401.6924. Bibcode:2014MPLA...2950046G. doi:10.1142/S0217732314500461.

[15] Hagen, C. R. (August 2014). "Obituaries - Gerald Stanford Guralnik". *Physics Today*. doi:10.1063/PT.3.2488.

[16] Kibble, T. W. B. (2009). "Englert–Brout–Higgs–Guralnik–Hagen–Kibble Mechanism (History)". *Scholarpedia* **4** (1): 8741. Bibcode:2009SchpJ...4.8741K. doi:10.4249/scholarpedia.8741.

[17] The Nobel Prize in Physics 2008 – official Nobel Prize website.

[18] Higgs, P. (24 November 2010). "My Life as a Boson" (PDF). Kings College London. Archived from the original (PDF) on 2014-05-01. – the original 2001 paper can be found at: Duff, M. J. and Liu; Liu, J. T., eds. (2003). *2001 A Spacetime Odyssey: Proceedings of the Inaugural Conference of the Michigan Center for Theoretical Physics*. World Scientific Publishing. pp. 86–88. ISBN 981-238-231-3.

[19] Anderson, P. (1963). "Plasmons, gauge invariance and mass". *Physical Review* **130**: 439. Bibcode:1963PhRv..130..439A. doi:10.1103/PhysRev.130.439.

[20] Klein, A.; Lee, B. (1964). "Does Spontaneous Breakdown of Symmetry Imply Zero-Mass Particles?". *Physical Review Letters* **12** (10): 266. Bibcode:1964PhRvL..12..266K. doi:10.1103/PhysRevLett.12.266.

[21] Higgs, Peter (1964). "Broken symmetries, massless particles and gauge fields".*Physics Letters***12**(2): 132–133.Bibcode:1H. doi:10.1016/0031-9163(64)91136-9.

[22] Higgs, Peter (2010-11-24). "My Life as a Boson" (PDF). Talk given by Peter Higgs at Kings College, London, Nov 24 2010. Retrieved 17 January 2013. Gilbert ... wrote a response to [Klein and Lee's paper] saying 'No, you cannot do that in a relativistic theory. You cannot have a preferred unit time-like vector like that.' This is where I came in, because the next month was when I responded to Gilbert's paper by saying 'Yes, you can have such a thing' but only in a gauge theory with a gauge field coupled to the current.

[23] G.S. Guralnik (2011). "Gauge Invariance and the Goldstone Theorem – 1965 Feldafing talk". *Modern Physics Letters A* **26** (19): 1381–1392. arXiv:1107.4592. Bibcode:2011MPLA...26.1381G. doi:10.1142/S0217732311036188.

[24] Higgs, Peter (1966). "Spontaneous Symmetry Breakdown without Massless Bosons". *Physical Review* **145** (4): 1156–1163. Bibcode:1966PhRv..145.1156H. doi:10.1103/PhysRev.145.1156.

[25] Kibble, Tom (1967). "Symmetry Breaking in Non-Abelian Gauge Theories".*Physical Review***155**(5): 1554–1561.Bibc4K. doi:10.1103/PhysRev.155.1554.

[26] "Guralnik, G S; Hagen, C R and Kibble, T W B (1967). Broken Symmetries and the Goldstone Theorem. Advances in Physics, vol. 2" (PDF).

[27] "Physical Review Letters – 50th Anniversary Milestone Papers". Physical Review Letters.

[28] S. Weinberg (1967). "A Model of Leptons". *Physical Review Letters* **19** (21): 1264–1266. Bibcode:1967PhRvL..19.1264W. doi:10.1103/PhysRevLett.19.1264.

[29] A. Salam (1968). N. Svartholm, ed. *Elementary Particle Physics: Relativistic Groups and Analyticity*. Eighth Nobel Symposium. Stockholm: Almquist and Wiksell. p. 367.

[30] S.L. Glashow (1961). "Partial-symmetries of weak interactions". *Nuclear Physics* **22** (4): 579–588. Bibcode:1961NucPh.G. doi:10.1016/0029-5582(61)90469-2.

[31] Ellis, John; Gaillard, Mary K.; Nanopoulos, Dimitri V. (2012). "A Historical Profile of the Higgs Boson". arXiv:1201.6045 [hep-ph].

[32] Politzer, David. "The Dilemma of Attribution". *Nobel Prize lecture, 2004*. Nobel Prize. Retrieved 22 January 2013. Sidney Coleman published in *Science* magazine in 1979 a citation search he did documenting that essentially no one paid any attention to Weinberg's Nobel Prize winning paper until the work of 't Hooft (as explicated by Ben Lee). In 1971 interest in Weinberg's paper exploded. I had a parallel personal experience: I took a one-year course on weak interactions from Shelly Glashow in 1970, and he never even mentioned the Weinberg–Salam model or his own contributions.

[33] Coleman, Sidney (1979-12-14). "The 1979 Nobel Prize in Physics". *Science* **206** (4424): 1290–1292. Bibcode:1979S .doi:10.1126/science.206.4424.1290. Retrieved 22 January 2013. – discussed by David Politzer in his 2004 Nobel speech. [32]

[34] Letters from the Past – A PRL Retrospective (50 year celebration, 2008)

[35] G.S. Guralnik, C.R. Hagen, T.W.B. Kibble (1968). "Broken Symmetries and the Goldstone Theorem". In R. L. Cool, R. E. Marshak. *Advances in Particle Physics* **2**. Interscience Publishers. pp. 567–708. ISBN 0-470-17057-3.

[36] American Physical Society – "J. J. Sakurai Prize for Theoretical Particle Physics".

[37] Merali, Zeeya (4 August 2010). "Physicists get political over Higgs". *Nature Magazine*. Retrieved 28 December 2011.

[38] Close, Frank (2011). *The Infinity Puzzle: Quantum Field Theory and the Hunt for an Orderly Universe*. Oxford: Oxford University Press. ISBN 978-0-19-959350-7.

[39] "LEP Electroweak Working Group".

[40] Bernstein, Jeremy (January 1974). "Spontaneous symmetry breaking, gauge theories, the Higgs mechanism and all that" (PDF). *Reviews of Modern Physics* **46** (1): 7–48. Bibcode:1974RvMP...46....7B. doi:10.1103/revmodphys.46.7. Retrieved 2012-12-10.

[41] José Luis Lucio and Arnulfo Zepeda (1987). *Proceedings of the II Mexican School of Particles and Fields, Cuernavaca-Morelos, 1986*. World Scientific. p. 29. ISBN 9971504340.

[42] Gunion, Dawson, Kane, and Haber (199). *The Higgs Hunter's Guide (1st ed.)*. pp. 11 (?). ISBN 9780786743186. – quoted as being in the first (1990) edition of the book by Peter Higgs in his talk "My Life as a Boson", 2001, ref#25.

30.6   Further reading

- Higgs, P. W. (1964). "Broken symmetries, massless particles and gauge fields". *Physics Letters* **12** (2): 132–201. Bibcode:1964PhL....12..132H. doi:10.1016/0031-9163(64)91136-9.

- Englert, F.; Brout, R. (1964). "Broken Symmetry and the Mass of Gauge Vector Mesons". *Physical Review Letters* **13** (9): 321. Bibcode:1964PhRvL..13..321E. doi:10.1103/PhysRevLett.13.321.

- Higgs, P. (1964). "Broken Symmetries and the Masses of Gauge Bosons". *Physical Review Letters* **13** (16): 508. Bibcode:1964PhRvL..13..508H. doi:10.1103/PhysRevLett.13.508.

- Guralnik, G.; Hagen, C.; Kibble, T. (1964). "Global Conservation Laws and Massless Particles". *Physical Review Letters* **13** (20): 585. Bibcode:1964PhRvL..13..585G. doi:10.1103/PhysRevLett.13.585.

- Higgs, P. (1966). "Spontaneous Symmetry Breakdown without Massless Bosons". *Physical Review* **145** (4): 1156. Bibcode:1966PhRv..145.1156H. doi:10.1103/PhysRev.145.1156.

- Nambu, Y.; Jona-Lasinio, G. (1961). "Dynamical Model of Elementary Particles Based on an Analogy with Superconductivity. I". *Physical Review* **122**: 345. Bibcode:1961PhRv..122..345N. doi:10.1103/PhysRev.122.345.

- Goldstone, J.; Salam, A.; Weinberg, S. (1962). "Broken Symmetries".*Physical Review***127**(3): 965.Bibcode:1G. doi:10.1103/PhysRev.127.965.

- Anderson, P. (1963). "Plasmons, Gauge Invariance, and Mass".*Physical Review***130**: 439.Bibcode:1963A. doi:10.1103/PhysRev.130.439.

- Klein, A.; Lee, B. (1964). "Does Spontaneous Breakdown of Symmetry Imply Zero-Mass Particles?". *Physical Review Letters* **12** (10): 266. Bibcode:1964PhRvL..12..266K. doi:10.1103/PhysRevLett.12.266.

- Gilbert, W. (1964). "Broken Symmetries and Massless Particles".*Physical Review Letters***12**(25): 713.Bibco13G. doi:10.1103/PhysRevLett.12.713.

- Guralnik, Gerald (2009). "The History of the Guralnik, Hagen and Kibble development of the Theory of Spontaneous Symmetry Breaking and Gauge Particles". *International Journal of Modern Physics A* **24** (14): 2601–2627. arXiv:0907.3466. Bibcode:2009IJMPA..24.2601G. doi:10.1142/S0217751X09045431., Guralnik, Gerald (2011). "The Beginnings of Spontaneous Symmetry Breaking in Particle Physics. Proceedings of the DPF-2011 Conference, Providence, RI, 8–13 August 2011". arXiv:1110.2253v1 [physics.hist-ph]., and Guralnik, Gerald (2013). "Heretical Ideas that Provided the Cornerstone for the Standard Model of Particle Physics". SPG MIT-TEILUNGEN March 2013, No. 39, (p. 14)

- Kobayashi, M.; Maskawa, T. (1973). "CP-Violation in the Renormalizable Theory of Weak Interaction". *Progress of Theoretical Physics* **49** (2): 652–657. Bibcode:1973PThPh..49..652K. doi:10.1143/PTP.49.652.

- 't Hooft, G.; Veltman, M. (1972). "Regularization and renormalization of gauge fields". *Nuclear Physics B* **44**: 189. Bibcode:1972NuPhB..44..189T. doi:10.1016/0550-3213(72)90279-9.

- G.S. Guralnik, C.R. Hagen, T.W.B. Kibble (1968). "Broken Symmetries and the Goldstone Theorem". In R. L. Cool, R. E. Marshak. *Advances in Particle Physics* **2**. Interscience Publishers. pp. 567–708. ISBN 0-470-17057-3.

30.7   External links

- *Physical Review Letters* - 50th Anniversary Milestone Papers
- American Physical Society - J. J. Sakurai Prize Winners
- Gerry Guralnik speaks at Brown University about the 1964 *PRL* papers
- In CERN Courier, Steven Weinberg reflects on spontaneous symmetry breaking
- Steven Weinberg on LHC
- Englert-Brout-Higgs-Guralnik-Hagen-Kibble Mechanism on Scholarpedia
- History of Englert-Brout-Higgs-Guralnik-Hagen-Kibble Mechanism on Scholarpedia
- "The History of the Guralnik, Hagen and Kibble development of the Theory of Spontaneous Symmetry Breaking and Gauge Particles"
- *International Journal of Modern Physics A*: "The History of the Guralnik, Hagen and Kibble development of the Theory of Spontaneous Symmetry Breaking and Gauge Particles"

- G.S. Guralnik (2011) "Gauge Invariance and the Goldstone Theorem - 1965 Feldafing talk". *International Journal of Modern Physics A*

- Spontaneous Symmetry Breaking in Gauge Theories: a Historical Survey

- CERN Courier Letter from GHK – December 2008

- God Particle

- 2010 Sakurai Prize Videos

- Brown University Celebration of 2010 Sakurai Prize - Videos

- The Hunt for the Higgs at Tevatron

- Physicists get political over Higgs

- Ian Sample on Controversy and Nobel Reform

- Massive by Ian Sample

- Blog Not Even Wrong, Review of Massive by Ian Sample

- Blog Not Even Wrong, Anderson-Higgs Mechanism

Chapter 31

Inflation (cosmology)

"Inflation model" and "Inflation theory" redirect here. For a general rise in the price level, see Inflation. For other uses, see Inflation (disambiguation).

In physical cosmology, **cosmic inflation**, **cosmological inflation**, or just **inflation** is a theory of exponential expansion of space in the early universe. The inflationary epoch lasted from 10^{-36} seconds after the Big Bang to sometime between 10^{-33} and 10^{-32} seconds. Following the inflationary period, the Universe continues to expand, but at a less rapid rate.[1]

Inflation was developed in the early 1980s. It explains the origin of the large-scale structure of the cosmos. Quantum fluctuations in the microscopic inflationary region, magnified to cosmic size, become the seeds for the growth of structure in the Universe (see galaxy formation and evolution and structure formation).[2] Many physicists also believe that inflation explains why the Universe appears to be the same in all directions (isotropic), why the cosmic microwave background radiation is distributed evenly, why the Universe is flat, and why no magnetic monopoles have been observed.

While the detailed particle physics mechanism responsible for inflation is not known, the basic picture makes a number of predictions that have been confirmed by observation.[3] The hypothetical field thought to be responsible for inflation is called the inflaton.[4]

In 2002, three of the original architects of the theory were recognized for their major contributions; physicists Alan Guth of M.I.T., Andrei Linde of Stanford and Paul Steinhardt of Princeton shared the prestigious Dirac Prize "for development of the concept of inflation in cosmology".[5]

31.1   Overview

Main article: Metric expansion of space

An expanding universe generally has a cosmological horizon, which, by analogy with the more familiar horizon caused by the curvature of the Earth's surface, marks the boundary of the part of the Universe that an observer can see. Light (or other radiation) emitted by objects beyond the cosmological horizon never reaches the observer, because the space in between the observer and the object is expanding too rapidly.

The observable universe is one *causal patch* of a much larger unobservable universe; other parts of the Universe cannot communicate with Earth yet. These parts of the Universe are outside our current cosmological horizon. In the standard hot big bang model, without inflation, the cosmological horizon moves out, bringing new regions into view. Yet as a local observer sees such a region for the first time, it looks no different from any other region of space the local observer has already seen: its background radiation is at nearly the same temperature as the background radiation of other regions, and its space-time curvature is evolving lock-step with the others. This presents a mystery: how did these new regions know what temperature and curvature they were supposed to have? They couldn't have learned it by getting signals, because they were not previously in communication with our past light cone.[9][10]

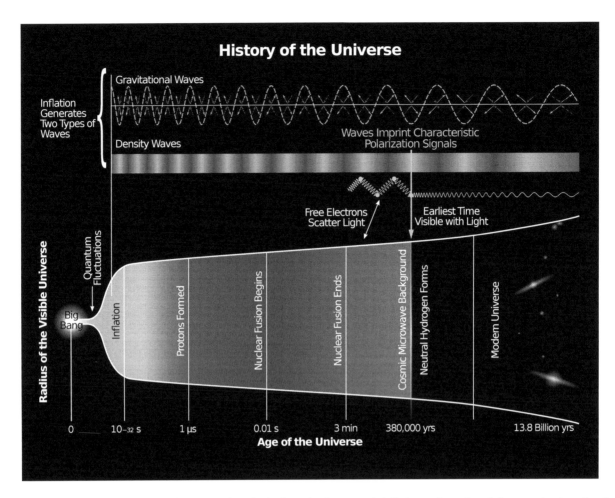

History of the Universe - gravitational waves are hypothesized to arise from cosmic inflation, a faster-than-light expansion just after the Big Bang (17 March 2014).[6][7][8]

Inflation answers this question by postulating that all the regions come from an earlier era with a big vacuum energy, or cosmological constant. A space with a cosmological constant is qualitatively different: instead of moving outward, the cosmological horizon stays put. For any one observer, the distance to the cosmological horizon is constant. With exponentially expanding space, two nearby observers are separated very quickly; so much so, that the distance between them quickly exceeds the limits of communications. The spatial slices are expanding very fast to cover huge volumes. Things are constantly moving beyond the cosmological horizon, which is a fixed distance away, and everything becomes homogeneous.

As the inflationary field slowly relaxes to the vacuum, the cosmological constant goes to zero and space begins to expand normally. The new regions that come into view during the normal expansion phase are exactly the same regions that were pushed out of the horizon during inflation, and so they are at nearly the same temperature and curvature, because they come from the same originally small patch of space.

The theory of inflation thus explains why the temperatures and curvatures of different regions are so nearly equal. It also predicts that the total curvature of a space-slice at constant global time is zero. This prediction implies that the total ordinary matter, dark matter and residual vacuum energy in the Universe have to add up to the critical density, and the evidence supports this. More strikingly, inflation allows physicists to calculate the minute differences in temperature of different regions from quantum fluctuations during the inflationary era, and many of these quantitative predictions have been confirmed.[11][12]

31.1.1 Space expands

To say that space expands exponentially means that two inertial observers are moving farther apart with accelerating velocity. In stationary coordinates for one observer, a patch of an inflating universe has the following polar metric:[13][14]

$$ds^2 = -\left(1 - \Lambda r^2\right) dt^2 + \frac{1}{1 - \Lambda r^2}\, dr^2 + r^2\, d\Omega^2.$$

This is just like an inside-out black hole metric—it has a zero in the dt component on a fixed radius sphere called the cosmological horizon. Objects are drawn away from the observer at $r = 0$ towards the cosmological horizon, which they cross in a finite proper time. This means that any inhomogeneities are smoothed out, just as any bumps or matter on the surface of a black hole horizon are swallowed and disappear.

Since the space–time metric has no explicit time dependence, once an observer has crossed the cosmological horizon, observers closer in take its place. This process of falling outward and replacement points closer in are always steadily replacing points further out—an exponential expansion of space–time.

This steady-state exponentially expanding spacetime is called a de Sitter space, and to sustain it there must be a cosmological constant, a vacuum energy proportional to Λ everywhere. In this case, the equation of state is $p = -\rho$. The physical conditions from one moment to the next are stable: the rate of expansion, called the Hubble parameter, is nearly constant, and the scale factor of the Universe is proportional to e^{Ht}. Inflation is often called a period of *accelerated expansion* because the distance between two fixed observers is increasing exponentially (i.e. at an accelerating rate as they move apart), while Λ can stay approximately constant (see deceleration parameter).

31.1.2 Few inhomogeneities remain

Cosmological inflation has the important effect of smoothing out inhomogeneities, anisotropies and the curvature of space. This pushes the Universe into a very simple state, in which it is completely dominated by the inflaton field, the source of the cosmological constant, and the only significant inhomogeneities are the tiny quantum fluctuations in the inflaton. Inflation also dilutes exotic heavy particles, such as the magnetic monopoles predicted by many extensions to the Standard Model of particle physics. If the Universe was only hot enough to form such particles *before* a period of inflation, they would not be observed in nature, as they would be so rare that it is quite likely that there are none in the observable universe. Together, these effects are called the inflationary "no-hair theorem"[15] by analogy with the no hair theorem for black holes.

The "no-hair" theorem works essentially because the cosmological horizon is no different from a black-hole horizon, except for philosophical disagreements about what is on the other side. The interpretation of the no-hair theorem is that the Universe (observable and unobservable) expands by an enormous factor during inflation. In an expanding universe, energy densities generally fall, or get diluted, as the volume of the Universe increases. For example, the density of ordinary "cold" matter (dust) goes down as the inverse of the volume: when linear dimensions double, the energy density goes down by a factor of eight; the radiation energy density goes down even more rapidly as the Universe expands since the wavelength of each photon is stretched (redshifted), in addition to the photons being dispersed by the expansion. When linear dimensions are doubled, the energy density in radiation falls by a factor of sixteen (see the solution of the energy density continuity equation for an ultra-relativistic fluid). During inflation, the energy density in the inflaton field is roughly constant. However, the energy density in everything else, including inhomogeneities, curvature, anisotropies, exotic particles, and standard-model particles is falling, and through sufficient inflation these all become negligible. This leaves the Universe flat and symmetric, and (apart from the homogeneous inflaton field) mostly empty, at the moment inflation ends and reheating begins.[16]

31.1.3 Duration

A key requirement is that inflation must continue long enough to produce the present observable universe from a single, small inflationary Hubble volume. This is necessary to ensure that the Universe appears flat, homogeneous and isotropic at the largest observable scales. This requirement is generally thought to be satisfied if the Universe expanded by a factor of at least 10^{26} during inflation.[17]

31.1.4 Reheating

Inflation is a period of supercooled expansion, when the temperature drops by a factor of 100,000 or so. (The exact drop is model dependent, but in the first models it was typically from 10^{27} K down to 10^{22} K.[18]) This relatively low temperature is maintained during the inflationary phase. When inflation ends the temperature returns to the pre-inflationary temperature; this is called *reheating* or thermalization because the large potential energy of the inflaton field decays into particles and fills the Universe with Standard Model particles, including electromagnetic radiation, starting the radiation dominated phase of the Universe. Because the nature of the inflation is not known, this process is still poorly understood, although it is believed to take place through a parametric resonance.[19][20]

31.2 Motivations

Inflation resolves several problems in Big Bang cosmology that were discovered in the 1970s.[21] Inflation was first proposed by Guth while investigating the problem of why no magnetic monopoles are seen today; he found that a positive-energy false vacuum would, according to general relativity, generate an exponential expansion of space. It was very quickly realised that such an expansion would resolve many other long-standing problems. These problems arise from the observation that to look like it does *today*, the Universe would have to have started from very finely tuned, or "special" initial conditions at the Big Bang. Inflation attempts to resolve these problems by providing a dynamical mechanism that drives the Universe to this special state, thus making a universe like ours much more likely in the context of the Big Bang theory.

31.2.1 Horizon problem

Main article: Horizon problem

The horizon problem is the problem of determining why the Universe appears statistically homogeneous and isotropic in accordance with the cosmological principle.[22][23][24] For example, molecules in a canister of gas are distributed homogeneously and isotropically because they are in thermal equilibrium: gas throughout the canister has had enough time to interact to dissipate inhomogeneities and anisotropies. The situation is quite different in the big bang model without inflation, because gravitational expansion does not give the early universe enough time to equilibrate. In a big bang with only the matter and radiation known in the Standard Model, two widely separated regions of the observable universe cannot have equilibrated because they move apart from each other faster than the speed of light and thus have never come into causal contact. In the early Universe, it was not possible to send a light signal between the two regions. Because they have had no interaction, it is difficult to explain why they have the same temperature (are thermally equilibrated). Historically, proposed solutions included the *Phoenix universe* of Georges Lemaître,[25] the related oscillatory universe of Richard Chase Tolman,[26] and the Mixmaster universe of Charles Misner. Lemaître and Tolman proposed that a universe undergoing a number of cycles of contraction and expansion could come into thermal equilibrium. Their models failed, however, because of the buildup of entropy over several cycles. Misner made the (ultimately incorrect) conjecture that the Mixmaster mechanism, which made the Universe *more* chaotic, could lead to statistical homogeneity and isotropy.[23][27]

31.2.2 Flatness problem

Main article: Flatness problem

The flatness problem is sometimes called one of the Dicke coincidences (along with the cosmological constant problem).[28][29] It became known in the 1960s that the density of matter in the Universe was comparable to the critical density necessary for a flat universe (that is, a universe whose large scale geometry is the usual Euclidean geometry, rather than a non-Euclidean hyperbolic or spherical geometry).[30]:61

Therefore, regardless of the shape of the universe the contribution of spatial curvature to the expansion of the Universe could not be much greater than the contribution of matter. But as the Universe expands, the curvature redshifts away more slowly than matter and radiation. Extrapolated into the past, this presents a fine-tuning problem because the contribution of curvature to the Universe must be exponentially small (sixteen orders of magnitude less than the density of radiation at big bang nucleosynthesis, for example). This problem is exacerbated by recent observations of the cosmic microwave background that have demonstrated that the Universe is flat to within a few percent.[31]

31.2.3 Magnetic-monopole problem

The magnetic monopole problem, sometimes called the exotic-relics problem, says that if the early universe were very hot, a large number of very heavy, stable magnetic monopoles would have been produced. This is a problem with Grand Unified Theories, which propose that at high temperatures (such as in the early universe) the electromagnetic force, strong, and weak nuclear forces are not actually fundamental forces but arise due to spontaneous symmetry breaking from a single gauge theory.[32] These theories predict a number of heavy, stable particles that have not been observed in nature. The most notorious is the magnetic monopole, a kind of stable, heavy "knot" in a magnetic field.[33][34] Monopoles are predicted to be copiously produced following Grand Unified Theories at high temperature,[35][36] and they should have persisted to the present day, to such an extent that they would become the primary constituent of the Universe.[37][38] Not only is that not the case, but all searches for them have failed, placing stringent limits on the density of relic magnetic monopoles in the Universe.[39] A period of inflation that occurs below the temperature where magnetic monopoles can be produced would offer a possible resolution of this problem: monopoles would be separated from each other as the Universe around them expands, potentially lowering their observed density by many orders of magnitude. Though, as cosmologist Martin Rees has written, "Skeptics about exotic physics might not be hugely impressed by a theoretical argument to explain the absence of particles that are themselves only hypothetical. Preventive medicine can readily seem 100 percent effective against a disease that doesn't exist!"[40]

31.3   History

31.3.1   Precursors

In the early days of General Relativity, Albert Einstein introduced the cosmological constant to allow a static solution, which was a three-dimensional sphere with a uniform density of matter. Later, Willem de Sitter found a highly symmetric inflating universe, which described a universe with a cosmological constant that is otherwise empty.[41] It was discovered that Einstein's universe is unstable, and that small fluctuations cause it to collapse or turn into a de Sitter universe.

In the early 1970s Zeldovich noticed the flatness and horizon problems of Big Bang cosmology; before his work, cosmology was presumed to be symmetrical on purely philosophical grounds. In the Soviet Union, this and other considerations led Belinski and Khalatnikov to analyze the chaotic BKL singularity in General Relativity. Misner's Mixmaster universe attempted to use this chaotic behavior to solve the cosmological problems, with limited success.

In the late 1970s, Sidney Coleman applied the instanton techniques developed by Alexander Polyakov and collaborators to study the fate of the false vacuum in quantum field theory. Like a metastable phase in statistical mechanics—water below the freezing temperature or above the boiling point—a quantum field would need to nucleate a large enough bubble of the new vacuum, the new phase, in order to make a transition. Coleman found the most likely decay pathway for vacuum decay and calculated the inverse lifetime per unit volume. He eventually noted that gravitational effects would be significant, but he did not calculate these effects and did not apply the results to cosmology.

In the Soviet Union, Alexei Starobinsky noted that quantum corrections to general relativity should be important for the early universe. These generically lead to curvature-squared corrections to the Einstein–Hilbert action and a form of $f(R)$ modified gravity. The solution to Einstein's equations in the presence of curvature squared terms, when the curvatures are large, leads to an effective cosmological constant. Therefore, he proposed that the early universe went through an inflationary de Sitter era.[42] This resolved the cosmology problems and led to specific predictions for the corrections to the microwave background radiation, corrections that were then calculated in detail.

In 1978, Zeldovich noted the monopole problem, which was an unambiguous quantitative version of the horizon problem, this time in a subfield of particle physics, which led to several speculative attempts to resolve it. In 1980 Alan Guth realized that false vacuum decay in the early universe would solve the problem, leading him to propose a scalar-driven inflation. Starobinsky's and Guth's scenarios both predicted an initial deSitter phase, differing only in mechanistic details.

31.3.2   Early inflationary models

Guth proposed inflation in January 1980 to explain the nonexistence of magnetic monopoles;[43][44] it was Guth who coined the term "inflation".[45] At the same time, Starobinsky argued that quantum corrections to gravity would replace the initial singularity of the Universe with an exponentially expanding deSitter phase.[46] In October 1980, Demosthenes Kazanas suggested that exponential expansion could eliminate the particle horizon and perhaps solve the horizon problem,[47] while Sato suggested that an exponential expansion could eliminate domain walls (another kind of exotic relic).[48] In 1981 Einhorn and Sato[49] published a model similar to Guth's and showed that it would resolve the puzzle of the magnetic monopole abundance in Grand Unified Theories. Like Guth, they concluded that such a model not only required fine tuning of the cosmological constant, but also would likely lead to a much too granular universe, i.e., to large density variations resulting from bubble wall collisions.

Guth proposed that as the early universe cooled, it was trapped in a false vacuum with a high energy density, which is much like a cosmological constant. As the very early universe cooled it was trapped in a metastable state (it was supercooled), which it could only decay out of through the process of bubble nucleation via quantum tunneling. Bubbles of true vacuum spontaneously form in the sea of false vacuum and rapidly begin expanding at the speed of light. Guth recognized that this model was problematic because the model did not reheat properly: when the bubbles nucleated, they did not generate any radiation. Radiation could only be generated in collisions between bubble walls. But if inflation lasted long enough to solve the initial conditions problems, collisions between bubbles became exceedingly rare. In any one causal patch it is likely that only one bubble would nucleate.

31.3.3   Slow-roll inflation

The bubble collision problem was solved by Linde[50] and independently by Andreas Albrecht and Paul Steinhardt[51] in a model named *new inflation* or *slow-roll inflation* (Guth's model then became known as *old inflation*). In this model, instead of tunneling out of a false vacuum state, inflation occurred by a scalar field rolling down a potential energy hill. When the field rolls very slowly compared to the expansion of the Universe, inflation occurs. However, when the hill becomes steeper, inflation ends and reheating can occur.

31.3.4   Effects of asymmetries

Eventually, it was shown that new inflation does not produce a perfectly symmetric universe, but that quantum fluctuations in the inflaton are created. These fluctuations form the primordial seeds for all structure created in the later universe.[52] These fluctuations were first calculated by Viatcheslav Mukhanov and G. V. Chibisov in analyzing Starobinsky's similar model.[53][54][55] In the context of inflation, they were worked out independently of the work of Mukhanov and Chibisov at the three-week 1982 Nuffield Workshop on the Very Early Universe at Cambridge University.[56] The fluctuations were calculated by four groups working separately over the course of the workshop: Stephen Hawking;[57] Starobinsky;[58] Guth and So-Young Pi;[59] and Bardeen, Steinhardt and Turner.[60]

31.4   Observational status

Inflation is a mechanism for realizing the cosmological principle, which is the basis of the standard model of physical cosmology: it accounts for the homogeneity and isotropy of the observable universe. In addition, it accounts for the observed flatness and absence of magnetic monopoles. Since Guth's early work, each of these observations has received

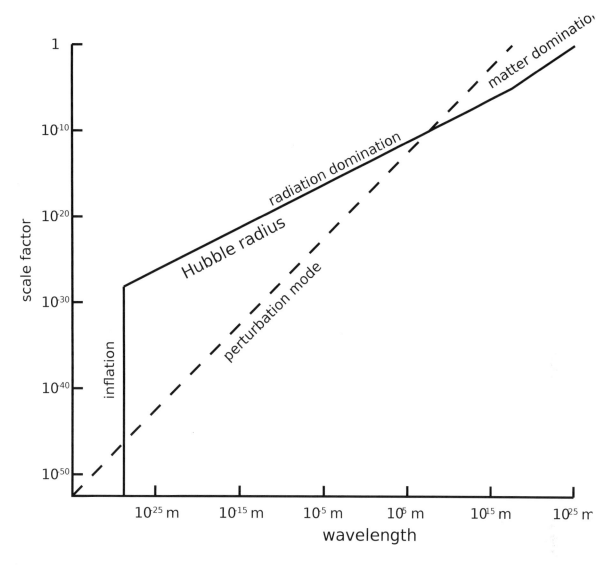

The physical size of the Hubble radius (solid line) as a function of the linear expansion (scale factor) of the universe. During cosmological inflation, the Hubble radius is constant. The physical wavelength of a perturbation mode (dashed line) is also shown. The plot illustrates how the perturbation mode grows larger than the horizon during cosmological inflation before coming back inside the horizon, which grows rapidly during radiation domination. If cosmological inflation had never happened, and radiation domination continued back until a gravitational singularity, then the mode would never have been outside the horizon in the very early universe, and no causal mechanism could have ensured that the universe was homogeneous on the scale of the perturbation mode.

further confirmation, most impressively by the detailed observations of the cosmic microwave background made by the Wilkinson Microwave Anisotropy Probe (**WMAP**) spacecraft.[11] This analysis shows that the Universe is flat to within at least a few percent, and that it is homogeneous and isotropic to one part in 100,000.

In addition, inflation predicts that the structures visible in the Universe today formed through the gravitational collapse of perturbations that were formed as quantum mechanical fluctuations in the inflationary epoch. The detailed form of the spectrum of perturbations called a nearly-scale-invariant Gaussian random field (or Harrison–Zel'dovich spectrum) is very specific and has only two free parameters, the amplitude of the spectrum and the *spectral index*, which measures the slight deviation from scale invariance predicted by inflation (perfect scale invariance corresponds to the idealized de Sitter universe).[61] Inflation predicts that the observed perturbations should be in thermal equilibrium with each other (these are called *adiabatic* or *isentropic* perturbations). This structure for the perturbations has been confirmed by the WMAP spacecraft and other cosmic microwave background (CMB) experiments,[11] and galaxy surveys, especially the ongoing Sloan Digital Sky Survey.[62] These experiments have shown that the one part in 100,000 inhomogeneities observed have exactly the form predicted by theory. Moreover, there is evidence for a slight deviation from scale invariance. The *spectral index*, n_s is equal to one for a scale-invariant spectrum. The simplest inflation models predict that this quantity is between 0.92 and 0.98.[63][64][65][66] From WMAP data it can be inferred that $n_s = 0.963 \pm 0.012$,[67] implying that it differs from one at the level of two standard deviations (2σ). This is considered an important confirmation of the theory of inflation.[11]

Various inflation theories have been proposed that make radically different predictions, but they generally have much more fine tuning than should be necessary.[63][64] As a physical model, however, inflation is most valuable in that it robustly predicts the initial conditions of the Universe based on only two adjustable parameters: the spectral index (that can only change in a small range) and the amplitude of the perturbations. Except in contrived models, this is true regardless of how inflation is realized in particle physics.

Occasionally, effects are observed that appear to contradict the simplest models of inflation. The first-year WMAP data suggested that the spectrum might not be nearly scale-invariant, but might instead have a slight curvature.[68] However, the third-year data revealed that the effect was a statistical anomaly.[11] Another effect remarked upon since the first cosmic microwave background satellite, the Cosmic Background Explorer is that the amplitude of the quadrupole moment of the CMB is unexpectedly low and the other low multipoles appear to be preferentially aligned with the ecliptic plane. Some have claimed that this is a signature of non-Gaussianity and thus contradicts the simplest models of inflation. Others have suggested that the effect may be due to other new physics, foreground contamination, or even publication bias.[69]

An experimental program is underway to further test inflation with more precise CMB measurements. In particular, high precision measurements of the so-called "B-modes" of the polarization of the background radiation could provide evidence of the gravitational radiation produced by inflation, and could also show whether the energy scale of inflation predicted by the simplest models (10^{15}–10^{16} GeV) is correct.[64][65] In March 2014, it was announced that B-mode CMB polarization consistent with that predicted from inflation had been demonstrated by a South Pole experiment.[6][7][8][70][71][72] However, on 19 June 2014, lowered confidence in confirming the findings was reported;[71][73][74] on 19 September 2014, a further reduction in confidence was reported[75][76] and, on 30 January 2015, even less confidence yet was reported.[77][78]

Other potentially corroborating measurements are expected from the Planck spacecraft, although it is unclear if the signal will be visible, or if contamination from foreground sources will interfere.[79] Other forthcoming measurements, such as those of 21 centimeter radiation (radiation emitted and absorbed from neutral hydrogen before the first stars turned on), may measure the power spectrum with even greater resolution than the CMB and galaxy surveys, although it is not known if these measurements will be possible or if interference with radio sources on Earth and in the galaxy will be too great.[80]

Dark energy is broadly similar to inflation and is thought to be causing the expansion of the present-day universe to accelerate. However, the energy scale of dark energy is much lower, 10^{-12} GeV, roughly 27 orders of magnitude less than the scale of inflation.

31.5   Theoretical status

In Guth's early proposal, it was thought that the inflaton was the Higgs field, the field that explains the mass of the elementary particles.[44] It is now believed by some that the inflaton cannot be the Higgs field[81] although the recent discovery of the Higgs boson has increased the number of works considering the Higgs field as inflaton.[82] One problem of this identification is the current tension with experimental data at the electroweak scale,[83] which is currently under study at the Large Hadron Collider (LHC). Other models of inflation relied on the properties of Grand Unified Theories.[51] Since the simplest models of grand unification have failed, it is now thought by many physicists that inflation will be included in a supersymmetric theory such as string theory or a supersymmetric grand unified theory. At present, while inflation is understood principally by its detailed predictions of the initial conditions for the hot early universe, the particle physics is largely *ad hoc* modelling. As such, although predictions of inflation have been consistent with the results of observational tests, many open questions remain.

31.5.1   Fine-tuning problem

One of the most severe challenges for inflation arises from the need for fine tuning. In new inflation, the *slow-roll conditions* must be satisfied for inflation to occur. The slow-roll conditions say that the inflaton potential must be flat (compared to the large vacuum energy) and that the inflaton particles must have a small mass.[84] New inflation requires the Universe to have a scalar field with an especially flat potential and special initial conditions. However, explanations for these fine-tunings have been proposed. For example, classically scale invariant field theories, where scale invariance is broken by quantum effects, provide an explanation of the flatness of inflationary potentials, as long as the theory can be studied through perturbation theory.[85]

Andrei Linde

Linde proposed a theory known as *chaotic inflation* in which he suggested that the conditions for inflation were actually satisfied quite generically. Inflation will occur in virtually any universe that begins in a chaotic, high energy state that has a scalar field with unbounded potential energy.[86] However, in his model the inflaton field necessarily takes values larger than one Planck unit: for this reason, these are often called *large field* models and the competing new inflation models are called *small field* models. In this situation, the predictions of effective field theory are thought to be invalid, as renormalization should cause large corrections that could prevent inflation.[87] This problem has not yet been resolved and some cosmologists argue that the small field models, in which inflation can occur at a much lower energy scale, are better models.[88] While inflation depends on quantum field theory (and the semiclassical approximation to quantum gravity) in an important way, it has not been completely reconciled with these theories.

Brandenberger commented on fine-tuning in another situation.[89] The amplitude of the primordial inhomogeneities produced in inflation is directly tied to the energy scale of inflation. This scale is suggested to be around 10^{16} GeV or 10^{-3} times the Planck energy. The natural scale is naïvely the Planck scale so this small value could be seen as another form of fine-tuning (called a hierarchy problem): the energy density given by the scalar potential is down by 10^{-12} compared to the Planck density. This is not usually considered to be a critical problem, however, because the scale of inflation corresponds naturally to the scale of gauge unification.

31.5.2 Eternal inflation

Main article: Eternal inflation

In many models, the inflationary phase of the Universe's expansion lasts forever in at least some regions of the Universe. This occurs because inflating regions expand very rapidly, reproducing themselves. Unless the rate of decay to the non-inflating phase is sufficiently fast, new inflating regions are produced more rapidly than non-inflating regions. In such models most of the volume of the Universe at any given time is inflating. All models of eternal inflation produce an infinite multiverse, typically a fractal.

Although new inflation is classically rolling down the potential, quantum fluctuations can sometimes lift it to previous levels. These regions in which the inflaton fluctuates upwards expand much faster than regions in which the inflaton has a lower potential energy, and tend to dominate in terms of physical volume. This steady state, which first developed by Vilenkin,[90] is called "eternal inflation". It has been shown that any inflationary theory with an unbounded potential is eternal.[91] It is a popular conclusion among physicists that this steady state cannot continue forever into the past.[92][93][94] Inflationary spacetime, which is similar to de Sitter space, is incomplete without a contracting region. However, unlike de Sitter space, fluctuations in a contracting inflationary space collapse to form a gravitational singularity, a point where densities become infinite. Therefore, it is necessary to have a theory for the Universe's initial conditions. Linde, however, believes inflation may be past eternal.[95]

In eternal inflation, regions with inflation have an exponentially growing volume, while regions that are not inflating don't. This suggests that the volume of the inflating part of the Universe in the global picture is always unimaginably larger than the part that has stopped inflating, even though inflation eventually ends as seen by any single pre-inflationary observer. Scientists disagree about how to assign a probability distribution to this hypothetical anthropic landscape. If the probability of different regions is counted by volume, one should expect that inflation will never end or applying boundary conditions that a local observer exists to observe it, that inflation will end as late as possible. Some physicists believe this paradox can be resolved by weighting observers by their pre-inflationary volume.

31.5.3 Initial conditions

Some physicists have tried to avoid the initial conditions problem by proposing models for an eternally inflating universe with no origin.[96][97][98][99] These models propose that while the Universe, on the largest scales, expands exponentially it was, is and always will be, spatially infinite and has existed, and will exist, forever.

Other proposals attempt to describe the ex nihilo creation of the Universe based on quantum cosmology and the following inflation. Vilenkin put forth one such scenario.[90] Hartle and Hawking offered the no-boundary proposal for the initial creation of the Universe in which inflation comes about naturally.[100]

Guth described the inflationary universe as the "ultimate free lunch":[101][102] new universes, similar to our own, are continually produced in a vast inflating background. Gravitational interactions, in this case, circumvent (but do not violate) the first law of thermodynamics (energy conservation) and the second law of thermodynamics (entropy and the arrow of time problem). However, while there is consensus that this solves the initial conditions problem, some have disputed this, as it is much more likely that the Universe came about by a quantum fluctuation. Don Page was an outspoken critic of inflation because of this anomaly.[103] He stressed that the thermodynamic arrow of time necessitates low entropy initial conditions, which would be highly unlikely. According to them, rather than solving this problem, the inflation theory aggravates it – the reheating at the end of the inflation era increases entropy, making it necessary for the initial state of the Universe to be even more orderly than in other Big Bang theories with no inflation phase.

Hawking and Page later found ambiguous results when they attempted to compute the probability of inflation in the Hartle-Hawking initial state.[104] Other authors have argued that, since inflation is eternal, the probability doesn't matter as long as it is not precisely zero: once it starts, inflation perpetuates itself and quickly dominates the Universe.[105][106]:223–225 However, Albrecht and Lorenzo Sorbo argued that the probability of an inflationary cosmos, consistent with today's observations, emerging by a random fluctuation from some pre-existent state is much higher than that of a non-inflationary cosmos. This is because the "seed" amount of non-gravitational energy required for the inflationary cosmos is so much less than that for a non-inflationary alternative, which outweighs any entropic considerations.[107]

Another problem that has occasionally been mentioned is the trans-Planckian problem or trans-Planckian effects.[108] Since the energy scale of inflation and the Planck scale are relatively close, some of the quantum fluctuations that have made up the structure in our universe were smaller than the Planck length before inflation. Therefore, there ought to be corrections from Planck-scale physics, in particular the unknown quantum theory of gravity. Some disagreement remains about the magnitude of this effect: about whether it is just on the threshold of detectability or completely undetectable.[109]

31.5.4 Hybrid inflation

Another kind of inflation, called *hybrid inflation*, is an extension of new inflation. It introduces additional scalar fields, so that while one of the scalar fields is responsible for normal slow roll inflation, another triggers the end of inflation: when inflation has continued for sufficiently long, it becomes favorable to the second field to decay into a much lower energy state.[110]

In hybrid inflation, one scalar field is responsible for most of the energy density (thus determining the rate of expansion), while another is responsible for the slow roll (thus determining the period of inflation and its termination). Thus fluctuations in the former inflaton would not affect inflation termination, while fluctuations in the latter would not affect the rate of expansion. Therefore, hybrid inflation is not eternal.[111][112] When the second (slow-rolling) inflaton reaches the bottom of its potential, it changes the location of the minimum of the first inflaton's potential, which leads to a fast roll of the inflaton down its potential, leading to termination of inflation.

31.5.5 Inflation and string cosmology

The discovery of flux compactifications opened the way for reconciling inflation and string theory.[113] *Brane inflation* suggests that inflation arises from the motion of D-branes[114] in the compactified geometry, usually towards a stack of anti-D-branes. This theory, governed by the *Dirac-Born-Infeld action*, is different from ordinary inflation. The dynamics are not completely understood. It appears that special conditions are necessary since inflation occurs in tunneling between two vacua in the string landscape. The process of tunneling between two vacua is a form of old inflation, but new inflation must then occur by some other mechanism.

31.5.6 Inflation and loop quantum gravity

When investigating the effects the theory of loop quantum gravity would have on cosmology, a loop quantum cosmology model has evolved that provides a possible mechanism for cosmological inflation. Loop quantum gravity assumes a quantized spacetime. If the energy density is larger than can be held by the quantized spacetime, it is thought to bounce back.[115]

31.6 Alternatives

Other models explain some of the observations explained by inflation. However none of these "alternatives" has the same breadth of explanation and still require inflation for a more complete fit with observation. They should therefore be regarded as adjuncts to inflation, rather than as alternatives.

31.6.1 Big bounce

The flatness and horizon problems are naturally solved in the Einstein-Cartan-Sciama-Kibble theory of gravity, without needing an exotic form of matter or free parameters.[116][117] This theory extends general relativity by removing a constraint of the symmetry of the affine connection and regarding its antisymmetric part, the torsion tensor, as a dynamical variable. The minimal coupling between torsion and Dirac spinors generates a spin-spin interaction that is significant in fermionic matter at extremely high densities. Such an interaction averts the unphysical Big Bang singularity, replacing it with a cusp-like bounce at a finite minimum scale factor, before which the Universe was contracting. The rapid expansion immediately after the Big Bounce explains why the present Universe at largest scales appears spatially flat, homogeneous and isotropic. As the density of the Universe decreases, the effects of torsion weaken and the Universe smoothly enters the radiation-dominated era.

31.6.2 String theory

String theory requires that, in addition to the three observable spatial dimensions, additional dimensions exist that are curled up or compactified (see also Kaluza–Klein theory). Extra dimensions appear as a frequent component of supergravity models and other approaches to quantum gravity. This raised the contingent question of why four space-time dimensions became large and the rest became unobservably small. An attempt to address this question, called *string gas cosmology*, was proposed by Robert Brandenberger and Cumrun Vafa.[118] This model focuses on the dynamics of the early universe considered as a hot gas of strings. Brandenberger and Vafa show that a dimension of spacetime can only expand if the strings that wind around it can efficiently annihilate each other. Each string is a one-dimensional object, and the largest number of dimensions in which two strings will generically intersect (and, presumably, annihilate) is three. Therefore, the most likely number of non-compact (large) spatial dimensions is three. Current work on this model centers on whether it can succeed in stabilizing the size of the compactified dimensions and produce the correct spectrum of primordial density perturbations.[119] Supporters admit that their model "does not solve the entropy and flatness problems of standard cosmology ..... and we can provide no explanation for why the current universe is so close to being spatially flat".[120]

31.6.3 Ekpyrotic and cyclic models

The ekpyrotic and cyclic models are also considered adjuncts to inflation. These models solve the horizon problem through an expanding epoch well *before* the Big Bang, and then generate the required spectrum of primordial density perturbations during a contracting phase leading to a Big Crunch. The Universe passes through the Big Crunch and emerges in a hot Big Bang phase. In this sense they are reminiscent of Richard Chace Tolman's oscillatory universe; in Tolman's model, however, the total age of the Universe is necessarily finite, while in these models this is not necessarily so. Whether the correct spectrum of density fluctuations can be produced, and whether the Universe can successfully navigate the Big Bang/Big Crunch transition, remains a topic of controversy and current research. Ekpyrotic models avoid the magnetic monopole problem as long as the temperature at the Big Crunch/Big Bang transition remains below the Grand Unified Scale, as this is the temperature required to produce magnetic monopoles in the first place. As things stand, there is no evidence of any 'slowing down' of the expansion, but this is not surprising as each cycle is expected to last on the order of a trillion years.

31.6.4 Varying C

Another adjunct, the varying speed of light model was offered by Jean-Pierre Petit in 1988, John Moffat in 1992 as well Albrecht and João Magueijo in 1999, instead of superluminal expansion the speed of light was 60 orders of magnitude faster than its current value solving the horizon and homogeneity problems in the early universe.

31.7 Criticisms

Since its introduction by Alan Guth in 1980, the inflationary paradigm has become widely accepted. Nevertheless, many physicists, mathematicians, and philosophers of science have voiced criticisms, claiming untestable predictions and a lack of serious empirical support.[105] In 1999, John Earman and Jesús Mosterín published a thorough critical review of inflationary cosmology, concluding, "we do not think that there are, as yet, good grounds for admitting any of the models of inflation into the standard core of cosmology."[121]

In order to work, and as pointed out by Roger Penrose from 1986 on, inflation requires extremely specific initial conditions of its own, so that the problem (or pseudo-problem) of initial conditions is not solved: "There is something fundamentally misconceived about trying to explain the uniformity of the early universe as resulting from a thermalization process. [...] For, if the thermalization is actually doing anything [...] then it represents a definite increasing of the entropy. Thus, the universe would have been even more special before the thermalization than after."[122] The problem of specific or "fine-tuned" initial conditions would not have been solved; it would have gotten worse. At a conference in 2015, Penrose said that "inflation isn't falsifiable, it's falsified. [...] BICEP did a wonderful service by bringing all the Inflation-ists out of their shell, and giving them a black eye."[123]

A recurrent criticism of inflation is that the invoked inflation field does not correspond to any known physical field, and that its potential energy curve seems to be an ad hoc contrivance to accommodate almost any data obtainable. Paul Steinhardt, one of the founding fathers of inflationary cosmology, has recently become one of its sharpest critics. He calls 'bad inflation' a period of accelerated expansion whose outcome conflicts with observations, and 'good inflation' one compatible with them: "Not only is bad inflation more likely than good inflation, but no inflation is more likely than either.... Roger Penrose considered all the possible configurations of the inflaton and gravitational fields. Some of these configurations lead to inflation ... Other configurations lead to a uniform, flat universe directly – without inflation. Obtaining a flat universe is unlikely overall. Penrose's shocking conclusion, though, was that obtaining a flat universe without inflation is much more likely than with inflation – by a factor of 10 to the googol (10 to the 100) power!"[105][106] Together with Anna Ijjas and Abraham Loeb, he wrote articles claiming that the inflationary paradigm is in trouble in view of the data from the Planck satellite.[124][125] Counter-arguments were presented by Alan Guth, David Kaiser, and Yasunori Nomura[126] and by Andrei Linde,[127] saying that "cosmic inflation is on a stronger footing than ever before".[126]

31.8 See also

- Brane cosmology

- Conservation of angular momentum

- Cosmology

- Dark flow

- Doughnut theory of the universe

- Hubble's law

- Non-minimally coupled inflation

- Nonlinear optics

- Varying speed of light

- Warm inflation

31.9 Notes

[1] "First Second of the Big Bang". *How The Universe Works 3*. 2014. Discovery Science.

[2] Tyson, Neil deGrasse and Donald Goldsmith (2004), *Origins: Fourteen Billion Years of Cosmic Evolution*, W. W. Norton & Co., pp. 84–5.

[3] Tsujikawa, Shinji (28 Apr 2003). "Introductory review of cosmic inflation". p. 4257. arXiv:hep-ph/0304257. Bibcode:T. In fact temperature anisotropies observed by the COBE satellite in 1992 exhibit nearly scale-invariant spectra as predicted by the inflationary paradigm. Recent observations of WMAP also show strong evidence for inflation.

[4] Guth, Alan H. (1997). *The Inflationary Universe: The Quest for a New Theory of Cosmic Origins*. Basic Books. pp. 233–234. ISBN 0201328402.

[5] "The Medallists: A list of past Dirac Medallists". *ictp.it*.

[6] Staff (17 March 2014). "BICEP2 2014 Results Release". *National Science Foundation*. Retrieved 18 March 2014.

[7] Clavin, Whitney (17 March 2014). "NASA Technology Views Birth of the Universe". *NASA*. Retrieved 17 March 2014.

[8] Overbye, Dennis (17 March 2014). "Space Ripples Reveal Big Bang's Smoking Gun". *The New York Times*. Retrieved 17 March 2014.

[9] Using Tiny Particles To Answer Giant Questions. Science Friday, 3 April 2009.

[10] See also Faster than light#Universal expansion.

[11] Spergel, D.N. (2006). "Three-year Wilkinson Microwave Anisotropy Probe (WMAP) observations: Implications for cosmology". WMAP... confirms the basic tenets of the inflationary paradigm...

[12] "Our Baby Universe Likely Expanded Rapidly, Study Suggests". *Space.com*.

[13] Melia, Fulvio (2007). "The Cosmic Horizon". *Monthly Notices of the Royal Astronomical Society* **382** (4): 1917–1921. arXiv:0711.4181. Bibcode:2007MNRAS.382.1917M. doi:10.1111/j.1365-2966.2007.12499.x.

[14] Melia, Fulvio; et al. (2009). "The Cosmological Spacetime". *International Journal of Modern Physics D* **18** (12): 1889–1901. arXiv:0907.5394. Bibcode:2009IJMPD..18.1889M. doi:10.1142/s0218271809015746.

[15] Kolb and Turner (1988).

[16] Barbara Sue Ryden (2003). *Introduction to cosmology*. Addison-Wesley. ISBN 978-0-8053-8912-8. Not only is inflation very effective at driving down the number density of magnetic monopoles, it is also effective at driving down the number density of every other type of particle, including photons.[202–207]

[17] This is usually quoted as 60 e-folds of expansion, where $e^{60} \approx 10^{26}$. It is equal to the amount of expansion since reheating, which is roughly $E_{\text{inflation}}/T_0$, where $T_0 = 2.7$ K is the temperature of the cosmic microwave background today. See, *e.g.* Kolb and Turner (1998) or Liddle and Lyth (2000).

[18] Guth, *Phase transitions in the very early universe*, in *The Very Early Universe*, ISBN 0-521-31677-4 eds Hawking, Gibbon & Siklos

[19] See Kolb and Turner (1988) or Mukhanov (2005).

[20] Kofman, Lev; Linde, Andrei; Starobinsky, Alexei (1994). "Reheating after inflation". *Physical Review Letters* **73** (5): 3195–3198. arXiv:hep-th/9405187. Bibcode:1986CQGra...3..811K. doi:10.1088/0264-9381/3/5/011.

[21] Much of the historical context is explained in chapters 15–17 of Peebles (1993).

[22] Misner, Charles W.; Coley, A A; Ellis, G F R; Hancock, M (1968). "The isotropy of the universe". *Astrophysical Journal* **151** (2): 431. Bibcode:1998CQGra..15..331W. doi:10.1088/0264-9381/15/2/008.

[23] Misner, Charles; Thorne, Kip S. and Wheeler, John Archibald (1973). *Gravitation*. San Francisco: W. H. Freeman. pp. 489–490, 525–526. ISBN 0-7167-0344-0.

[24] Weinberg, Steven (1971). *Gravitation and Cosmology*. John Wiley. pp. 740, 815. ISBN 0-471-92567-5.

[25] Lemaître, Georges (1933). "The expanding universe". *Annales de la Société Scientifique de Bruxelles* **47A**: 49., English in *Gen. Rel. Grav.* **29**:641–680, 1997.

[26] R. C. Tolman (1934). *Relativity, Thermodynamics, and Cosmology*. Oxford: Clarendon Press. ISBN 0-486-65383-8. LCCN 34032023. Reissued (1987) New York: Dover ISBN 0-486-65383-8.

[27] Misner, Charles W.; Leach, P G L (1969). "Mixmaster universe".*Physical Review Letters***22**(15): 1071–74.Bibcode:2008. doi:10.1088/1751-8113/41/15/155201.

[28] Dicke, Robert H. (1970). *Gravitation and the Universe*. Philadelphia: American Philosopical Society.

[29] Dicke, Robert H.; P. J. E. Peebles (1979). "The big bang cosmology – enigmas and nostrums". In ed. S. W. Hawking and W. Israel. *General Relativity: an Einstein Centenary Survey*. Cambridge University Press.

[30] Alan P. Lightman (1 January 1993). *Ancient Light: Our Changing View of the Universe*. Harvard University Press. ISBN 978-0-674-03363-4.

[31] "WMAP- Content of the Universe". *nasa.gov*.

[32] Since supersymmetric Grand Unified Theory is built into string theory, it is still a triumph for inflation that it is able to deal with these magnetic relics. See, *e.g.* Kolb and Turner (1988) and Raby, Stuart (2006). ed. Bruce Hoeneisen, ed. "Grand Unified Theories". arXiv:hep-ph/0608183.

[33] 't Hooft, Gerard (1974). "Magnetic monopoles in Unified Gauge Theories".*Nuclear Physics B***79**(2): 276–84.Bibcode:. doi:10.1016/0550-3213(74)90486-6.

[34] Polyakov, Alexander M. (1974). "Particle spectrum in quantum field theory".*JETP Letters***20**: 194–5.Bibcode:1974JETPL.

[35] Guth, Alan; Tye, S. (1980). "Phase Transitions and Magnetic Monopole Production in the Very Early Universe". *Physical Review Letters* **44** (10): 631–635; Erratum *ibid.*,**44**:963, 1980. Bibcode:1980PhRvL..44..631G. doi:10.1103/PhysRevLett.44.631.

[36] Einhorn, Martin B; Stein, D. L.; Toussaint, Doug (1980). "Are Grand Unified Theories Compatible with Standard Cosmology?". *Physical Review D* **21** (12): 3295–3298. Bibcode:1980PhRvD..21.3295E. doi:10.1103/PhysRevD.21.3295.

[37] Zel'dovich, Ya.; Khlopov, M. Yu. (1978). "On the concentration of relic monopoles in the universe". *Physics Letters B* **79** (3): 239–41. Bibcode:1978PhLB...79..239Z. doi:10.1016/0370-2693(78)90232-0.

[38] Preskill, John (1979). "Cosmological production of superheavy magnetic monopoles". *Physical Review Letters* **43** (19): 1365–1368. Bibcode:1979PhRvL..43.1365P. doi:10.1103/PhysRevLett.43.1365.

[39] See, *e.g.* Yao, W.–M.; Amsler, C.; Asner, D.; Barnett, R. M.; Beringer, J.; Burchat, P. R.; Carone, C. D.; Caso, C.; Dahl, O.; d'Ambrosio, G.; De Gouvea, A.; Doser, M.; Eidelman, S.; Feng, J. L.; Gherghetta, T.; Goodman, M.; Grab, C.; Groom, D. E.; Gurtu, A.; Hagiwara, K.; Hayes, K. G.; Hernández-Rey, J. J.; Hikasa, K.; Jawahery, H.; Kolda, C.; Kwon, Y.; Mangano, M. L.; Manohar, A. V.; Masoni, A.; et al. (2006). "Review of Particle Physics". *J. Phys. G* **33** (1): 1–1232. arXiv:astro-ph/0601168. Bibcode:2006JPhG...33....1Y. doi:10.1088/0954-3899/33/1/001.

[40] Rees, Martin. (1998). *Before the Beginning* (New York: Basic Books) p. 185 ISBN 0-201-15142-1

[41] de Sitter, Willem (1917). "Einstein's theory of gravitation and its astronomical consequences. Third paper". *Monthly Notices of the Royal Astronomical Society* **78**: 3–28. Bibcode:1917MNRAS..78....3D. doi:10.1093/mnras/78.1.3.

[42] Starobinsky, A. A. (December 1979). "Spectrum Of Relict Gravitational Radiation And The Early State Of The Universe". *Journal of Experimental and Theoretical Physics Letters* **30**: 682. Bibcode:1979JETPL..30..682S.; Starobinskii, A. A. (December 1979). "Spectrum of relict gravitational radiation and the early state of the universe". *Pisma Zh. Eksp. Teor. Fiz. (Soviet Journal of Experimental and Theoretical Physics Letters)* **30**: 719. Bibcode:1979ZhPmR..30..719S.

[43] SLAC seminar, "10^{-35} seconds after the Big Bang", 23 January 1980. see Guth (1997), pg 186

[44] Guth, Alan H. (1981). "Inflationary universe: A possible solution to the horizon and flatness problems" (PDF). *Physical Review D* **23** (2): 347–356. Bibcode:1981PhRvD..23..347G. doi:10.1103/PhysRevD.23.347.

[45] Chapter 17 of Peebles (1993).

[46] Starobinsky, Alexei A. (1980). "A new type of isotropic cosmological models without singularity". *Physics Letters B* **91**: 99–102. Bibcode:1980PhLB...91...99S. doi:10.1016/0370-2693(80)90670-X.

[47] Kazanas, D. (1980). "Dynamics of the universe and spontaneous symmetry breaking". *Astrophysical Journal* **241**: L59–63. Bibcode:1980ApJ...241L..59K. doi:10.1086/183361.

[48] Sato, K. (1981). "Cosmological baryon number domain structure and the first order phase transition of a vacuum". *Physics Letters B* **33**: 66–70. Bibcode:1981PhLB...99...66S. doi:10.1016/0370-2693(81)90805-4.

[49] Einhorn, Martin B; Sato, Katsuhiko (1981). "Monopole Production In The Very Early Universe In A First Order Phase Transition". *Nuclear Physics B* **180** (3): 385–404. Bibcode:1981NuPhB.180..385E. doi:10.1016/0550-3213(81)90057-2.

[50] Linde, A (1982). "A new inflationary universe scenario: A possible solution of the horizon, flatness, homogeneity, isotropy and primordial monopole problems". *Physics Letters B* **108** (6): 389–393. Bibcode:1982PhLB..108..389L. doi:10.1016/0370-2693(82)91219-9.

[51] Albrecht, Andreas; Steinhardt, Paul (1982). "Cosmology for Grand Unified Theories with Radiatively Induced Symmetry Breaking" (PDF). *Physical Review Letters* **48** (17): 1220–1223. Bibcode:1982PhRvL..48.1220A. doi:10.1103/PhysRevLe0.

[52] J.B. Hartle (2003). *Gravity: An Introduction to Einstein's General Relativity* (1st ed.). Addison Wesley. p. 411. ISBN 0-8053-8662-9

[53] See Linde (1990) and Mukhanov (2005).

[54] Chibisov, Viatcheslav F.; Chibisov, G. V. (1981). "Quantum fluctuation and "nonsingular" universe". *JETP Letters* **33**: 532–5. Bibcode:1981JETPL..33..532M.

[55] Mukhanov, Viatcheslav F. (1982). "The vacuum energy and large scale structure of the universe". *Soviet Physics JETP* **56**: 258–65.

[56] See Guth (1997) for a popular description of the workshop, or *The Very Early Universe*, ISBN 0-521-31677-4 eds Hawking, Gibbon & Siklos for a more detailed report

[57] Hawking, S.W. (1982). "The development of irregularities in a single bubble inflationary universe". *Physics Letters B* **115** (4): 295–297. Bibcode:1982PhLB..115..295H. doi:10.1016/0370-2693(82)90373-2.

[58] Starobinsky, Alexei A. (1982). "Dynamics of phase transition in the new inflationary universe scenario and generation of perturbations". *Physics Letters B* **117** (3–4): 175–8. Bibcode:1982PhLB..117..175S. doi:10.1016/0370-2693(82)90541-X.

[59] Guth, A.H. (1982). "Fluctuations in the new inflationary universe". *Physical Review Letters* **49**(15): 1110–3. Bibcode:G. doi:10.1103/PhysRevLett.49.1110.

[60] Bardeen, James M.; Steinhardt, Paul J.; Turner, Michael S. (1983). "Spontaneous creation Of almost scale-free density perturbations in an inflationary universe". *Physical Review D* **28** (4): 679–693. Bibcode:1983PhRvD..28..679B. doi:10.1103/Ph.

[61] Perturbations can be represented by Fourier modes of a wavelength. Each Fourier mode is normally distributed (usually called Gaussian) with mean zero. Different Fourier components are uncorrelated. The variance of a mode depends only on its wavelength in such a way that within any given volume each wavelength contributes an equal amount of power to the spectrum of perturbations. Since the Fourier transform is in three dimensions, this means that the variance of a mode goes as k^{-3} to compensate for the fact that within any volume, the number of modes with a given wavenumber k goes as k^3.

[62] Tegmark, M.; Eisenstein, Daniel J.; Strauss, Michael A.; Weinberg, David H.; Blanton, Michael R.; Frieman, Joshua A.; Fukugita, Masataka; Gunn, James E.; et al. (August 2006). "Cosmological constraints from the SDSS luminous red galaxies". *Physical Review D* **74** (12). arXiv:astro-ph/0608632. Bibcode:2006PhRvD..74l3507T. doi:10.1103/PhysRevD.74.123507.

[63] Steinhardt, Paul J. (2004). "Cosmological perturbations: Myths and facts". *Modern Physics Letters A* **19** (13 & 16): 967–82. Bibcode:2004MPLA...19..967S. doi:10.1142/S0217732304014252.

[64] Boyle, Latham A.; Steinhardt, PJ; Turok, N (2006). "Inflationary predictions for scalar and tensor fluctuations reconsidered". *Physical Review Letters* **96** (11): 111301. arXiv:astro-ph/0507455. Bibcode:2006PhRvL..96k1301B. doi:10.1103/P. PMID 16605810.

[65] Tegmark, Max (2005). "What does inflation really predict?". *JCAP* **0504** (4): 001. arXiv:astro-ph/0410281.Bibco. doi:10.1088/1475-7516/2005/04/001.

[66] This is known as a "red" spectrum, in analogy to redshift, because the spectrum has more power at longer wavelengths.

[67] Komatsu, E.; Smith, K. M.; Dunkley, J.; Bennett, C. L.; Gold, B.; Hinshaw, G.; Jarosik, N.; Larson, D.; et al. (January 2010). "Seven-Year Wilkinson Microwave Anisotropy Probe (WMAP) Observations: Cosmological Interpretation". *The Astrophysical Journal Supplement Series* **192** (2): 18. arXiv:1001.4538. Bibcode:2011ApJS..192...18K. doi:10.1088/0067-0049/192/2/18.

[68] Spergel, D. N.; Verde, L.; Peiris, H. V.; Komatsu, E.; Nolta, M. R.; Bennett, C. L.; Halpern, M.; Hinshaw, G.; et al. (2003). "First year Wilkinson Microwave Anisotropy Probe (WMAP) observations: determination of cosmological parameters". *Astrophysical Journal Supplement Series* **148** (1): 175–194. arXiv:astro-ph/0302209. Bibcode:2003ApJS..148..175S. doi:10.1086/377226.

[69] See cosmic microwave background#Low multipoles for details and references.

[70] Overbye, Dennis (24 March 2014). "Ripples From the Big Bang". *New York Times*. Retrieved 24 March 2014.

[71] Ade, P.A.R. (BICEP2 Collaboration); et al. (19 June 2014). "Detection of B-Mode Polarization at Degree Angular Scales by BICEP2". *Physical Review Letters* **112** (24): 241101. arXiv:1403.3985. Bibcode:2014PhRvL.112x1101A. doi:10.1103/Phys1. PMID 24996078.

[72] Woit, Peter (13 May 2014). "BICEP2 News". *Not Even Wrong*. Columbia University. Retrieved 19 January 2014.

[73] Overbye, Dennis (19 June 2014). "Astronomers Hedge on Big Bang Detection Claim". *New York Times*. Retrieved 20 June 2014.

[74] Amos, Jonathan (19 June 2014). "Cosmic inflation: Confidence lowered for Big Bang signal". *BBC News*. Retrieved 20 June 2014.

[75] Planck Collaboration Team (19 September 2014). "Planck intermediate results. XXX. The angular power spectrum of polarized dust emission at intermediate and high Galactic latitudes". *ArXiv*. arXiv:1409.5738. Bibcode:2014arXiv1409.5738P. Retrieved 22 September 2014.

[76] Overbye, Dennis (22 September 2014). "Study Confirms Criticism of Big Bang Finding". *New York Times*. Retrieved 22 September 2014.

[77] Clavin, Whitney (30 January 2015). "Gravitational Waves from Early Universe Remain Elusive". *NASA*. Retrieved 30 January 2015.

[78] Overbye, Dennis (30 January 2015). "Speck of Interstellar Dust Obscures Glimpse of Big Bang". *New York Times*. Retrieved 31 January 2015.

[79] Rosset, C.; PLANCK-HFI collaboration (2005). "Systematic effects in CMB polarization measurements". *Exploring the universe: Contents and structures of the universe (XXXIXth Rencontres de Moriond)*.

[80] Loeb, A.; Zaldarriaga, M (2004). "Measuring the small-scale power spectrum of cosmic density fluctuations through 21 cm tomography prior to the epoch of structure formation". *Physical Review Letters* **92** (21): 211301. arXiv:astro-ph/0312134. Bibcode:2004PhRvL..92u1301L. doi:10.1103/PhysRevLett.92.211301. PMID 15245272.

[81] Guth, Alan (1997). *The Inflationary Universe*. Addison–Wesley. ISBN 0-201-14942-7.

[82] Choi, Charles (Jun 29, 2012). "Could the Large Hadron Collider Discover the Particle Underlying Both Mass and Cosmic Inflation?". Scientific American. Retrieved Jun 25, 2014."The virtue of so-called Higgs inflation models is that they might explain inflation within the current Standard Model of particle physics, which successfully describes how most known particles and forces behave. Interest in the Higgs is running hot this summer because CERN, the lab in Geneva, Switzerland, that runs the LHC, has said it will announce highly anticipated findings regarding the particle in early July."

[83] Salvio, Alberto (2013-08-09). "Higgs Inflation at NNLO after the Boson Discovery". *Phys.Lett. B727 (2013) 234-239* **727**: 234–239. arXiv:1308.2244. Bibcode:2013PhLB..727..234S. doi:10.1016/j.physletb.2013.10.042.

[84] Technically, these conditions are that the logarithmic derivative of the potential, $\epsilon = (1/2)(V'/V)^2$ and second derivative $\eta = V''/V$ are small, where V is the potential and the equations are written in reduced Planck units. See, *e.g.* Liddle and Lyth (2000), pg 42-43.

[85] Salvio, Strumia (2014-03-17). "Agravity". *JHEP 1406 (2014) 080* **2014**. arXiv:1403.4226. Bibcode:2014JHEP...06..080S. doi:10.1007/JHEP06(2014)080.

[86]Linde, Andrei D. (1983). "Chaotic inflation".*Physics Letters B***129**(3): 171–81.Bibcode:1983PhLB..129..177L.doi:10.10-2693(83)90837-7.

[87] Technically, this is because the inflaton potential is expressed as a Taylor series in $\varphi/m P_l$, where φ is the inflaton and $m P_l$ is the Planck mass. While for a single term, such as the mass term $m_\varphi^4 (\varphi/m P_l)^2$, the slow roll conditions can be satisfied for φ much greater than $m P_l$, this is precisely the situation in effective field theory in which higher order terms would be expected to contribute and destroy the conditions for inflation. The absence of these higher order corrections can be seen as another sort of fine tuning. See *e.g.* Alabidi, Laila; Lyth, David H (2006). "Inflation models and observation". *JCAP* **0605** (5): 016. arXiv:astro-ph/0510441. Bibcode:2006JCAP...05..016A. doi:10.1088/1475-7516/2006/05/016.

[88] See, *e.g.* Lyth, David H. (1997). "What would we learn by detecting a gravitational wave signal in the cosmic microwave background anisotropy?". *Physical Review Letters* **78** (10): 1861–3. arXiv:hep-ph/9606387. Bibcode:1997PhRvL..78.1861L. doi:10.1103/PhysRevLett.78.1861.

[89] Brandenberger, Robert H. (November 2004). "Challenges for inflationary cosmology". arXiv:astro-ph/0411671.

[90]Vilenkin, Alexander (1983). "The birth of inflationary universes".*Physical Review D***27**(12): 2848–2855.Bibcode:1983Ph. doi:10.1103/PhysRevD.27.2848.

[91]A. Linde (1986). "Eternal chaotic inflation".*Modern Physics Letters A***1**(2): 81–85.Bibcode:1986MPLA....1...81L.do000129. A. Linde (1986). "Eternally existing self-reproducing chaotic inflationary universe" (PDF). *Physics Letters B* **175** (4): 395–400. Bibcode:1986PhLB..175..395L. doi:10.1016/0370-2693(86)90611-8.

[92] A. Borde, A. Guth and A. Vilenkin (2003). "Inflationary space-times are incomplete in past directions". *Physical Review Letters* **90** (15): 151301. arXiv:gr-qc/0110012. Bibcode:2003PhRvL..90o1301B. doi:10.1103/PhysRevLett.90.151301. PMID 12732026.

[93] A. Borde (1994). "Open and closed universes, initial singularities and inflation". *Physical Review D* **50** (6): 3692–702. arXiv:gr-qc/9403049. Bibcode:1994PhRvD..50.3692B. doi:10.1103/PhysRevD.50.3692.

[94] A. Borde and A. Vilenkin (1994). "Eternal inflation and the initial singularity". *Physical Review Letters* **72** (21): 3305–9. arXiv:gr-qc/9312022. Bibcode:1994PhRvL..72.3305B. doi:10.1103/PhysRevLett.72.3305.

[95] Linde (2005, §V).

[96] Carroll, Sean M.; Chen, Jennifer (2005). "Does inflation provide natural initial conditions for the universe?". *Gen. Rel. Grav.* **37** (10): 1671–4. arXiv:gr-qc/0505037. Bibcode:2005GReGr..37.1671C. doi:10.1007/s10714-005-0148-2.

[97] Carroll, Sean M.; Jennifer Chen (2004). "Spontaneous inflation and the origin of the arrow of time". arXiv:hep-th/0410270.

[98] Aguirre, Anthony; Gratton, Steven (2003). "Inflation without a beginning: A null boundary proposal". *Physical Review D* **67** (8): 083515. arXiv:gr-qc/0301042. Bibcode:2003PhRvD..67h3515A. doi:10.1103/PhysRevD.67.083515.

[99] Aguirre, Anthony; Gratton, Steven (2002). "Steady-State Eternal Inflation". *Physical Review D* **65** (8): 083507. arXiv:astro-ph/0111191. Bibcode:2002PhRvD..65h3507A. doi:10.1103/PhysRevD.65.083507.

[100]Hartle, J.; Hawking, S. (1983). "Wave function of the universe".*Physical Review D***28**(12): 2960–2975.Bibcode:1983PH. doi:10.1103/PhysRevD.28.2960.; See also Hawking (1998).

[101] Hawking (1998), p. 129.

[102] Wikiquote

[103] Page, Don N. (1983). "Inflation does not explain time asymmetry". *Nature* **304** (5921): 39–41. Bibcode:1983Natur.304...39P. doi:10.1038/304039a0.; see also Roger Penrose's book The Road to Reality: A Complete Guide to the Laws of the Universe.

[104] Hawking, S. W.; Page, Don N. (1988). "How probable is inflation?".*Nuclear Physics B***298**(4): 789–809.Bibcode:H. doi:10.1016/0550-3213(88)90008-9.

[105] Steinhardt, Paul J. (2011). "The inflation debate: Is the theory at the heart of modern cosmology deeply flawed?" (*Scientific American*, April; pp. 18-25).

[106] Paul J. Steinhardt; Neil Turok (2007). *Endless Universe: Beyond the Big Bang*. Broadway Books. ISBN 978-0-7679-1501-4.

[107] Albrecht, Andreas; Sorbo, Lorenzo (2004). "Can the universe afford inflation?". *Physical Review D* **70** (6): 063528. arXiv:hep-th/0405270. Bibcode:2004PhRvD..70f3528A. doi:10.1103/PhysRevD.70.063528.

[108] Martin, Jerome; Brandenberger, Robert (2001). "The trans-Planckian problem of inflationary cosmology". *Physical Review D* **63** (12): 123501. arXiv:hep-th/0005209. Bibcode:2001PhRvD..63l3501M. doi:10.1103/PhysRevD.63.123501.

[109] Martin, Jerome; Ringeval, Christophe (2004). "Superimposed Oscillations in the WMAP Data?". *Physical Review D* **69** (8): 083515. arXiv:astro-ph/0310382. Bibcode:2004PhRvD..69h3515M. doi:10.1103/PhysRevD.69.083515.

[110] Robert H. Brandenberger, "A Status Review of Inflationary Cosmology", proceedings Journal-ref: BROWN-HET-1256 (2001), (available from arXiv:hep-ph/0101119v1 11 January 2001)

[111] Andrei Linde, "Prospects of Inflation", *Physica Scripta Online* (2004) (available from arXiv:hep-th/0402051)

[112] Blanco-Pillado et al., "Racetrack inflation", (2004) (available from arXiv:hep-th/0406230)

[113] Kachru, Shamit; Kallosh, Renata; Linde, Andrei; Maldacena, Juan; McAllister, Liam; Trivedi, Sandip P (2003). "Towards inflation in string theory". *JCAP* **0310** (10): 013. arXiv:hep-th/0308055. Bibcode:2003JCAP...10..013K. doi:10.1088/1475-7516/2003/10/013.

[114] G. R. Dvali, S. H. Henry Tye, *Brane inflation, Phys.Lett.* **B450**, 72-82 (1999), arXiv:hep-ph/9812483.

[115] Bojowald, Martin (October 2008). "Big Bang or Big Bounce?: New Theory on the Universe's Birth". Retrieved 2015-08-31.

[116] Poplawski, N. J. (2010). "Cosmology with torsion: An alternative to cosmic inflation". *Physics Letters B* **694** (3): 181–185. arXiv:1007.0587. Bibcode:2010PhLB..694..181P. doi:10.1016/j.physletb.2010.09.056.

[117] Poplawski, N. (2012). "Nonsingular, big-bounce cosmology from spinor-torsion coupling". *Physical Review D* **85** (10): 107502. arXiv:1111.4595. Bibcode:2012PhRvD..85j7502P. doi:10.1103/PhysRevD.85.107502.

[118] Brandenberger, R; Vafa, C. (1989). "Superstrings in the early universe".*Nuclear Physics B***316**(2): 391–410.Bibcode:1. doi:10.1016/0550-3213(89)90037-0.

[119] Battefeld, Thorsten; Watson, Scott (2006). "String Gas Cosmology". *Reviews Modern Physics* **78** (2): 435–454. arXiv:hep-th/0510022. Bibcode:2006RvMP...78..435B. doi:10.1103/RevModPhys.78.435.

[120] Brandenberger, Robert H.; Nayeri, ALI; Patil, Subodh P.; Vafa, Cumrun (2007). "String Gas Cosmology and Structure Formation". *International Journal of Modern Physics A* **22** (21): 3621–3642. arXiv:hep-th/0608121. Bibcode:2007IJMPA..22.3621B. doi:10.1142/S0217751X07037159.

[121] Earman, John; Mosterín, Jesús (March 1999). "A Critical Look at Inflationary Cosmology". *Philosophy of Science* **66**: 1–49. doi:10.2307/188736 (inactive 2015-01-14). JSTOR 188736.

[122] Penrose, Roger (2004). *The Road to Reality: A Complete Guide to the Laws of the Universe*. London: Vintage Books, p. 755. See also Penrose, Roger (1989). "Difficulties with Inflationary Cosmology". *Annals of the New York Academy of Sciences* **271**: 249–264. Bibcode:1989NYASA.571..249P. doi:10.1111/j.1749-6632.1989.tb50513.x.

[123] Hložek, Renée (12 June 2015). "CMB@50 day three". Retrieved 15 July 2015.
This is a collation of remarks from the third day of the "Cosmic Microwave Background @50" conference held at Princeton, 10–12 June 2015.

[124] Ijjas, Anna; Steinhardt, Paul J.; Loeb, Abraham. "Inflationary paradigm in trouble after Planck2013". *Physics Letters* **B723**: 261–266. arXiv:1304.2785. Bibcode:2013PhLB..723..261I. doi:10.1016/j.physletb.2013.05.023.

[125] Ijjas, Anna; Steinhardt, Paul J.; Loeb, Abraham. "Inflationary schism after Planck2013". *Physics Letters* **B736**: 142–146. Bibcode:2014PhLB..736..142I. doi:10.1016/j.physletb.2014.07.012.

[126] Guth, Alan H.; Kaiser, David I.; Nomura, Yasunori. "Inflationary paradigm after Planck 2013". *Physics Letters* **B733**: 112–119. arXiv:1312.7619. Bibcode:2014PhLB..733..112G. doi:10.1016/j.physletb.2014.03.020.

[127] Linde, Andrei. "Inflationary cosmology after Planck 2013". arXiv:1402.0526. Bibcode:2014arXiv1402.0526L.

31.10 References

- Guth, Alan (1997). *The Inflationary Universe: The Quest for a New Theory of Cosmic Origins*. Perseus. ISBN 0-201-32840-2.

- Hawking, Stephen (1998). *A Brief History of Time*. Bantam. ISBN 0-553-38016-8.

- Hawking, Stephen; Gary Gibbons (1983). *The Very Early Universe*. Cambridge University Press. ISBN 0-521-31677-4.

- Kolb, Edward; Michael Turner (1988). *The Early Universe*. Addison-Wesley. ISBN 0-201-11604-9.

- Linde, Andrei (1990). *Particle Physics and Inflationary Cosmology*. Chur, Switzerland: Harwood. arXiv:hep-th/0503203. ISBN 3-7186-0490-6.

- Linde, Andrei (2005) "Inflation and String Cosmology", *eConf* **C040802** (2004) L024; *J. Phys. Conf. Ser.* **24** (2005) 151–60; arXiv:hep-th/0503195 v1 2005-03-24.

- Liddle, Andrew; David Lyth (2000). *Cosmological Inflation and Large-Scale Structure*. Cambridge. ISBN 0-521-57598-2.

- Lyth, David H.; Riotto, Antonio (1999). "Particle physics models of inflation and the cosmological density perturbation". *Phys. Rept.* **314** (1–2): 1–146. arXiv:hep-ph/9807278. Bibcode:1999PhR...314....1L. doi:10.1016/S0370-1573(98)00128-8.

- Mukhanov, Viatcheslav (2005). *Physical Foundations of Cosmology*. Cambridge University Press. ISBN 0-521-56398-4.

- Vilenkin, Alex (2006). *Many Worlds in One: The Search for Other Universes*. Hill and Wang. ISBN 0-8090-9523-8.

- Peebles, P. J. E. (1993). *Principles of Physical Cosmology*. Princeton University Press. ISBN 0-691-01933-9.

31.11 External links

- Was Cosmic Inflation The 'Bang' Of The Big Bang?, by Alan Guth, 1997

- An Introduction to Cosmological Inflation by Andrew Liddle, 1999

- update 2004 by Andrew Liddle

- hep-ph/0309238 Laura Covi: Status of observational cosmology and inflation

- hep-th/0311040 David H. Lyth: Which is the best inflation model?

- The Growth of Inflation *Symmetry*, December 2004

- Guth's logbook showing the original idea

- WMAP Bolsters Case for Cosmic Inflation, March 2006

- NASA March 2006 WMAP press release

- Max Tegmark's *Our Mathematical Universe* (2014), "Chapter 5: Inflation"

Chapter 32

Coupling (physics)

In physics, two systems are **coupled** if they are interacting with each other. Of special interest is the **coupling** of two (or more) vibratory systems (e.g. pendula or resonant circuits) by means of springs or magnetic fields, etc. Characteristic for a coupled oscillation is the effect of beat.

The concept of coupling is particularly important in physical cosmology, in which various forms of matter gradually decouple and recouple between each other.

Coupling is also important in physics for the generation of plasmas. In electrical discharges, the coupling of an exciting field and a medium creates plasmas. The quality of the coupling of an exciting field of given frequency to a charged particle depends on resonance.

32.1 See also

- Coupling constant

- Frequency classification of plasmas

- Spin spin coupling

- Spin-orbit coupling

- Self-coupling

Chapter 33

Cosmological constant

In cosmology, the **cosmological constant** (usually denoted by the Greek capital letter lambda: Λ) is the value of the energy density of the vacuum of space. It was originally introduced by Albert Einstein in 1917[1] as an addition to his theory of general relativity to "hold back gravity" and achieve a static universe, which was the accepted view at the time. Einstein abandoned the concept after Hubble's 1929 discovery that all galaxies outside the Local Group (the group that contains the Milky Way Galaxy) are moving away from each other, implying an overall expanding universe. From 1929 until the early 1990s, most cosmology researchers assumed the cosmological constant to be zero.

Since the 1990s, several developments in observational cosmology, especially the discovery of the accelerating universe from distant supernovae in 1998 (in addition to independent evidence from the cosmic microwave background and large galaxy redshift surveys), have shown that around 70% of the mass–energy density of the universe can be attributed to dark energy. While dark energy is poorly understood at a fundamental level, the main required properties of dark energy are that it functions as a type of anti-gravity, it dilutes much more slowly than matter as the universe expands, and it clusters much more weakly than matter, or perhaps not at all. The cosmological constant is the simplest possible form of dark energy since it is constant in both space and time, and this leads to the current standard model of cosmology known as the Lambda-CDM model, which provides a good fit to many cosmological observations as of 2014.

33.1   Equation

The cosmological constant Λ appears in Einstein's field equation in the form of

$$R_{\mu\nu} - \frac{1}{2} R \, g_{\mu\nu} + \Lambda \, g_{\mu\nu} = \frac{8\pi G}{c^4} T_{\mu\nu},$$

where R and g describe the structure of spacetime, T pertains to matter and energy affecting that structure, and G and c are conversion factors that arise from using traditional units of measurement. When Λ is zero, this reduces to the original field equation of general relativity. When T is zero, the field equation describes empty space (the vacuum).

The cosmological constant has the same effect as an intrinsic energy density of the vacuum, ϱ_{vac} (and an associated pressure). In this context, it is commonly moved onto the right-hand side of the equation, and defined with a proportionality factor of 8π: $\Lambda = 8\pi\varrho_{vac}$, where unit conventions of general relativity are used (otherwise factors of G and c would also appear, i.e. $\Lambda = 8\pi \, (G/c^2)\varrho_{vac} = \kappa \, \varrho_{vac}$, where κ is Einstein's constant). It is common to quote values of energy density directly, though still using the name "cosmological constant", with convention $8\pi \, G = 1$. (In fact, the true dimension of Λ is a length^{-2} and it has the value of ~$1 \cdot 10^{-52}$ m^{-2} or in reduced Planck units : ~$3 \cdot 10^{-122}$, calculated with the best present (2015) values of $\Omega\Lambda = 0.6911 \pm 0.0062$ and $H_o = 67.74 \pm 0.46$ km/s / Mpc $= 2.195 \pm 0.015 \cdot 10^{-18}$ s^{-1}).

A positive vacuum energy density resulting from a cosmological constant implies a negative pressure, and vice versa. If the energy density is positive, the associated negative pressure will drive an accelerated expansion of the universe, as observed. (See dark energy and cosmic inflation for details.)

239

33.1.1   ΩΛ (Omega Lambda)

Instead of the cosmological constant itself, cosmologists often refer to the ratio between the energy density due to the cosmological constant and the critical density of the universe, the tipping point for a sufficient density to stop the universe from expanding forever. This ratio is usually denoted ΩΛ, and is estimated to be 0.6911 ± 0.0062, according to the recent Planck results released in 2015.[2] In a flat universe ΩΛ is the fraction of the energy of the universe due to the cosmological constant, i.e., what we would intuitively call the fraction of the universe that is made up of dark energy. Note that this value changes over time: the critical density changes with cosmological time, but the energy density due to the cosmological constant remains unchanged throughout the history of the universe: the amount of dark energy increases as the universe grows, while the amount of matter does not.

33.1.2   Equation of state

Another ratio that is used by scientists is the equation of state, usually denoted w, which is the ratio of pressure that dark energy puts on the universe to the energy per unit volume.[3] This ratio is $w = -1$ for a true cosmological constant, and is generally different for alternative time-varying forms of vacuum energy such as quintessence.

33.2   History

Einstein included the cosmological constant as a term in his field equations for general relativity because he was dissatisfied that otherwise his equations did not allow, apparently, for a static universe: gravity would cause a universe that was initially at dynamic equilibrium to contract. To counteract this possibility, Einstein added the cosmological constant.[4] However, soon after Einstein developed his static theory, observations by Edwin Hubble indicated that the universe appears to be expanding; this was consistent with a cosmological solution to the *original* general relativity equations that had been found by the mathematician Friedmann, working on the Einstein equations of general relativity. Einstein later reputedly referred to his failure to accept the validation of his equations—when they had predicted the expansion of the universe in theory, before it was demonstrated in observation of the cosmological red shift—as the "biggest blunder" of his life.[5][6]

In fact, adding the cosmological constant to Einstein's equations does not lead to a static universe at equilibrium because the equilibrium is unstable: if the universe expands slightly, then the expansion releases vacuum energy, which causes yet more expansion. Likewise, a universe that contracts slightly will continue contracting. [7]:59

However, the cosmological constant remained a subject of theoretical and empirical interest. Empirically, the onslaught of cosmological data in the past decades strongly suggests that our universe has a positive cosmological constant.[4] The explanation of this small but positive value is an outstanding theoretical challenge (*see the section below*).

Finally, it should be noted that some early generalizations of Einstein's gravitational theory, known as classical unified field theories, either introduced a cosmological constant on theoretical grounds or found that it arose naturally from the mathematics. For example, Sir Arthur Stanley Eddington claimed that the cosmological constant version of the vacuum field equation expressed the "epistemological" property that the universe is "self-gauging", and Erwin Schrödinger's pure-affine theory using a simple variational principle produced the field equation with a cosmological term.

33.3   Positive value

Observations announced in 1998 of distance–redshift relation for Type Ia supernovae[8][9] indicated that the expansion of the universe is accelerating. When combined with measurements of the cosmic microwave background radiation these implied a value of $\Omega_\Lambda \simeq 0.7$,[10] a result which has been supported and refined by more recent measurements. There are other possible causes of an accelerating universe, such as quintessence, but the cosmological constant is in most respects the simplest solution. Thus, the current standard model of cosmology, the Lambda-CDM model, includes the cosmological constant, which is measured to be on the order of 10^{-52} m^{-2}, in metric units. Multiplied by other constants that appear in the equations, it is often expressed as 10^{-52} m^{-2}, 10^{-35} s^{-2}, 10^{-47} GeV^4, 10^{-29} g/cm^3.[11] In terms of Planck units, and as a natural dimensionless value, the cosmological constant, λ, is on the order of 10^{-122}.[12]

As was only recently seen, by works of 't Hooft, Susskind[13] and others, a positive cosmological constant has surprising consequences, such as a finite maximum entropy of the observable universe (see the holographic principle).

33.4 Predictions

33.4.1 Quantum field theory

See also: Vacuum catastrophe

A major outstanding problem is that most quantum field theories predict a huge value for the quantum vacuum. A common assumption is that the quantum vacuum is equivalent to the cosmological constant. Although no theory exists that supports this assumption, arguments can be made in its favor.[14]

Such arguments are usually based on dimensional analysis and effective field theory. If the universe is described by an effective local quantum field theory down to the Planck scale, then we would expect a cosmological constant of the order of M_{pl}^4. As noted above, the measured cosmological constant is smaller than this by a factor of 10^{-120}. This discrepancy has been called "the worst theoretical prediction in the history of physics!".[15]

Some supersymmetric theories require a cosmological constant that is exactly zero, which further complicates things. This is the *cosmological constant problem*, the worst problem of fine-tuning in physics: there is no known natural way to derive the tiny cosmological constant used in cosmology from particle physics.

33.4.2 Anthropic principle

One possible explanation for the small but non-zero value was noted by Steven Weinberg in 1987 following the anthropic principle.[16] Weinberg explains that if the vacuum energy took different values in different domains of the universe, then observers would necessarily measure values similar to that which is observed: the formation of life-supporting structures would be suppressed in domains where the vacuum energy is much larger. Specifically, if the vacuum energy is negative and its absolute value is substantially larger than it appears to be in the observed universe (say, a factor of 10 larger), holding all other variables (e.g. matter density) constant, that would mean that the universe is closed; furthermore, its lifetime would be shorter than the age of our universe, possibly too short for intelligent life to form. On the other hand, a universe with a large positive cosmological constant would expand too fast, preventing galaxy formation. According to Weinberg, domains where the vacuum energy is compatible with life would be comparatively rare. Using this argument, Weinberg predicted that the cosmological constant would have a value of less than a hundred times the currently accepted value.[17] In 1992, Weinberg refined this prediction of the cosmological constant to 5 to 10 times the matter density.[18]

This argument depends on a lack of a variation of the distribution (spatial or otherwise) in the vacuum energy density, as would be expected if dark energy were the cosmological constant. There is no evidence that the vacuum energy does vary, but it may be the case if, for example, the vacuum energy is (even in part) the potential of a scalar field such as the residual inflaton (also see quintessence). Another theoretical approach that deals with the issue is that of multiverse theories, which predict a large number of "parallel" universes with different laws of physics and/or values of fundamental constants. Again, the anthropic principle states that we can only live in one of the universes that is compatible with some form of intelligent life. Critics claim that these theories, when used as an explanation for fine-tuning, commit the inverse gambler's fallacy.

In 1995, Weinberg's argument was refined by Alexander Vilenkin to predict a value for the cosmological constant that was only ten times the matter density,[19] i.e. about three times the current value since determined.

33.4.3 Cyclic model

More recent work has suggested the problem may be indirect evidence of a cyclic universe possibly as allowed by string theory. With every cycle of the universe (Big Bang then eventually a Big Crunch) taking about a trillion (10^{12}) years, "the amount of matter and radiation in the universe is reset, but the cosmological constant is not. Instead, the cosmological constant gradually diminishes over many cycles to the small value observed today."[20] Critics respond that, as the authors acknowledge in their paper, the model "entails ... the same degree of tuning required in any cosmological model".[21]

33.5   See also

- Higgs mechanism

- Lambdavacuum solution

- Naturalness (physics)

- Quantum electrodynamics

- de Sitter relativity

- Unruh effect

33.6   Further reading

- Michael, E., University of Colorado, Department of Astrophysical and Planetary Sciences, "The Cosmological Constant"

- Ferguson, Kitty (1991). *Stephen Hawking: Quest For A Theory of Everything*, Franklin Watts. ISBN 0-553-29895-X.

- John D. Barrow and John K. Webb (June 2005). "Inconstant Constants". *Scientific American*.

- *Beyond the Cosmological Standard Model*[22] (2014)

33.7   References

[1] Einstein, A (1917). "Kosmologische Betrachtungen zur allgemeinen Relativitaetstheorie". *Sitzungsberichte der Königlich Preussischen Akademie der Wissenschaften Berlin*. part 1: 142–152.

[2] Collaboration, Planck, PAR Ade, N Aghanim, C Armitage-Caplan, M Arnaud, et al., Planck 2015 results. XIII. Cosmological parameters. arXiv preprint 1502.1589v2 , 6 Feb 2015.

[3] Hogan, Jenny (2007). "Welcome to the Dark Side".*Nature***448**(7151): 240–245.Bibcode:2007Natur.448..240H.doi:10.1a. PMID 17637630.

[4] Urry, Meg (2008). *The Mysteries of Dark Energy*. Yale Science. Yale University.

[5] Gamov, George (1970). *My World Line*. Viking Press. p. 44. ISBN 978-0670503766

[6] Rosen, Rebecca J. "Einstein Likely Never Said One of His Most Oft-Quoted Phrases". *The Atlantic*. The Atlantic Media Company. Retrieved 10 August 2013.

[7] Barbara Sue Ryden (2003). *Introduction to cosmology*. Addison-Wesley. ISBN 978-0-8053-8912-8.

[8] Riess, A.; et al. (September 1998). "Observational Evidence from Supernovae for an Accelerating Universe and a Cosmological Constant". *The Astronomical Journal* **116** (3): 1009–1038. arXiv:astro-ph/9805201. Bibcode:1998AJ....116.1009R. doi:10.1086/300499.

[9] Perlmutter, S.; et al. (June 1999). "Measurements of Omega and Lambda from 42 High-Redshift Supernovae". *The Astrophysical Journal* **517** (2): 565–586. arXiv:astro-ph/9812133. Bibcode:1999ApJ...517..565P. doi:10.1086/307221.

[10] See e.g. Baker, Joanne C.; et al. (1999). "Detection of cosmic microwave background structure in a second field with the Cosmic Anisotropy Telescope". *Monthly Notices of the Royal Astronomical Society* **308** (4): 1173–1178. arXiv:astro-ph/9904415. Bibcode:1999MNRAS.308.1173B. doi:10.1046/j.1365-8711.1999.02829.x.

[11] Tegmark, Max; et al. (2004). "Cosmological parameters from SDSS and WMAP". *Physical Review D* **69** (103501): 103501. arXiv:astro-ph/0310723. Bibcode:2004PhRvD..69j3501T. doi:10.1103/PhysRevD.69.103501.

[12] John D. Barrow The Value of the Cosmological Constant

[13] Lisa Dyson, Matthew Kleban, Leonard Susskind: "Disturbing Implications of a Cosmological Constant"

[14] Rugh, S; Zinkernagel, H. (2001). "The Quantum Vacuum and the Cosmological Constant Problem". *Studies in History and Philosophy of Modern Physics* **33** (4): 663–705. doi:10.1016/S1355-2198(02)00033-3.

[15] MP Hobson, GP Efstathiou & AN Lasenby (2006). *General Relativity: An introduction for physicists* (Reprinted with corrections 2007 ed.). Cambridge University Press. p. 187. ISBN 978-0-521-82951-9.

[16] Weinberg, S (1987). "Anthropic Bound on the Cosmological Constant". *Phys. Rev. Lett.* **59** (22): 2607–2610. Bibcode:1W. doi:10.1103/PhysRevLett.59.2607. PMID 10035596.

[17] Alexander Vilenkin, *Many Worlds in One: The Search for Other Universes*, ISBN 978-0-8090-9523-0, pp. 138–9

[18] Weinberg, Steven (1993). *Dreams of a Final Theory: the search for the fundamental laws of nature*. Vintage Press. p. 182. ISBN 0-09-922391-0.

[19] Alexander Vilenkin, *Many Worlds in One: The Search for Other universes*, ISBN 978-0-8090-9523-0, p. 146, which references Vilenkin' *Predictions from quantum cosmology*, Physical Review Letters, vol 74, p. 846 (1995)

[20] 'Cyclic universe' can explain cosmological constant, NewScientistSpace, 4 May 2006

[21] Steinhardt, P. J.; Turok, N. (2002-04-25). "A Cyclic Model of the Universe". *Science* **296** (5572): 1436–1439. arXiv:hep-th/0111030v2. Bibcode:2002Sci...296.1436S. doi:10.1126/science.1070462. PMID 11976408. Retrieved 2012-04-29.

[22] Austin Joyce, Bhuvnesh Jain, Justin Khoury, Mark Trodden (2014). "Beyond the Cosmological Standard Model". *Arxiv*.

33.8 External links

- Cosmological constant (astronomy) at *Encyclopædia Britannica*

- Carroll, Sean M., *"The Cosmological Constant"* (short), *"The Cosmological Constant"*(extended).

- 'Cyclic universe' can explain cosmological constant.

- News story: More evidence for dark energy being the cosmological constant

- Cosmological constant article from Scholarpedia

- Copeland, Ed; Merrifield, Mike. "Λ – Cosmological Constant". *Sixty Symbols*. Brady Haran for the University of Nottingham.

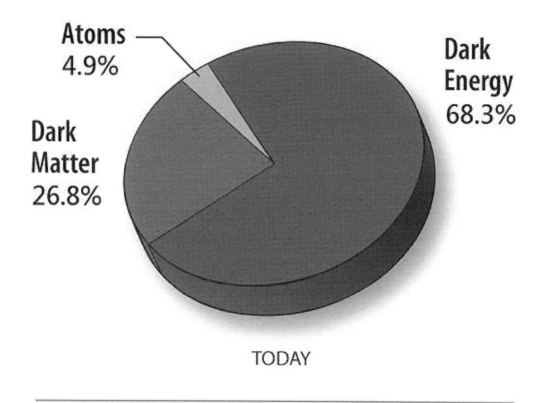

TODAY

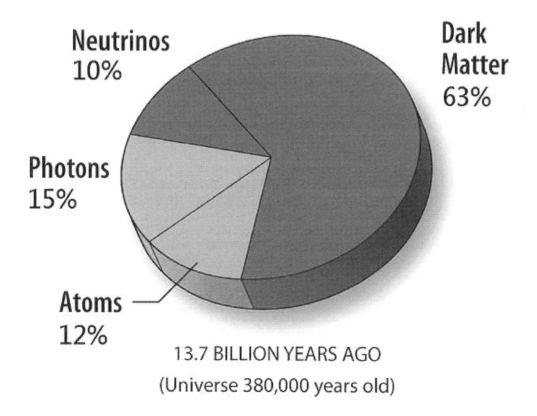

13.7 BILLION YEARS AGO
(Universe 380,000 years old)

Estimated ratios of dark matter and dark energy (which may be the cosmological constant) in the universe. According to current theories of physics, dark energy now dominates as the largest source of energy of the universe, in contrast to earlier epochs when it was insignificant.

33.9 Text and image sources, contributors, and licenses

33.9.1 Text

- **Higgs boson** *Source:* https://en.wikipedia.org/wiki/Higgs_boson?oldid=684961993 *Contributors:* AxelBoldt, CYD, ClaudeMuncey, Bryan Derksen, Manning Bartlett, Roadrunner, David spector, Heron, Ewen, Stevertigo, Edward, Boud, TeunSpaans, Dante Alighieri, Ixfd64, Gaurav, TakuyaMurata, CesarB, Anders Feder, Mgimpel~enwiki, Bueller 007, Mark Foskey, Kaihsu, Samw, Cherkash, Lee M, Mxn, Ehn, Timwi, Dcoetzee, Wikiborg, Kbk, Tpbradbury, Phys, Bevo, Topbanana, JonathanDP81, AnonMoos, Bcorr, Jerzy, BenRG, Slawojarek, Phil Boswell, Donarreiskoffer, Robbot, Josh Cherry, ChrisO~enwiki, Owain, Iwpg, Goethean, Altenmann, Nurg, Lowellian, Merovingian, Rursus, Caknuck, Hadal, Alba, Mattflaschen, David Gerard, M-Falcon, Giftlite, Graeme Bartlett, Harp, ShaneCavanaugh, Lethe, Herbee, Jrquinlisk, Xerxes314, Ds13, Fleminra, Dratman, Muzzle, Varlaam, Jason Quinn, Foobar, DÁ„ugosz, Golbez, Bodhitha, Mmm~enwiki, Aughtandzero, Quadell, Selva, Kaldari, Fred Stober, Johnflux, RetiredUser2, Thincat, Elektron, Bbbl67, Icairns, J0m1eisler, Cructacean, Tdent, TJSwoboda, John-Armagh, Safety Cap, ProjeX, Njh@bandsman.co.uk, Mike Rosoft, Chris Howard, Jkl, Discospinster, Rich Farmbrough, FT2, Qutezuce, Vsmith, Pie4all88, Kooo, David Schaich, Xgenei, Mal~enwiki, Dbachmann, Mani1, Bender235, ESkog, RJHall, Ylee, Pt, El C, Lycurgus, Lars~enwiki, Laurascudder, Art LaPella, Bookofjude, Brians, TheMile, Dragon76, Smalljim, C S, Reuben, La goutte de pluie, Rangelov, Sasquatch, Bawolff, Tritium6, Eritain, HasharBot~enwiki, Jumbuck, Yoweigh, Alansohn, Andrew Gray, JohnAlbertRigali, Axl, Sligocki, Kocio, Mlm42, Tom12519, Chuckupd, Atomicthumbs, Wtmitchell, KapilTagore, Endersdouble, Dirac1933, DrGaellon, Falcorian, Itinerant, DarTar, Joriki, Reinoutr, Linas, Mindmatrix, Jamsta, Sburke, Benbest, Jonburchel, Thruston, TotoBaggins, GregorB, J M Rice, CharlesC, Waldir, Christopher Thomas, Karam.Anthony.K, Tevatron~enwiki, RichardWeiss, Ashmoo, Fleisher, Kbdank71, GrundyCamellia, Drbogdan, Rjwilmsi, Nightscream, Koavf, Strait, XP1, Martaf, BlueMoonlet, MZMcBride, Mike Peel, NeonMerlin, R.e.b., Jehochman, Bubba73, Afterwriting, A Man In Black, Splarka, RobertG, Nihiltres, Norvy, Itinerant1, Gurch, Mark J, Nimur, Shawn@garbett.org, ElfQrin, Danny-DaWriter, Goudzovski, Diza, Consumed Crustacean, Srleffler, Sbove, Chobot, DVdm, Bgwhite, Zentropa, Bambaiah, Wester, Hairy Dude, Huw Powell, Wikky Horse, Pip2andahalf, RussBot, Jacques Antoine, Bhny, JabberWok, Hellbus, Archelon, Eleassar, Rsrikanth05, Salsb, Big Brother 1984, NawlinWiki, Folletto, Buster79, Trovatore, Neutron, SCZenz, Daniel Mietchen, Gadget850, Bota47, Karl Andrews, Dnawebmaster, Jezzabr, Thor Waldsen, Crisco 1492, Deeday-UK, Daniel C, WAS 4.250, Paul Magnussen, Closedmouth, D'Agosta, Bondegezou, Netrapt, Egumtow, LeonardoRob0t, Ilmari Karonen, NeilN, Kgf0, Maryhit, Dragon of the Pants, SmackBot, Nahald, Moeron, Ashley thomas80, Slashme, InverseHypercube, Melchoir, Cinkcool, Baad, Jagged 85, Nickst, Frymaster, AnOddName, ZerodEgo, Giandrea, Gilliam, Ohnoitsjamie, Skizzik, Carl.bunderson, Aurimas, Dauto, JCSantos, TimBentley, RevenDS, Jprg1966, Rick7425, Cadmasteradam, Roscelese, Epastore, DHN-bot~enwiki, Sbharris, Eusebeus, Scwlong, Modest Genius, Famspear, V1adis1av, Rhodesh, Fiziker, Lantrix, Grover cleveland, Jmnbatista, Wen D House, Flyguy649, Jgwacker, Daqu, Mesons, Rezecib, Martijn Hoekstra, Pulu, BullRangifer, Andrew c, Gildir, Kendrick7, Marcus Brute, Vina-iwbot~enwiki, Yevgeny Kats, Frglee, TriTertButoxy, CIS, SashatoBot, Lambiam, Mukadderat, Hi2lok, Kuru, Khazar, Shirifan, Eikern, Tktktk, JorisvS, DMurphy, Mgiganteus1, Bjankuloski06en~enwiki, IronGargoyle, Aardvark23, Loadmaster, Smith609, Deceglie, Hvn0413, Xiphoris, Norm mit, Keith-264, Kencf0618, Britannica~enwiki, Paul venter, Newone, Twas Now, GDallimore, Benplowman, Airstrike~enwiki, Chetvorno, DKqwerty, Harold f, JForget, Laplacian, Er ouz, Jtuggle, Banedon, Ruslik0, Krioni, McVities, Keithh, Rotiro, Yaris678, Slazenger, Cydebot, Martinthoegersen, Gogo Dodo, Anonymi, Lewisxxxusa, Mat456, Jlmorgan, Hippypink, Michael C Price, Quibik, AndersFeder, Raoul NK, PKT, Thijs!bot, Keraunos, Anupam, Headbomb, Marek69, Kathovo, James086, Hcobb, D.H, Logicat, Jomoal99, Northumbrian, Oreo Priest, JitendraS, -dennis-~enwiki, Widefox, Seaphoto, Orionus, QuiteUnusual, Readro, Hsstr8, Tlabshier, Tim Shuba, Yellowdesk, TuvicBot, JAnDbot, Asmeurer, Tigga, Jde123, Roman à clef, Zekemurdock, Mcorazao, Mozart998, Kborland, Bongwarrior, NeverWorker, Ronstew, Marcel Kosko, Jpod2, J mcandrews, Walter Wpg, Trugster, JMBryant, Vanished user ty12kl89jq10, CodeCat, Allstarecho, Brian Fenton, JaGa, GermanX, Alangarr, WLU, TimidGuy, Mr Shark, Pagw, Andre.holzner, Sigmundg, Ben Mac-Dui, David Nicoson, Anaxial, JTiago, CommonsDelinker, Leyo, Gah4, Fconaway, Oddz, Tgeairn, J.delanoy, Fatka, Pharaoh of the Wizards, Maurice Carbonaro, Stephanwehner, Foober, Aveh8, McSly, Memory palace, NewEnglandYankee, Policron, 83d40m, Usp, Lamp90, Austinian, Izno, SoCalSuperEagle, Robprain, Cuzkatzimhut, Deor, Schucker, VolkovBot, Off-shell, ABF, Eliga~enwiki, JohnBlackburne, AlnoktaBOT, Tburket, Davidwr, Philip Trueman, Spemble, TXiKiBoT, Quatschman, The Original Wildbear, Gwib, Fatram, Kipb9, Andrius.v, Matan568, Nxavar, Nafhan, Photonh2o, Impunv, Peterbullockismyname, Cerebellum, Martin451, Praveen pillay, LoverOfArt, Abdullais4u, Justinrossetti, Cgwaldman, Bcody80, BotKung, Tennisnutt92, Dirkbb, Antixt, Francis Flinch, Moose-32, Ptrslv72, TheBendster, Masterofpsi, Jonbutterworth, Adrideba, SieBot, StAnselm, Manyugarg, PlanetStar, Jor63, Meldor, OlliffeObscurity, Jdcanfield, Yintan, Abhishikt, Flyer22, Graycrow, Infestor, Hrishirise, Cablehorn, Arthur Smart, Aperseghin, Mattmeskill, Gobbledygeek, Cthomas3, Steven Crossin, Nskillen, Sunrise, Afernand74, Jimtpat, Iknowyourider, StaticGull, Jfromcanada, MvL1234, Sphilbrick, Nergaal, Denisarona, Escape Orbit, Quinling, Martarius, PhysicsGrad2013, ClueBot, Victor Chmara, The Thing That Should Not Be, TomRed, Alyjack, Infrasonik, Mx3, Master1228, Drmies, Frmorrison, Loves martyr, Polyamorph, Nobaddude, Sjdunn9, Kitsunegami, Ktr101, Excirial, Joeyfjj, Wmlschlotterer, Pawan ctn, Lartoven, Artur80, Sun Creator, BobertWABC, PeterTheWall, Nondisclosure, M.O.X, SchreiberBike, JasonAQuest, Another Believer, Scf1984, 1ForTheMoney, Anoopan, Wnt, Darkicebot, TimothyRias, XLinkBot, Rreagan007, Resonance cascade, JinJian, Jabberwoch, Hess88, Hybirdd, Tayste, Addbot, Proofreader77, Mortense, Jacopo Werther, DBGustavson, DOI bot, Betterusername, Ocdnctx, OttRider, Cgd8d, Leszek Jańczuk, WikiUserPedia, NjardarBot, Download, LaaknorBot, AndersBot, Favonian, AgadaUrbanit, HandThatFeeds, Tide rolls, Lightbot, ScAvenger, SPat, Zorrobot, Jarble, ScienceApe, Dnamanish, Luckas-bot, Yobot, Chreod, EchetusXe, Nsbinsnj, Evans1982, Amble, Now dance, fu.cker, dance!, Anypodetos, Nallimbot, Trinitrix, SkepticalPoet, Pulickkal, Fernandosmission, Apollo reactor, Csmallw, AnomieBOT, Novemberrain94, 1exec1, Jim1138, JackieBot, Gc9580, LlywelynII, Materialscientist, Citation bot, Brightgalrs, Onesius, ArthurBot, Northryde, LilHelpa, Xqbot, Konor org, Noonehasthisnameithink, Engineering Guy, Yutenite, Newzebras, Universalsuffrage, DeadlyMETAL, Tomdo08, Professor J Lawrence, Br77rino, Srich32977, Arni.leibovits, StevenVerstoep, ProtectionTaggingBot, Vdkdaan, Omnipaedista, RibotBOT, Kyng, Waleswatcher, WissensDürster, Ace111, Kristjan.Jonasson, MerlLinkBot, Ernsts, Chaheel Riens, A. di M., A.amitkumar, Markdavid2000, 蝦蝦, Dave3457, FrescoBot, Weyesr1, Paine Ellsworth, Kenneth Dawson, Cdw1952, CamB424, CamB4242, Steve Quinn, N4tur4le, Jc odcsmf, Cannolis, Dolyn, Citation bot 1, Openmouth, Gil987, OriumX, Biker Biker, Gautier lebon, Pinethicket, Edderso, Boson15, Jonesey95, Three887, CarsonsDad, Calmer Waters, Jusses2, RedBot, BiObserver, Aknochel, Meier99, Trappist the monk, Puzl bustr, Proffsl, Higgshunter, Mary at CERN, Periglas, Zanhe, Lotje, Callanecc, Comet Tuttle, Jdigitalbath, Vrenator, SeoMac, ErikvanB, Tbhotch, Minimac, Coolpranjal, Mean as custard, RjwilmsiBot, TjBot, Olegrog, 123Mike456Winston789, Weaselpit, Newty23125, Techhead7890, Tesseract2, Skamecrazy123, Northern Arrow, Mukogodo, J36miles, EmausBot, John of Reading, Wikitanvir-

Bot, Stryn, Dadaist6174, Nuujinn, Montgolfière, Fotoni, GoingBatty, RA0808, Bengt Nyman, Bt8257, Gimmetoo, KHamsun, Swfarnsworth, LHC Tommy, Slightsmile, NikiAnna, TeeTylerToe, Dekker451, Hhhippo, Evanh2008, JSquish, Kkm010, ZéroBot, John Cline, Liquidmetalrob, Fæ, Bollyjeff, Érico Júnior Wouters, StringTheory11, Stevengoldfarb, Sgerbic, Opkdx, Quondum, AndrewN, Tbushman, Makecat, Timetraveler3.14, Foonle77, Tolly4bolly, Wiggles007, Irenan, Nobleacuff, Brandmeister, Baseballrocks538, Chris81w, Inswoon, Maschen, Donner60, Ontyx, Angelo souti, ChuispastonBot, ChiZeroOne, Ninjalectual, Exsmokey, Herk1955, I hate whitespace, Rocketrod1960, Whoop whoop pull up, Ajuvr, Petrb, Grapple X, ClueBot NG, Perfectlight, Aaron Booth, Gareth Griffith-Jones, Siswick, MelbourneStar, Gilderien, PhysicsAboveAll, Manu.ajm, Muon, Parcly Taxel, O.Koslowski, Widr, Mohd. Toukir Hamid, Diyar se, Helpful Pixie Bot, Popcornduff, Aesir.le, Bibcode Bot, 2001:db8, Lowercase sigmabot, BG19bot, Scottaleger, Mcarmier, Jibu8, Loupatriz67, Dave4478, Frze, Ervin Goldfain, Reader505, Mark Arsten, Lovetrivedi, BarbaraMervin, Silvrous, Drcooljoe, Cadiomals, Joydeep, Altaïr, Piet De Pauw, Jeancey, Sovereign8, Visuall, Ownedroad9, Brainssturm, Jw2036, Writ Keeper, DPL bot, Nickni28, Philpill691, Lee.boston, Scientist999, Benjiboy187, Duxwing, Cengime, Skiret girdet njozet, GRighta, Downtownclaytonbrown, Diasjordan, Ghsetht, Marioedesouza, BattyBot, 1narendran, LORDCOTTINGHAM2, NO SOPA, Tchaliburton, Wijnburger, StarryGrandma, Mdann52, Dilaton, Magikal Samson, Samuelled, Dja1979, Georgegroom, BecurSansnow, EuroCarGT, MSUGRA, Rhlozier, Pscott558, Turullulla, Blueprinteditor, Misterharris~enwiki, AstroDoc, Bigbear213, Dexbot, Randomizer3, Daggerbot, DoctorLazarusLong, Caroline1981, Nitpicking polisher, SoledadKabocha, Gsmanu007, Windows.dll, Mogism, Prabal123koirala, Abitoby, Clidog, Rongended, Darryl from Mars, Cerabot~enwiki, MuonRay, CuriousMind01, TheTruth72, Capt. Mohan Kuruvilla, Gatheringstorm2, Jason7898, Mumbai999999, SkepticalKid, Cjean42, Nmrzuk, Lugia2453, Mafuee, Frosty, SFK2, Thegodparticlebook, Rijensky, Mishra866868, Rockstar999999999, Toddbeck911, Nilaykumar07, Thepalerider2012, WikiPhysTech, The Anonymouse, Ahmar Saeed, Pincrete, Apidium23, Prahas.wiki, Exenola, Pdotpwns, Epicgenius, Fireballninja, Greengreengreenred, ⁇⁇, Technogeek101, NicoPosner, Apurva Godghase, Durfyy, Soumya Mittal, American In Brazil, SaifAli13, Qwerkysteve, Spatiandas, Retroherb, Tango303, Hoppeduppeanut, Redplain, Shaelote, Quadrum, AntiguanAcademic, Simpsonojsim, Agyeyaankur, DavidLeighEllis, Ethanthevelociraptor, Qfang12, Comp.arch, Eletro1903, E8xE8, HeineBOB, Kahtar, Depthdiver, JAaron95, Mfb, Anrnusna, Stamptrader, Man of Steel 85, Cteirmn, AiraCobra, MyNameIsn'tElvis, Meganlock8, Sxxximf, Drsoumyadeepb, 22merlin, Ndidi Okonkwo Nwuneli, Monkbot, Dialga5555, Fred1810, Akro7, Pewpewpewpapapa, BradNorton1979, 21bhargav, Whistlemethis, Thinking Skeptically, Amk365, Gagnonlg, Knowledgebattle, L21234, TheNextMessiah, Naterealm224, Joey van Helsing, Adrian Lamplighter, Arnab santra, Gemadi, BATMAN1021, Isambard Kingdom, Mercedes321, DrKitts, KasparBot, JJMC89, GBjun3, TheRoamer64, Firstcause, Seventhorbitday, RobeDM, Wordfunk and Anonymous: 963

- **Elementary particle** *Source:* https://en.wikipedia.org/wiki/Elementary_particle?oldid=684282599 *Contributors:* CYD, Mav, Bryan Derksen, XJaM, Heron, Stevertigo, Patrick, Fbjon, Looxix~enwiki, Александър, Julesd, Glenn, AugPi, Mxn, Timwi, Reddi, Tpbradbury, Furrykef, Bevo, Donarreiskoffer, Robbot, Craig Stuntz, Nurg, Papadopc, Wikibot, Jimduck, Anthony, Ancheta Wis, Giftlite, DavidCary, Mikez, Haselhurst, Monedula, Xerxes314, Alison, Guanaco, Greydream, Anythingyouwant, Bodnotbod, Kate, Brianjd, Mormegil, Urvabara, Rich Farmbrough, Guanabot, Qutezuce, Hidaspal, Dmr2, Goplat, RJHall, RoyBoy, Robotje, Neonumbers, ליאור, Dirac1933, DV8 2XL, Azmaverick623, Blaxthos, Kay Dekker, Joriki, Simetrical, TomTheHand, Mpatel, Isnow, Ggonnell, Palica, Strait, Miserlou, Ligulem, Naraht, DannyWilde, Lmatt, Srleffler, Chobot, Cactus.man, Roboto de Ajvol, YurikBot, Hairy Dude, NTBot~enwiki, Ohwilleke, Albert Einsteins pipe, Stephenb, Chaos, Vibritannia, SCZenz, Edwardlalone, Larsobrien, Bota47, BraneJ, Dna-webmaster, Arthur Rubin, Oyvind, GrinBot~enwiki, SmackBot, Mrcoolbp, Bomac, GrGBL~enwiki, Chris the speller, MalafayaBot, George Rodney Maruri Game, Silly rabbit, Complexica, MovGP0, Fmalan, Scwlong, Amazins490, Cybercobra, EPM, Garry Denke, Drphilharmonic, Sadi Carnot, ArglebargleIV, Tktktk, NongBot~enwiki, WhiteHatLurker, Jonhall, Dekaels~enwiki, Jynus, Newone, Courcelles, Laplace's Demon, SchmittM, J Milburn, Fordmadoxfraud, Cydebot, Bvcrist, Kozuch, Thijs!bot, Lord Hawk, Headbomb, MichaelMaggs, Escarbot, Ssr, JAnDbot, Eurobas, Acroterion, VoABot II, Appraiser, BatteryIncluded, R'n'B, Sgreddin, MikeBaharmast, Lk69, Acalamari, DraakUSA, TomasBat, Joshua Issac, Kenneth M Burke, Ken g6, Idioma-bot, VolkovBot, SarahLawrence Scott, Nxavar, JhsBot, Abdullais4u, Lejarrag, Antixt, PGWG, SieBot, Timb66, Sonicology, PlanetStar, Bamkin, Dhatfield, Byrialbot, Svick, Perfectapproach, Thorncrag, Big55e, ClueBot, Jmorris84, Maxtitan, Alexbot, Dekisugi, Paradoxalterist, Saintlucifer2008, Cockshut12345, Rreagan007, RP459, Truthnlove, Addbot, Yakiv Gluck, Draco 2k, Mac Dreamstate, Funky Fantom, CarsracBot, HerculeBot, Legobot, Blah28948, Luckas-bot, Zhitelew, KamikazeBot, Kulmalukko, Orion11M87, AnomieBOT, Girl Scout cookie, Templatehater, Icalanise, Citation bot, Onesius, Vuerqex, Bci2, ArthurBot, Rightly, Xqbot, Phazvmk, Kirin13, FrescoBot, Delphinus1997, Steve Quinn, Robo37, SuperJew, HRoestBot, Sthyne, Hellknowz, Yahia.barie, Skyerise, Tobi - Tobsen, FoxBot, Physics therapist, Think!97, Bj norge, RjwilmsiBot, Beyond My Ken, EmausBot, John of Reading, Mnkyman, GoingBatty, Mthorndill, ZéroBot, Bollyjeff, StringTheory11, Markinvancouver, Quantumor, Maschen, RolteVolte, Negovori, NTox, I hate whitespace, ClueBot NG, CocuBot, Widr, Micah.yannatos1, Helpful Pixie Bot, Guzman.c, Bibcode Bot, BG19bot, Spaceawesome, Rainbot, Leaverward, Let'sBuildTheFuture, Eduardofeld, Sha-256, Dr.RobertTweed, ZX95, Joeinwiki, Mark viking, Cephas Atheos, Yo butt, Snakeboy666, Psyruby42, Haminoon, Sardeth42, TaiSakuma, LadyCailin, Morph dtlr, Delbert7, Karam adel, Liance, Isambard Kingdom, KasparBot, Are you freaking kidding me, Kurousagi and Anonymous: 187

- **Standard Model** *Source:* https://en.wikipedia.org/wiki/Standard_Model?oldid=684586063 *Contributors:* AxelBoldt, Derek Ross, CYD, Bryan Derksen, The Anome, Ed Poor, Andre Engels, Roadrunner, David spector, Isis~enwiki, Youandme, Ram-Man, Stevertigo, Edward, Patrick, Boud, Michael Hardy, SebastianHelm, Looxix~enwiki, Julesd, Glenn, AugPi, Mxn, Raven in Orbit, Reddi, Phr, Tpbradbury, Populus, Haoherb428, Phys, Floydian, Bevo, Pierre Boreal, AnonMoos, BenRG, Jeffq, Dmytro, Drxenocide, Robbot, Nurg, Securiger, Texture, Roscoe x, Fuelbottle, Mattflaschen, Tobias Bergemann, Alan Liefting, Ancheta Wis, Giftlite, Dbenbenn, Harp, Herbee, Monedula, LeYaYa, Xerxes314, Dratman, Alison, JeffBobFrank, Dmmaus, Pharotic, Brockert, Bodhitha, Andycjp, Sonjaaa, HorsePunchKid, APH, Icairns, AmarChandra, Gscshoyru, Kate, Arivero, FT2, Rama, Vsmith, David Schaich, Xezbeth, D-Notice, Dfan, Bender235, Pt, El C, Laurascudder, Shanes, Drhex, Fogger~enwiki, Brim, Rbj, Jeodesic, Jumbuck, Alansohn, Gary, ChristopherWillis, Guy Harris, Axl, Sligocki, Kocio, Stillnotelf, Alinor, Wtmitchell, Egg, TenOfAllTrades, H2g2bob, Killing Vector, Linas, Mindmatrix, Benbest, Dodiad, Mpatel, Faethon, TPickup, Faethon34, Palica, Dysepsion, Faethon36, Qwertyca, Drbogdan, Rjwilmsi, Zbxgscqf, Macumba, Strangethingintheland, Dstudent, R.e.b., Bubba73, Drrngrvy, Agasicles, FlaBot, Naraht, Agasides, DannyWilde, Dave1g, Itinerant1, Gparker, Jrtayloriv, Goudzovski, Chobot, Bgwhite, FrankTobia, YurikBot, Bambaiah, Ohwilleke, VoxMoose, Bhny, JabberWok, Bovineone, Krbabu, SCZenz, JulesH, Davemck, Lomn, E2mb0t~enwiki, Dna-webmaster, Jrf, Dv82matt, Tetracube, Hirak 99, Arthur Rubin, Netrapt, JLaTondre, Caco de vidro, RG2, GrinBot~enwiki, That Guy, From That Show!, Hal peridol, SmackBot, YellowMonkey, Tom Lougheed, Melchoir, Bazza 7, KocjoBot~enwiki, Jagged 85, Thunderboltz, Setanta747 (locked), Skizzik, Dauto, Chris the speller, Bluebot, TimBentley, Sirex98, Silly rabbit, Complexica, Metacomet, DHN-bot~enwiki, MovGP0, QFT, Kittybrewster, Addshore, Jmnbatista, Cybercobra, Jgwacker, BullRangifer, Soarhead77, Daniel.Cardenas, Yevgeny Kats, Byelf2007, TriTertButoxy, Craig Bolon, Ajnosek, Ekjon Lok, Bjankuloski06, Tarcieri, Waggers, JarahE, Michaelbusch, Lottamiata, Newone,

Twas Now, IanOfNorwich, Srain, Patrickwooldridge, J Milburn, Mosaffa, Gatortpk, Vessels42, Geremia, Van helsing, Harrigan, Phatom87, Cydebot, David edwards, Verdy p, Michael C Price, Xantharius, Crum375, JamesAM, Thijs!bot, Epbr123, Headbomb, Phy1729, Stannered, Tariqhada, Seaphoto, Orionus, Voyaging, Gnixon, Jbaranao, Jrw@pobox.com, Len Raymond, Narssarssuaq, Bakken, CattleGirl, Davidoaf, Vanished user ty12kl89jq10, Lvwarren, Taborgate, Leyo, HEL, J.delanoy, Hans Dunkelberg, Stephanwehner, Wbellido, Aoosten, Jackson-walters, The Transliterator, DadaNeem, Student7, Joshmt, WJBscribe, Jozwolf, Hexane2000, BernardZ, Awren, Sheliak, Physicist brazuca, Schucker, Goop Goop, Fences and windows, Dextrose, Mcewan, Swamy g, TXiKiBoT, Sharikkamur, Thrawn562, Voorlandt, Escalona, Setreset, PDFbot, Pleroma, UnitedStatesian, Piyush Sriva, Kacser, Billinghurst, Francis Flinch, Moose-32, Ptrslv72, David Barnard, SieBot, ShiftFn, Robdunst, Jim E. Black, SheepNotGoats, Gerakibot, Nozzer42, Mr swordfish, Wing gundam, Bamkin, Likebox, Arthur Smart, HungarianBarbarian, Commutator, KathrynLybarger, Iomesus, C0nanPayne, Crazz bug 5, ClueBot, Superwj5, Wwheaton, Garyzx, SuperHamster, Elsweyn, Maldmac, DragonBot, Djr32, Diagramma Della Verita, Nymf, Eeekster, Brews ohare, NuclearWarfare, PhySusie, Ordovico, Mastertek, DumZiBoT, BodhisattvaBot, Guarracino, Mitch Ames, Truthnlove, Stephen Poppitt, Tayste, Addbot, Deepmath, Eric Drexler, DWHalliday, Mjamja, Leszek Jańczuk, NjardarBot, Mwoldin, Bassbonerocks, Barak Sh, AgadaUrbanit, Lightbot, Smeagol 17, Abjiklam, Ve744, Luckas-bot, Yobot, Orion11M87, AnomieBOT, JackieBot, Icalanise, Citation bot, ArthurBot, Northryde, LilHelpa, Xqbot, Sionus, Professor J Lawrence, Tomwsulcer, Edsegal, GrouchoBot, Trongphu, QMarion II, Ernsts, A. di M., Bytbox, FrescoBot, Paine Ellsworth, Aliotra, Steve Quinn, Citation bot 1, Rameshngbot, MJ94, RedBot, MastiBot, Aknochel, Sijothankam, Puzl bustr, Beta Orionis, Physics therapist, Bj norge, Innotata, Jesse V., RjwilmsiBot, Mathewsyriac, Afteread, EmausBot, Bookalign, WikitanvirBot, Wilhelm-physiker, Bdijkstra, DerNeedle, Kenmint, Dbraize, Tanner Swett, HeptishHotik, بەهار مەهنەشین, Suslindisambiguator, Quondum, Webbeh, UniversumExNihilo, Vanished user fijw983kjaslkekfhj45, Maschen, RockMagnetist, Stormymountain, $Z\varepsilon\tau\alpha\,\zeta$, Whoop whoop pull up, Isocliff, ClueBot NG, Smtchahal, Snotbot, Tonypak, O.Koslowski, CharleyQuinton, Dsperlich, Theopolisme, ZakMarksbury, Helpful Pixie Bot, Bibcode Bot, BG19bot, Tirebiter78, AvocatoBot, Lukys~enwiki, Stapletongrey, Ownedroad9, Chip123456, ChrisGualtieri, Khazar2, Billyfesh399, Rhlozier, JYBot, Dexbot, Doom636, Rongended, Cerabot~enwiki, CuriousMind01, Cjean42, Jayanta mallick, Joeinwiki, Kowtje, JPaestpreornJeolhlna, Eyesnore, Euan Richard, Nigstomper, Particle physicist, Prokaryotes, Jernahthern, Ginsuloft, Dimension10, JNrgbKLM, Krabaey, 1codesterS, FelixRosch, Monkbot, Delbert7, BradNorton1979, Lathamboyle, Tetra quark, KasparBot, Buckbill10, S3rr8s and Anonymous: 357

- **Quantum** *Source:* https://en.wikipedia.org/wiki/Quantum?oldid=683220301 *Contributors:* The Anome, Stevertigo, Ahoerstemeier, Smack, RodC, Charles Matthews, Jitse Niesen, Topbanana, Robbot, Sverdrup, Academic Challenger, Pengo, Graeme Bartlett, Lethe, Alison, Bensaccount, Finn-Zoltan, Jaan513, Smartcowboy, Raylu, Tsemii, Andreas Kaufmann, Freakofnurture, EugeneZelenko, Masudr, Vsmith, Too Old, El C, Laurascudder, RoyBoy, Omoo, Truthflux, Marco Polo, John Vandenberg, Viriditas, Dungodung, Hujaza, Jag123, Sam Korn, Nsaa, Anthony Appleyard, Atlant, Andrewpmk, Ricky81682, BryanD, Spangineer, Wtmitchell, Dalillama, Deathphoenix, Ceyockey, JHolman, Yougotavirus, Ashmoo, Bilbo1507, Qwertyus, Dennis Estenson II, Platypus222, Scorpionman, FlaBot, Nihiltres, Jeff02, Whodunit, Fresheneesz, Wiki-WikiPhil, Salvatore Ingala, DaGizza, Roboto de Ajvol, YurikBot, Mhocker, 4C~enwiki, Qwertzy2, Wimt, NawlinWiki, DragonHawk, Vanished user 1029384756, Mikeblas, Hosterweis, Orthografer, Geoffrey.landis, Garion96, Mebden, RG2, SmackBot, Melchoir, Vald, Bluebot, Complexica, Bbq332, Jfsamper, Scwlong, Stevenmitchell, Gragox, LoveEncounterFlow, Sadi Carnot, Fjjf, Mental Blank, Chymicus, SashatoBot, JorisvS, Bjankuloski06en~enwiki, Andypandy.UK, TastyPoutine, Laplace's Demon, Tawkerbot2, Filelakeshoe, Bupper, Wafulz, Mariodivece, GavinMorley, Dragon's Blood, Bicala, Xxanthippe, Thijs!bot, Epbr123, Davidhorman, Nick Number, Oreo Priest, AntiVandalBot, Drakonicon, Fdmt, JAnDbot, Db099221, Burakburak, Bongwarrior, VoABot II, Kinston eagle, Soulbot, Zamb, Crunchy Numbers, MartinBot, AlexiusHoratius, Juventas, C. Trifle, Maurice Carbonaro, NerdyNSK, Acalamari, Openforbusiness, Stewartrfc, Ontarioboy, Pundit, Pdcook, Leebo, DavidBrahm, TXiKiBoT, Djkrajnik, Kilmer-san, Yungjui, Synthebot, Lova Falk, Falcon8765, Grinq, HiDrNick, SieBot, Dannyeder, BotMultichill, Pengyanan, JerrySteal, Tiptoety, Oxymoron83, OKBot, Ngexpert5, Francvs, Seanruiz, FlamingSilmaril, Gratedparmesan, ClueBot, Abhinav, Alexbot, KnowledgeBased, Quantumpundit, SilvonenBot, NHJG, Truthnlove, NCDane, Addbot, War sharks, Favonian, Tide rolls, Lightbot, Deasmumhain, Zorrobot, Luckas-bot, Yobot, Nutfortuna, Julia W, Amirobot, Vortico, AnomieBOT, DemocraticLuntz, Greenbreen, AdjustShift, Jalexsmith1991, Materialscientist, Citation bot, ArthurBot, Parkyere, Capricorn42, Programming gecko, GrouchoBot, ProtectionTaggingBot, LtBert44, Nacefe, Chjoaygame, FrescoBot, Citation bot 1, Guruspiritual, Jsjunkie, Robo Cop, Lightlowemon, FoxBot, ඒ෪ෙ, Vrenator, Darsie42, Raidon Kane, Orphan Wiki, Booknotes, Wikipelli, K.zaman1710, Loggin12354, Grapeguy7, H3llBot, Makecat, Wayne Slam, Flightx52, Harishng, Kurt hueston, DASHBotAV, ClueBot NG, Dr Miles Long, Joefromrandb, Gexmeansgecko, Calabe1992, Bibcode Bot, Iisthphir, J991, Ajmah 200, Achowat, Vishakh24, BrightStarSky, Vogone, Reatlas, Ginsuloft, Kingnyancatjack, Monkbot, Jrafner, Farhan babra, Llllll2009, QuantaRaj39, Shubhamlanje1, KasparBot and Anonymous: 200

- **Mass generation** *Source:* https://en.wikipedia.org/wiki/Mass_generation?oldid=676957128 *Contributors:* Rjwilmsi, Bhny, JorisvS, Niceguyedc, Ktr101, Tom.Reding, ClueBot NG, Accedie, Bibcode Bot, Itchmean, Evensteven, Prestigiouzman, Ambermae2002 and Anonymous: 10

- **Symmetry (physics)** *Source:* https://en.wikipedia.org/wiki/Symmetry_(physics)?oldid=684906899 *Contributors:* The Anome, Heron, Stevertigo, Patrick, Michael Hardy, Bloodshedder, Rorro, Giftlite, BenFrantzDale, Netoholic, NetBot, I9Q79oL78KiL0QTFHgyc, Physicistjedi, Danski14, Mattpickman, Reaverdrop, Oleg Alexandrov, Woohookitty, StradivariusTV, Commander Keane, Mpatel, BD2412, Eyu100, Mathbot, Nihiltres, Jrtayloriv, X42bn6, Bhny, Archelon, Paul D. Anderson, Sbyrnes321, SmackBot, Incnis Mrsi, Complexica, Mets501, Quodfui, JRSpriggs, Myasuda, AndrewHowse, Cydebot, Michael C Price, Christian75, Divey, YK Times, PhilKnight, Email4mobile, Homunq, CodeCat, Janus Shadowsong, YohanN7, Paradoctor, Fratrep, ClueBot, Ottre, Bob108, Brews ohare, SchreiberBike, Thamuzino, Rror, Bradv, Addbot, Fyrael, Stylus881, PV=nRT, Luckas-bot, Yobot, Hotbody, Manganite, Rubinbot, Point-set topologist, Locobot, A. di M., FrescoBot, HRoestBot, MastiBot, 8af4bf06611c, ZéroBot, Maschen, Shivsagardharam, Vkpd11, Muskid, Dexbot, Hemlisp, Augustus Leonhardus Cartesius, Rabbitflyer, KHEname, BioticPixels, Matli, Ryanexler and Anonymous: 40

- **Weak interaction** *Source:* https://en.wikipedia.org/wiki/Weak_interaction?oldid=679571094 *Contributors:* AxelBoldt, Chenyu, Sodium, Bryan Derksen, Tarquin, AstroNomer~enwiki, Andre Engels, XJaM, Heron, JohnOwens, Gdarin, Delirium, Andrewa, Andres, Emperorbma, Timwi, Fibonacci, Phys, Phil Boswell, Lowellian, Mayooranathan, Tobias Bergemann, Giftlite, Sj, Herbee, Xerxes314, Jcobb, Mckaysalisbury, Munkee, Toby Woodwark, Bbbl67, Icairns, AmarChandra, Lumidek, Jørgen Friis Bak, Discospinster, ArnoldReinhold, Roybb95~enwiki, Gianluigi, Joanjoc~enwiki, Shanes, AJP, AtomicDragon, Danski14, Alansohn, Arthena, Axl, SidneySM, Hwefhasvs, DV8 2XL, Nightstallion, Kazvor-pal, Linas, StradivariusTV, Benbest, Bbatsell, Palica, Tevatron~enwiki, Graham87, BD2412, Ketiltrout, Rjwilmsi, Strait, Erkcan, The wub, FlaBot, Naraht, Itinerant1, Srleffler, Chobot, Krishnavedala, YurikBot, Borgx, Bambaiah, Hairy Dude, Jimp, Sillybilly, Conscious, Epolk, Jab-berWok, Gaius Cornelius, Shaddack, SCZenz, Irishguy, Shimei, Willtron, RG2, Phr en, That Guy, From That Show!, Luk, SmackBot, DavidKernow, Tom Lougheed, WookieInHeat, Dauto, Chris the speller, Philosopher, Moshe Constantine Hassan Al-Silverburg, Complexica, DHN-bot~enwiki, Zirconscot, BIL, Wen D House, "alyosha", Maxwahrhaftig, Akriasas, Vina-iwbot~enwiki, Bdushaw, TTE, SashatoBot, Fontenello,

Herr apa, Condem, Tony Fox, MottyGlix, JRSpriggs, Heartofgoldfish, Calmargulis, Green caterpillar, Joelholdsworth, Cydebot, Michael C Price, Mtpaley, Thijs!bot, ChKa, Kichwa Tembo, Headbomb, Hcobb, Icep, Escarbot, AntiVandalBot, Jimeree, Steelpillow, JAnDbot, Magioladitis, Swpb, باسل, Wormcast, DAGwyn, Giggy, Khalid Mahmood, Gah4, Tarotcards, 2help, Lighted Match, DorganBot, Halmstad, Idiomabot, VolkovBot, Jcuadros, Hilarious Bookbinder, TXiKiBoT, Rei-bot, CaptinJohn, Awl, Shenanegins, BotKung, Wingedsubmariner, Antixt, Xxxlilbritxxx, Ptrslv72, Monty845, AlleborgoBot, SieBot, Paolo.dL, Skyentist, Ptr123, ClueBot, Bondchic007, SuperHamster, Erudecorp, Rotational, Jackey0105, Alexbot, Cenarium, Zomno, Zahnrad, He6kd, TimothyRias, InternetMeme, Timo Metzemakers, Stephen Poppitt, Addbot, Some jerk on the Internet, Markdman, ChenzwBot, Ehrenkater, Tide rolls, Luckas-bot, Yobot, Les boys, Kilom691, THEN WHO WAS PHONE?, Rifter0x0000, Duping Man, Dickdock, Magog the Ogre, AnomieBOT, Materialscientist, Citation bot, Quebec99, Kreigiron, Xqbot, Drilnoth, BurntSynapse, GrouchoBot, Omnipaedista, RibotBOT, Workanode, Jaz1305, Mnmngb, Dave3457, FrescoBot, Charles.walker, LucienBOT, Ionutzmovie, Grandiose, Pinethicket, Boulaur, Rameshngbot, RedBot, 23790AD, Tea with toast, Jauhienij, FoxBot, Earthandmoon, RjwilmsiBot, Itamarhason, Newty23125, EmausBot, WikitanvirBot, GA bot, GoingBatty, Splibubay, StringTheory11, Braswiki, Git2010, Wayne Slam, Jsayre64, Maschen, ChuispastonBot, ClueBot NG, VinculumMan, Physics is all gnomes, Fjpyanez, Mouse20080706, Helpful Pixie Bot, Geo7777, Bibcode Bot, Junaid2754, Bolatbek, Phbarnacle, Neutral current, Glevum, Idenshi, Marioedesouza, Dexbot, Spray787, Reatlas, CsDix, Jamesmcmahon0, Ihatedirac2k13, Kharkiv07, Jwratner1, YimmyYohnson, Monkbot, BalderdashVonDrivel, ASCarretero, Malerisch, Lachlan Newland, Tetra quark, KasparBot and Anonymous: 155

- **Higgs mechanism** *Source:* https://en.wikipedia.org/wiki/Higgs_mechanism?oldid=685001216 *Contributors:* CYD, Roadrunner, Ubiquity, Michael Hardy, Julesd, Palfrey, Charles Matthews, Doradus, Tpbradbury, Phys, Sbisolo, David Gerard, Ancheta Wis, Giftlite, Herbee, Edcolins, Lumidek, Ukexpat, Benzh~enwiki, Chris Howard, FT2, Mat cross, David Schaich, Saintswithin, Mal~enwiki, Bender235, Viriditas, BDD, Oleg Alexandrov, Kelly Martin, Linas, BoLingua, Duncan.france, Christopher Thomas, BD2412, Rjwilmsi, Koavf, Strait, R.e.b., Jehochman, FlaBot, Goudzovski, Markdroberts, Gareth E Kegg, Chobot, Algebraist, YurikBot, Bambaiah, Wester, Darsie, AVM, Bhny, Stephenb, Długosz, Dna-webmaster, Tetracube, Caco de vidro, Finell, Triple333, SmackBot, Maksim-e~enwiki, ZerodEgo, Chris the speller, Sbharris, Jmnbatista, Lambiam, JorisvS, Ckatz, Meco, Newone, Benabik, MarsRover, Myasuda, Xxanthippe, Michael C Price, Quibik, Ldussan, Difty, Thijs!bot, Epbr123, Headbomb, West Brom 4ever, Mattfiller, D.H, RogierBrussee, VoABot II, Bakken, Jpod2, RickyCayley, JohnWilliams, Hekerui, Rif Winfield, MartinBot, Haydarhan, Gillleke, Cuzkatzimhut, VolkovBot, Off-shell, LokiClock, TXiKiBoT, Calwiki, Moose-32, Ptrslv72, Coffee, Gerakibot, Likebox, JacquesPHI, Henry Delforn (old), Pac72, Mr. Stradivarius, LoserJoke, ClueBot, General Epitaph, Wwheaton, Drmies, Auntof6, Brews ohare, M.O.X, Crowsnest, XLinkBot, Scvblwxq, Addbot, Eric Drexler, SpBot, Bob K31416, Barak Sh, Tide rolls, Yoavd, Luckas-bot, Yobot, Ptbotgourou, Fraggle81, Galaxydraem, AnomieBOT, Ciphers, Rubinbot, ArthurBot, LilHelpa, Xqbot, TheAMmollusc, Capricorn42, DSisyphBot, RibotBOT, Waleswatcher, Benzen, FrescoBot, BenzolBot, XeBot, Citation bot 1, Benji1986, O.anatinus, RedBot, MastiBot, Aknochel, Beth Ann Lindstrom, Felix0411, Meier99, Mary at CERN, WildBot, EmausBot, WikitanvirBot, Japs 88, LHC Tommy, Slawekb, JSquish, Quondum, L Kensington, Cerlbar, Zueignung, BabbaQ, CBuiltother, PhysicsAboveAll, Giuseppe Vitiello, Jj1236, Parthdu, Curb Chain, Bibcode Bot, Tirebiter78, Ownedroad9, ChrisGualtieri, Dexbot, Abits52, Konbini, CuriousMind01, Ajsal.ea, Itchmean, Cjean42, Crigeos, Crbeals, Jwratner1, Atotalstranger, Jzampardi, KasparBot and Anonymous: 146

- **Branching fraction** *Source:* https://en.wikipedia.org/wiki/Branching_fraction?oldid=570243404 *Contributors:* Michael Hardy, Giftlite, Harp, Xerxes314, Jonathunder, Hooperbloob, Ifdef, RobPlatt, Physchim62, Limulus, Zoidberg~enwiki, Shreshth91, Chris the speller, Headbomb, Jotempe, TXiKiBoT, Addbot, Timeroot, Minivip, Double sharp, ZéroBot and Anonymous: 6

- **Excited state** *Source:* https://en.wikipedia.org/wiki/Excited_state?oldid=675596412 *Contributors:* XJaM, Andres, Smack, Giftlite, Bensaccount, Vogon, RetiredUser2, Icairns, Vsmith, Jag123, Passw0rd, Alansohn, Andrewwall, Andrew Gray, Rjwilmsi, Phantom784, Fresheneesz, Mushin, Conscious, Grafen, Andreaskem, Wknight94, Pb30, The Photon, Xaosflux, Armeria, Pieter Kuiper, Bduke, OrangeDog, Behaafarid, Voyajer, DMacks, Spiritia, Anlace, John, Buchanan-Hermit, Atomobot, Rozzychan~enwiki, ShelfSkewed, Thijs!bot, Hazmat2, Headbomb, RogueNinja, Alphachimpbot, JAnDbot, Loonymonkey, Dell9300, Uvainio, Phasechange, Warut, Bieusinh92, ARTE, Jorfer, Equazcion, Bovineboy2008, Nomaan8, Bartosik, TXiKiBoT, A4bot, Ronningt, Spinningspark, AlleborgoBot, YonaBot, Kropotkine 113, Behtis, Cm176, Djr32, Pifreak94, Boleyn, SkyLined, Prowikipedians, Addbot, Chzz, Frozenguild, Numbo3-bot, Zorrobot, Luckas-bot, Yobot, Jnivekk, Materialscientist, Citation bot, JohnnyB256, Acebulf, Djalover, Mnmngb, Rushbugled13, Armando-Martin, Enchirion, Hezimmerman, Tommy2010, Chemprofguy, ZéroBot, AManWithNoPlan, SBaker43, Wwwmrzkwww, ClueBot NG, Bibcode Bot, Snow Blizzard, Johnhgagon, Thepoopinator, Vinoodle, Mpj7 and Anonymous: 66

- **Scalar boson** *Source:* https://en.wikipedia.org/wiki/Scalar_boson?oldid=653639224 *Contributors:* AugPi, Phys, Giftlite, Linas, Mpatel, RussBot, Bhny, SmackBot, Sergio.ballestrero, QFT, WLevine, Magioladitis, Tdadamemd, Sigmundur, Antixt, Jim E. Black, JL-Bot, ClueBot, Naradawickramage, Addbot, Dr. Universe, Anypodetos, J04n, Erik9bot, Fortdj33, Tomville219, RedBot, GoingBatty, Carbosi, ZéroBot, RolteVolte, Darine Of Manor, Parcly Taxel, Beaumont877, Mogism and Anonymous: 15

- **Spin (physics)** *Source:* https://en.wikipedia.org/wiki/Spin_(physics)?oldid=684695863 *Contributors:* AxelBoldt, CYD, The Anome, Larry_Sanger, Andre Engels, XJaM, David spector, Stevertigo, Xavic69, Michael Hardy, Tim Starling, Dominus, Cyp, Stevenj, Glenn, AugPi, Rossami, Nikai, Andres, Med, Mxn, Charles Matthews, Timwi, Kbk, 4lex, Reina riemann, E23~enwiki, Phys, Wtrmute, Bevo, Elwoz, Robbot, Gandalf61, Blainster, DHN, Hadal, Papadopc, Jheise, Anthony, Diberri, Xanzzibar, Giftlite, Smjg, Lethe, Lupin, MathKnight, Xerxes314, Average Earth-man, AlistairMcMillan, Ato, Andycjp, Gzuckier, Beland, Karol Langner, Spiralhighway, Elroch, B.d.mills, Tsemii, Frau Holle, Mike Rosoft, Igorivanov~enwiki, FT2, MuDavid, Paul August, Pt, Susvolans, Army1987, Wood Thrush, SpeedyGonsales, Physicistjedi, Obradovic Goran, Neonumbers, Keenan Pepper, Count Iblis, Egg, Linas, Palica, Torquil~enwiki, Ashmoo, Grammarbot, Zoz, Rjwilmsi, Zbxgscqf, Drrngrvy, FlaBot, Mathbot, TheMidnighters, Itinerant1, Ewlyahoocom, Gurch, Fresheneesz, Srleffler, Kri, Chobot, DVdm, YurikBot, Bambaiah, Hairy Dude, JabberWok, Rsrikanth05, NawlinWiki, Buster79, Hwasungmars, Kkmurray, Werdna, Djdaedalus, Simen, Netrapt, Mpjohans, KSevcik, GrinBot~enwiki, Joshronsen, Bo Jacoby, Sbyrnes321, DVD R W, Shanesan, KasugaHuang, That Guy, From That Show!, SmackBot, Unyoyega, Bluebot, Complexica, DHN-bot~enwiki, Sergio.ballestrero, Vladis1av, QFT, Voyajer, Terryeo, Ryanluck, Radagast83, Jgrahamc, Michael-Billington, Richard001, DMacks, Daniel.Cardenas, Bidabadi~enwiki, Sadi Carnot, Bdushaw, Andrei Stroe, Tesseran, SashatoBot, Leo C Stein, Vanished user 9i39j3, UberCryxic, Jonas Ferry, Vgy7ujm, Loodog, Jaganath, Terry Bollinger, Wierdw123, Inquisitus, Beefyt, Jc37, Dreftymac, Newone, RokasT~enwiki, Jaksmata, Aepryus, JRSpriggs, Joostvandeputte~enwiki, CRGreathouse, David s graff, Ahmes, JasonHise, Eric Le Bigot, Bmk, Myasuda, Mct mht, FilipeS, Cydebot, A876, Thijs!bot, Barticus88, Headbomb, Brichcja, Davidhorman, Oreo Priest, Wide-fox, Orionus, Tlabshier, Accordionman, Astavats, JAnDbot, Em3ryguy, MER-C, Igodard, PhilKnight, .anacondabot, Sangak, Magioladitis, Swpb, LorenzoB, Monurkar~enwiki, TechnoFaye, Brilliand, R'n'B, CommonsDelinker, Victor Blacus, J.delanoy, Numbo3, Sackm, Maurice

Carbonaro, Klatkinson, Cmichael, Uberdude85, Craklyn, CardinalDan, VolkovBot, Error9312, JohnBlackburne, Bolzano~enwiki, TXiKiBoT, Hqb, Anonymous Dissident, Costela, BotKung, Kganjam, Petergans, Kbrose, SieBot, BotMultichill, The way, the truth, and the light, RadicalOne, Jasondet, Paolo.dL, R J Sutherland, Lightmouse, JackSchmidt, Martarius, ClueBot, JonnybrotherJr, Warbler271, Mild Bill Hiccup, David Trochos, Outerrealm, Sbian, Peachypoh, SchreiberBike, Ant59, Crowsnest, XLinkBot, Addbot, Mathieu Perrin, Narayansg, Imeriki al-Shimoni, Sriharsha.karnati, Numbo3-bot, Tide rolls, Lightbot, Luckas-bot, Yobot, Nallimbot, AnomieBOT, Lendtuffz, Citation bot, Nepahwin, ArthurBot, Obersachsebot, Xqbot, Sionus, WandringMinstrel, Francine Rogers, Pradameinhoff, Tom1936, Ernsts, A. di M., NoldorinElf, Daleang, Baz.77.243.99.32, LucienBOT, Paine Ellsworth, Tobby72, Freddy78, Craig Pemberton, C.Bluck, Jondn, Pokyrek, Citation bot 1, I dream of horses, Adlerbot, Casimir9999, Kallikanzarid, Jkforde, Michael9422, Miracle Pen, Sgravn, 8af4bf06611c, Garuh knight, EmausBot, Beatnik8983, GoingBatty, JustinTime55, Zhenyok 1, Atomicann, JSquish, ZéroBot, Harddk, Neh0000, Quondum, Jacksccsi, Maschen, Zueignung, Rasinj, RockMagnetist, Eg-T2g, ClueBot NG, Paolo328, Gilderien, Frietjes, Widr, PhiMAP, Helpful Pixie Bot, Bibcode Bot, BG19bot, PUECH P.-F., Mark Arsten, Yudem, F=q(E+v^B), Blaspie55, Halfb1t, Robertwilliams2011, Dexbot, Foreverascone, ScitDei, Mark viking, Pedantchemist, W. P. Uzer, Francois-Pier, Mathphysman, Aidan Clark, Brotter121, KasparBot and Anonymous: 230

- **Electric charge** *Source:* https://en.wikipedia.org/wiki/Electric_charge?oldid=685481752 *Contributors:* AxelBoldt, Mav, Andre Engels, Roadrunner, Peterlin~enwiki, Heron, JohnOwens, Michael Hardy, Ixfd64, Delirium, Looxix~enwiki, Ellywa, Mdebets, Glenn, Rossami, Nikai, Andres, Raven in Orbit, Reddi, Omegatron, Gakrivas, Lumos3, Rogper~enwiki, Gentgeen, Robbot, Fredrik, Dukeofomnium, Wikibot, Fuelbottle, Wjbeaty, Giftlite, DavidCary, Herbee, Snowdog, Dratman, Valen~enwiki, RScheiber, Jason Quinn, Brockert, OldakQuill, Manuel Anastácio, LiDaobing, Karol Langner, Icairns, Iantresman, GNU, Vincom2, Discospinster, Guanabot, Jpk, Dbachmann, ZeroOne, Laurascudder, Bobo192, Rbj, Giraffedata, Kjkolb, Scentoni, Mdd, Alansohn, Atlant, ABCD, Velella, Wtshymanski, HenkvD, Mikeo, DV8 2XL, Gene Nygaard, HenryLi, Oleg Alexandrov, Nuno Tavares, Cimex, Rocastelo, StradivariusTV, Oliphaunt, BillC, Eleassar777, Cyberman, Palica, BD2412, Demonuk, Edison, SMC, Krash, Dougluce, FlaBot, Psyphen, Nivix, Alfred Centauri, Gurch, Kri, Gdrbot, Manscher, YurikBot, Bambaiah, Lucinos~enwiki, Stephenb, Manop, Pseudomonas, JDoorjam, TDogg310, Chichui, Kkmurray, Wknight94, Light current, Enormousdude, Johndburger, Tcsetattr, Pinikas, Reyk, Canley, Geoffrey.landis, JDspeeder1, GrinBot~enwiki, Mejor Los Indios, Sbyrnes321, Marquez~enwiki, Moeron, Vald, Thunderboltz, Dmitry sychov, HalfShadow, Gilliam, Oscarthecat, Andy M. Wang, Chris the speller, Lenko, DHN-bot~enwiki, Dual Freq, Hallenrm, Rrburke, The tooth, MichaelBillington, Hgilbert, Drphilharmonic, Daniel.Cardenas, Springnuts, Yevgeny Kats, Andrei Stroe, DJIndica, Naui~enwiki, Nmnogueira, SashatoBot, Richard L. Peterson, Slowmover, Cronholm144, Mgiganteus1, Bjankuloski06en~enwiki, Nonsuch, Ben Moore, RandomCritic, MarkSutton, Stikonas, Dicklyon, Levineps, Igoldste, Tawkerbot2, Chetvorno, JForget, CmdrObot, Kehrli, Jsd, Myasuda, Cydebot, Fl, Bvcrist, Meno25, Gogo Dodo, WISo, Christian75, Ssilvers, Thijs!bot, Epbr123, Barticus88, N5iln, Mojo Hand, Headbomb, Gerry Ashton, Escarbot, Aadal, AntiVandalBot, Seaphoto, Prolog, DarkAudit, Lyricmac, Tim Shuba, WikifingHelper, Asgrrr, JAnDbot, Acroterion, Bongwarrior, VoABot II, J2thawiki, Sstolper, Jjurik, Bubba hotep, User A1, DerHexer, InvertRect, Robin S, MartinBot, M. Bilal Shafiq, LedgendGamer, Pharaoh of the Wizards, Numbo3, Hans Dunkelberg, NightFalcon90909, Uncle Dick, Ginsengbomb, Katalaveno, DarkFalls, NewEnglandYankee, QuickClown, Juliancolton, ACBest, Treisijs, Lseixas, Jefferson Anderson, Sheliak, Philip Trueman, TXiKiBoT, The Original Wildbear, Ayan2289, Nickipedia 008, LuizBalloti, Monty845, Jpalpant, Biscuittin, Demmy100, SieBot, Gerakibot, Caltas, Gastin, Wing gundam, Msadaghd, JerrySteal, Jojalozzo, Oxymoron83, Faradayplank, Avnjay, Anchor Link Bot, Neo., Loren.wilton, ClueBot, The Thing That Should Not Be, Arakunem, Termine, Mild Bill Hiccup, Stephaninator, LeoFrank, Excirial, Kocher2006, Jusdafax, Brews ohare, Cenarium, Jotterbot, PhySusie, SchreiberBike, Wuzur, JDPhD, Versus22, Thinking Stone, Rror, Cernms, Truthnlove, Addbot, Some jerk on the Internet, CanadianLinuxUser, NjardarBot, LaaknorBot, Scottyferguson, LinkFA-Bot, Naidevinci, Ocwaldron, Tide rolls, Lightbot, JDSperling, Legobot, Luckas-bot, Yobot, CinchBug, Duping Man, AnomieBOT, DemocraticLuntz, Sertion, Jim1138, IRP, Pyrrhus16, Kingpin13, Bluerasberry, Materialscientist, Geek1337~enwiki, ImperatorExercitus, Xqbot, TheAMmollusc, Phazvmk, Addihockey10, Capricorn42, Nnivi, ProtectionTaggingBot, RibotBOT, Srr712, A. di M., Constructive editor, Frozenevolution, Ryryrules100, Jc3s5h, Drunauthorized, Mithrandir, Steve Quinn, Davidteng, Fast kartwheels, BenzolBot, DivineAlpha, AstaBOTh15, Pinethicket, I dream of horses, Jivee Blau, Calmer Waters, Tinton5, MastiBot, Serols, Meaghan, Lalrang2007, Logical Gentleman, FoxBot, TobeBot, SchreyP, Jonkerz, Ndkartik, Vrenator, Taytaylisious09, Ammodramus, Jamietw, DARTH SIDIOUS 2, Eshmate, Irfanyousufdar, EmausBot, John of Reading, GoingBatty, K6ka, Darkfight, Hhhippo, JSquish, Harddk, Stephen C Wells, Liam McM, Sonygal, L Kensington, Donner60, Peter Karlsen, Sven Manguard, Planetscared, ClueBot NG, Jack Greenmaven, Cking1414, Ihwood, Ulflund, CocuBot, MelbourneStar, O.Koslowski, Brickmack, AvocatoBot, Ushakaron, Rm1271, Altaïr, F=q(E+v^B), Snow Blizzard, Brad7777, Bhaskarandpm, Eduardofeld, GoShow, Dexbot, JoshyyP, Brandonsmacgregor, Reeceyboii, Frosty, Reatlas, I am One of Many, Eyesnore, Tentinator, Germeten, Nablacdy, Spyglasses, Freddyboi69, 20M030810, SpecialPiggy, Marizperoj, Peterfreed, Rigid hexagon, Jiteshkumar727464, Dyeith, Podayeruma, Oleaster, Layfi, BlueDecker, GeneralizationsAreBad, Pritam kumar Barik, KasparBot, Ramprakashsfc and Anonymous: 428

- **Color charge** *Source:* https://en.wikipedia.org/wiki/Color_charge?oldid=643420544 *Contributors:* CYD, Daniel Mahu, Looxix~enwiki, Theresaknott, Glenn, AugPi, Sethmahoney, Coren, The Anomebot, DJ Clayworth, Phys, Robbot, Wereon, M-Falcon, Herbee, Xerxes314, Dratman, Alison, JeffBobFrank, Jason Quinn, Eequor, RetiredUser2, 4pq1injbok, Lovelac7, MuDavid, Bender235, Jlcooke, Gauge, Ben Webber, El C, Euyyn, Wiki-Ed, Matt McIrvin, Fwb22, Njaard, Keenan Pepper, DrGaellon, Velho, Linas, Cruccone, Southwest, BD2412, Crazynas, Jeff02, Frank Tobia, Bambaiah, Gaius Cornelius, E2mb0t~enwiki, Finell, KasugaHuang, Felisse, SmackBot, Incnis Mrsi, Bluebot, Ahnood, UNV, Al-phathon, VMS Mosaic, Daniel bg, Dcamp314, Michael Bednarek, Keisetsu, CRGreathouse, Joelholdsworth, ChrisKennedy, TicketMan, Ker-aunos, Headbomb, West Brom 4ever, Cultural Freedom, JAnDbot, Yill577, Shim'on, Melamed katz, KylieTastic, Liometopum, Aaron Roten-berg, Sie Bot, WereSpielChequers, RadicalOne, BartekChom, Beofluff, Muhends, Dstebbins, ClueBot, Drmies, Manishearth, Andrewjmcneil, RexxS, Truthnlove, Addbot, Download, CarsracBot, AgadaUrbanit, Mrmister99823, Luckas-bot, Yobot, Amirobot, Götz, Jim1138, Citationbot, Arthur Bot, Obersachsebot, Xqbot, BenzolBot, TobeBot, Dinamik-bot, Scgtrp, Serketan, Hhhippo, AvicBot, JSquish, Maschen, Kate460, Helpful Pixie Bot, Super Duper Fixer Upper, BattyBot, Qashqaiilove, Cjean42, Graphium, Haezlgrace, Samproctor125 and Anonymous: 61

- **Doublet state** *Source:* https://en.wikipedia.org/wiki/Doublet_state?oldid=557497461 *Contributors:* Lupin, Acjelen, Anthony Appleyard, Linas, JarlaxleArtemis, Rjwilmsi, MarSch, Srleffler, Bambaiah, That Guy, From That Show!, SmackBot, The way, the truth, and the light, Addbot, Luckas-bot, Erik9bot, LucienBOT, Steve Quinn, Drift chambers, Fylbecatulous and Anonymous: 4

- **Weak isospin** *Source:* https://en.wikipedia.org/wiki/Weak_isospin?oldid=679239808 *Contributors:* Xavic69, Charles Matthews, Giftlite, RScheiber, Hidaspal, Ian Pitchford, Chobot, Roboto de Ajvol, YurikBot, Bambaiah, Jimp, RussBot, Paul D. Anderson, Jheriko, Bbabba, Cydebot, Michael C Price, Headbomb, Igodard, Tokei-so, Andre.holzner, Pamputt, AlleborgoBot, Muhends, L.smithfield, Tvine, Addbot, Icalanise, ArthurBot, Ernsts, Puzl bustr, John of Reading and Anonymous: 17

- **Tachyonic field** *Source:* https://en.wikipedia.org/wiki/Tachyonic_field?oldid=666604614 *Contributors:* Stevertigo, Giftlite, FT2, Sumanch, Długosz, Ashley thomas80, BernardZ, JohnBlackburne, Lamro, Tide rolls, AnomieBOT, Waleswatcher, FrescoBot, Tom.Reding, RjwilmsiBot, Newty23125, ClueBot NG, Bibcode Bot, Utvik, Yamaha5, Astra731, Monkbot and Anonymous: 5

- **Tachyon condensation** *Source:* https://en.wikipedia.org/wiki/Tachyon_condensation?oldid=666465490 *Contributors:* Julesd, Bogdangiusca, Jengod, Charles Matthews, BenRG, Giftlite, Xerxes314, Anville, Lumidek, FT2, Lycurgus, Jag123, Joke137, Marudubshinki, Rjwilmsi, Leeyc0, Roboto de Ajvol, Salsb, Trovatore, CecilWard, SamuelRiv, FyzixFighter, SmackBot, Timeshifter, Grokmoo, QFT, Doczilla, Lavateraguy, Michael C Price, Pstanton, Jpod2, MarkJefferys, Rominandreu, TXiKiBoT, Antixt, Addbot, DOI bot, MagnusA.Bot, AnomieBOT, Waleswatcher, Citation bot 1, Merongb10, Gil987, Ruthhaworth, Whoop whoop pull up, Bibcode Bot, Zgstehdyp and Anonymous: 22

- **Gauge boson** *Source:* https://en.wikipedia.org/wiki/Gauge_boson?oldid=682924361 *Contributors:* Bryan Derksen, Andre Engels, Michael Hardy, Ahoerstemeier, Bueller 007, LouI, Phys, Robbot, Gwrede, Rholton, Rursus, DavidI9999, Giftlite, Xerxes314, Alison, JeffBobFrank, Chinasaur, Andris, Garth 187, Beland, Setokaiba, Icairns, AmarChandra, Lumidek, Vsmith, Roybb95~enwiki, Mal~enwiki, La goutte de pluie, Nk, Kusma, Ringbang, Mpatel, Nakos2208~enwiki, Tevatron~enwiki, Kbdank71, Chobot, Roboto de Ajvol, Hairy Dude, Salsb, StuRat, ArielGold, RG2, InverseHypercube, Niels Olson, Sadi Carnot, TriTertButoxy, Ekjon Lok, Bjankuloski06en~enwiki, Phatom87, Headbomb, Tyco.skinner, Knotwork, Swpb, Maurice Carbonaro, Gombang, TXiKiBoT, Odellus, Antixt, AlleborgoBot, SieBot, Jim E. Black, Homonihilis, BOTarate, DumZiBoT, SilvonenBot, Addbot, Bertman600, NjardarBot, Numbo3-bot, Lightbot, Zorrobot, Luckas-bot, Yobot, Citation bot, ArthurBot, A. di M., Rameshngbot, RedBot, RobinK, Mary at CERN, TjBot, EmausBot, ZéroBot, StringTheory11, Mentibot, Dsperlich, CeraBot, Galactic Messiah, DerekWinters, Fisherv, KasparBot and Anonymous: 42

- **W and Z bosons** *Source:* https://en.wikipedia.org/wiki/W_and_Z_bosons?oldid=676803444 *Contributors:* AxelBoldt, Sodium, Mav, Bryan Derksen, The Anome, Ap, Andre Engels, Danny, Roadrunner, DrBob, Michael Hardy, Tim Starling, Karada, Egil, Ahoerstemeier, Ryan Cable, Julesd, Mxn, Charles Matthews, Ike9898, Saltine, Phys, Topbanana, BenRG, Finlay McWalter, Twang, Phil Boswell, Donarreiskoffer, Robbot, Pigsonthewing, Nurg, DHN, Xanzzibar, M-Falcon, Giftlite, Tremolo, Harp, Herbee, Xerxes314, Jeremy Henty, Bodhitha, LiDaobing, RetiredUser2, Icairns, Mike Rosoft, Vsmith, Gianluigi, Kjoonlee, Drhex, Obradovic Goran, Jérôme, Fkbreitl, Cameron.simpson, Gene Nygaard, Linas, LoopZilla, Graham87, Kbdank71, Rjwilmsi, Strait, Mike Peel, Lmatt, Goudzovski, Chobot, FrankTobia, Roboto de Ajvol, Ugha, Mushin, Bambaiah, Wester, Hairy Dude, Hellbus, Salsb, Seb35, Długosz, Turbolinux999, Ravedave, Scottfisher, Dna-webmaster, Modify, Argo Navis, Teply, Sbyrnes321, SmackBot, Tom Lougheed, Jagged 85, ZerodEgo, Dauto, Bluebot, Shaggorama, Sbharris, Niels Olson, Radagast83, Acdx, John, Lottamiata, Happy-melon, Tubezone, MightyWarrior, Joelholdsworth, Tangobot, Michael C Price, Quibik, Dchristle, Realjanuary, Headbomb, Davidhorman, Nosirrom, Certain, Gökhan, JAnDbot, Tigga, Omeganian, Brimofinsanity, TheEditrix2, Trapezoidal, Magioladitis, ThoHug, Leyo, Lilac Soul, HEL, Rod57, Y2H, HiEv, Adam Zivner, Madblueplanet, Sheliak, Dextrose, Anonymous Dissident, Synthebot, Antixt, Coronellian~enwiki, SieBot, STANMAR725, Jim E. Black, Gerakibot, Martin Kealey, CutOffTies, Fratrep, ClueBot, Mild Bill Hiccup, Alexbot, Carsrac, SkyLined, Dieppu, Stephen Poppitt, Addbot, Eric Drexler, Toyokuni3, Mjamja, Ronkonkaman, Download, CarsracBot, ChenzwBot, Lightbot, M sotirov, Luckas-bot, Yobot, Jim1138, MehrdadAfshari, ArthurBot, Ernsts, A. di M., Howard McCay, FrescoBot, Paine Ellsworth, D'ohBot, Citation bot 1, Gil987, Tom.Reding, Swallerick, FoxBot, Earthandmoon, Tm1729, TjBot, Антон Глiнiсты, Newty23125, EmausBot, Mnkyman, StringTheory11, Quondum, MisterDub, WaterCrane, Whoop whoop pull up, ClueBot NG, Helpful Pixie Bot, Bibcode Bot, BG19bot, Bakkedal, JYBot, Mamaphyskerin, Anrnusna, MartinNicklin, Boidal-Quantized and Anonymous: 137

- **Alternatives to the Standard Model Higgs** *Source:* https://en.wikipedia.org/wiki/Alternatives_to_the_Standard_Model_Higgs?oldid=681205 883 *Contributors:* Dante Alighieri, Charles Matthews, Twang, K igor k, Mike Rosoft, Δ, FT2, Pjacobi, David Schaich, Dbachmann, Ben-der235, Physicistjedi, Kocio, Hdeasy, Count Iblis, Siafu, Linas, LoopZilla, Cscott, Aodhd, Rjwilmsi, Tone, Hillman, Conscious, Cybrspunk, Dogcow, SmackBot, Baad, Njerseyguy, Jmnbatista, JorisvS, Vagary, Will314159, Ruslik0, Rotiro, Phatom87, Raoul NK, Andyjsmith, Head-bomb, Second Quantization, DagosNavy, Mollwollfumble, Omar.zanusso, VolkovBot, Ptrslv72, TimothyRias, Dthomsen8, WikHead, Ad-dbot, Mjamja, Yobot, Wireader, Materialscientist, Citation bot, Mcoupal, Omnipaedista, Senouf, A. di M., N4tur4le, Citation bot 1, FranBarnard, RjwilmsiBot, Chricho, Partouf, Wisconsinbadger, Helpful Pixie Bot, Bibcode Bot, Orentago, Ervin Goldfain, Higgs4all, Brainssturm, Lee.boston, Policarpo Y. Ulianov, Impsswoon, Jackson Reality and Anonymous: 33

- **Fermion** *Source:* https://en.wikipedia.org/wiki/Fermion?oldid=683048587 *Contributors:* AxelBoldt, Chenyu, Derek Ross, CYD, Mav, Bryan Derksen, The Anome, Ben-Zin~enwiki, Alan Peakall, Dominus, Dcljr, Looxix~enwiki, Glenn, Nikai, Andres, Wikiborg, David Latapie, Phys, Bevo, Stormie, Olathe, Donarreiskoffer, Robbot, Merovingian, Rorro, Wikibot, HaeB, Giftlite, Fropuff, Xerxes314, Vivektewary, JoJan, Karol Langner, Tothebarricades.tk, Icairns, Hidaspal, Vsmith, Laurascudder, Lysdexia, Ashlux, Graham87, Magister Mathematicae, Kbdank71, Syndicate, Strait, Protez, Drrngrvy, FlaBot, Srleffler, Chobot, YurikBot, RobotE, Jimp, Bhny, Captaindan, SpuriousQ, Salsb, Lomn, Enormousdude, CharlesHBennett, Federalist51, Tom Lougheed, Unyoyega, Jrockley, MK8, BabuBhatt, Complexica, Zachorious, Shalom Yechiel, QFT, Garry Denke, Daniel.Cardenas, SashatoBot, Flipperinu, Dan Gluck, LearningKnight, Happy-melon, Paulfriedman7, Cydebot, Meno25, Zalgo, Thijs!bot, Mbell, Headbomb, Nick Number, Orionus, Shlomi Hillel, CosineKitty, NE2, Mwarren us, ZPM, Vanished user ty12k189jq10, Joshua Davis, R'n'B, Tensegrity, Rod57, Dgiraffes, Alpvax, VolkovBot, TXiKiBoT, Red Act, Anonymous Dissident, Abdullais4u, בל יכול, Tanhueiming, Antixt, Haiviet~enwiki, EmxBot, Kbrose, SieBot, Likebox, Jojalozzo, Dhatfield, Oxymoron83, TubularWorld, ClueBot, Seervoitek, Rodhullandemu, Jorisverbiest, Feebas factor, ChandlerMapBot, Nilradical, Wikeepedian, Stephen Poppitt, Addbot, Vectorboson, Luckas-bot, Yobot, Planlips, Dickdock, AnomieBOT, Icalanise, Materialscientist, Xqbot, Br77rino, Balaonair, ⁇, Paine Ellsworth, Blackoutjack, Kikeku, Rameshngbot, Tom.Reding, RedBot, Alarichus, Michael9422, Silicon-28, TjBot, EmausBot, WikitanvirBot, Quazar121, Solomonfromfinland, JSquish, Fimin, Quondum, AManWithNoPlan, EdoBot, ClueBot NG, PBot1, EthanChant, Bibcode Bot, BG19bot, Petermahlzahn, KingKhan85, ChrisGualtieri, BoethiusUK, DerekWinters, Tentinator, JNrgbKLM, Mohit rajpal, KasparBot, Jiswin1992 and Anonymous: 120

- **Yukawa interaction** *Source:* https://en.wikipedia.org/wiki/Yukawa_interaction?oldid=679514336 *Contributors:* Phys, Alan Liefting, Giftlite, Xerxes314, Lumidek, MuDavid, Bender235, Jag123, LutzL, Natalya, Linas, MarSch, Roboto de Ajvol, YurikBot, Ugha, Bambaiah, Welsh, RG2, SmackBot, Droll, Sbharris, Akriasas, Skittleys, Difty, ChKa, Headbomb, WVhybrid, Yonidebot, Sheliak, MystBot, SkyLined, Addbot, Luckas-bot, AnomieBOT, LilHelpa, Suslindisambiguator, Maschen, ChuispastonBot, Makecat-bot, Wicklet and Anonymous: 16

- **Boson** *Source:* https://en.wikipedia.org/wiki/Boson?oldid=684807698 *Contributors:* CYD, The Anome, Xaonon, Aldie, Enchanter, Roadrunner, Ben-Zin~enwiki, Lisiate, Michael Hardy, Tim Starling, Kroose, Looxix~enwiki, Andrewa, Glenn, Andres, Kaihsu, Samw, Panoramix,

Schneelocke, CAkira, Wikiborg, The Anomebot, Saltine, Phys, Drxenocide, Robbot, Altenmann, Bkalafut, Merovingian, Rorro, Hadal, Robinh, VanishedUser kfljdfjsg33k, Giftlite, Fropuff, Pharotic, Isidore, Alexf, Beland, Jossi, Icairns, Zfr, Cructacean, Ornil, Mormegil, DanielCD, Noisy, Discospinster, Rich Farmbrough, Guanabot, Hidaspal, Sunborn, Kbh3rd, Jensbn, Alxndr, La goutte de pluie, Anthony Appleyard, Jlandahl, Leoadec, H2g2bob, Rocastelo, Benbest, Mpatel, Nakos2208~enwiki, GregorB, SDC, Palica, Ashmoo, Graham87, Kbdank71, Zzedar, Drbogdan, Strait, Master Justin, Wragge, FlaBot, Srleffler, Chobot, YurikBot, Bhny, The1physicist, Salsb, NawlinWiki, Welsh, Pyg, Dna-webmaster, Enormousdude, NeilN, Finell, Hal peridol, SmackBot, Incnis Mrsi, Ashley thomas80, Melchoir, Gilliam, MK8, MalafayaBot, Complexica, Epastore, DHN-bot~enwiki, Scienz Guy, Sbharris, QFT, Voyajer, Grover cleveland, Philvarner, Bradenripple, SashatoBot, Lambiam, Turbothy, T-dot, MagnaMopus, Candamir, WhiteHatLurker, Dicklyon, Treyp, Focomoso, Dan Gluck, UltraHighVacuum, Iridescent, Mathninja, Buckyboy314, Ianji, Cydebot, Stebbins, W.F.Galway, VashiDonsk, Tenbergen, Ward3001, Abtvctkto61, Thijs!bot, Barticus88, Mbell, Frozenport, Headbomb, MichaelMaggs, Escarbot, Orionus, Shan23, Alomas, JAnDbot, Deflective, CosineKitty, Pkoppenb, TheEditrix2, Magioladitis, VoABot II, Inertiatic076, Vanished user ty12kl89jq10, CodeCat, MartinBot, STBot, R'n'B, Tarotcards, Uberdude85, RuneSylvester, The Wild Falcon, Asnr 6, TXiKiBoT, Hqb, Anonymous Dissident, Abdullais4u, Bertrem, Moutane, Dirkbb, Antixt, Jeraaldo, BriEnBest, SieBot, Jim E. Black, Gerakibot, RadicalOne, Flyer22, Radon210, Sunayanaa, Jojalozzo, Tpvibes, Nsajjansajja, Owhanow~enwiki, Mike2vil, Mgurgan, VanishedUser sdu9aya9fs787sads, Danthewhale, PipepBot, Rodhullandemu, ChandlerMapBot, Excirial, PixelBot, Nilradical, Cenarium, Wikeepedian, Ouchitburns, Addbot, Bwr6, Minami Kana, Aboctok, Numbo3-bot, OlEnglish, David0811, WikiDreamer Bot, Jack who built the house, Luckas-bot, Yobot, Ptbotgourou, Senator Palpatine, AnomieBOT, Jim1138, JackieBot, Kingpin13, Materialscientist, Xqbot, Gravitivistically, Daners, Tomwsulcer, GrouchoBot, MeDrewNotYou, ⁇⁇, Ace of Spades, Alarics, Paul Laroque, Rameshngbot, RedBot, FoxBot, DixonDBot, Michael9422, Weedwhacker128, Tbhotch, TjBot, Ripchip Bot, EmausBot, JSquish, Kkm010, HiW-Bot, ZéroBot, StringTheory11, Lagomen, Robhenry9, Tls60, RockMagnetist, ClueBot NG, Raghavankl, GioGziro95, HBook, Helpful Pixie Bot, Bibcode Bot, 2001:db8, AvocatoBot, Nickni28, Minsbot, Blogger 20, Protomaestro, Abitoby, Darryl from Mars, NoRwEgIaNbAcTeRiUm, Jason7898, Valluvan888, Ov.kulkarni, Crpandya, Enamex, Lugia2453, Graphium, Federicoaolivieri, 77Mike77, Rltb, 314Username, Dllaughingwang, Codeusirae, Sometree, DR ROBERT HALT, KasparBot, Jiswin1992 and Anonymous: 224

- **Gluon** *Source:* https://en.wikipedia.org/wiki/Gluon?oldid=684614468 *Contributors:* AxelBoldt, CYD, Bryan Derksen, Gdarin, TakuyaMurata, Card~enwiki, Looxix~enwiki, Ellywa, Ahoerstemeier, Med, Schneelocke, Phys, Phil Boswell, Donarreiskoffer, Fredrik, Merovingian, Hadal, Giftlite, Herbee, Xerxes314, Eequor, Darrien, Keith Edkins, RetiredUser2, Icairns, Mike Rosoft, AlexChurchill, HedgeHog, Kenny TM~~enwiki, David Schaich, Ioliver, Mashford, El C, Kwamikagami, Ardric47, Obradovic Goran, Alansohn, Guy Harris, Dachannien, Ricky81682, Batmanand, Velella, Kazvorpal, April Arcus, Forteblast, Mpatel, Palica, BD2412, Kbdank71, Rjwilmsi, Macumba, Strait, Mike Peel, Bubba73, Klortho, FlaBot, Srleffler, Chobot, YurikBot, Wavelength, Bambaiah, Hairy Dude, Jimp, JabberWok, Zelmerszoetrop, Salsb, SCZenz, Randolf Richardson, Ravedave, Danlaycock, Bota47, LeonardoRob0t, Anclation~enwiki, Physicsdavid, Erudy, GrinBot~enwiki, Kgf0, SmackBot, Melchoir, Cessator, Benjaminevans82, Abtal, MK8, Colonies Chris, Can't sleep, clown will eat me, Decltype, Qcdmaestro, Edconrad, Darkpoison99, FredrickS, Omsharan, Pegasusbot, Gregbard, ProfessorPaul, Thijs!bot, Headbomb, Rriegs, Oreo Priest, AntiVandalBot, Shambolic Entity, Deflective, Mujokan, Yill577, Happycool, Mother.earth, Martynas Patasius, WiiWillieWiki, HEL, Hans Dunkelberg, Gombang, Inwind, Sheliak, Jonthaler, VolkovBot, TXiKiBoT, Davehi1, Kriak, Anonymous Dissident, Imasleepviking, AlleborgoBot, EJF, SieBot, Steven Crossin, OKBot, ClueBot, Wwheaton, Qsaw, Nucularphysicist, Ottava Rima, Gordon Ecker, Rhododendrites, Brews ohare, Cacadril, RexxS, JKeck, Against the current, SkyLined, Addbot, DOI bot, Lightbot, Skippy le Grand Gourou, Luckas-bot, Planlips, AnomieBOT, Jim1138, JackieBot, Citation bot, Bci2, ArthurBot, Xqbot, Neil95, Triclops200, Omnipaedista, TorKr, ⁇⁇, Paine Ellsworth, Ivoras, Citation bot 1, Pekayer11, Rameshngbot, PNG, RjwilmsiBot, TjBot, Lilcal89012, EmausBot, Socob, JSquish, StringTheory11, Quondum, TyA, Maschen, RolteVolte, ClueBot NG, Timothy jordan, Maplelanefarm, Bibcode Bot, BG19bot, Gravitoweak, Cadiomals, Tropcho, Fraulein451, DrHjmHam, Rhlozier, D.shinkaruk, Yaara dildaara, BronzeRatio, Monkbot, Yikkayaya, KasparBot and Anonymous: 143

- **Tau (particle)** *Source:* https://en.wikipedia.org/wiki/Tau_(particle)?oldid=685074264 *Contributors:* Bryan Derksen, Iluvcapra, Ahoerstemeier, Bueller 007, Schneelocke, Dysprosia, Donarreiskoffer, Merovingian, Rorro, Davidl9999, Millosh, Harp, Herbee, Codepoet, Xerxes314, Bodhitha, CryptoDerk, Icairns, Rich Farmbrough, Pjacobi, Martpol, Sunborn, Kjoonlee, El C, Reuben, JellyWorld, RobPlatt, RJFJR, Falcorian, Dmitry Brant, Christopher Thomas, Palica, Rjwilmsi, Strait, Mike Peel, FlaBot, DannyWilde, Goudzovski, Chobot, RobotE, Bambaiah, AcidHelmNun, JabberWok, Eleassar, Salsb, SCZenz, Zwobot, Ospalh, PS2pcGAMER, Bota47, Someones life, Poulpy, Physicsdavid, Incnis Mrsi, Dauto, Pieter Kuiper, Loodog, JorisvS, MTSbot~enwiki, WISo, Q43, Thijs!bot, Headbomb, Davidhorman, Hcobb, Escarbot, RogueNinja, Yill577, Soulbot, Kostisl, STBotD, Sheliak, Joyko~enwiki, VolkovBot, Fences and windows, TXiKiBoT, Awl, Jba138, SieBot, OKBot, ImageRemovalBot, Plastikspork, Djr32, Alexbot, TimothyRias, Assosiation, BodhisattvaBot, SkyLined, J Hazard, Addbot, Eric Drexler, Ronhjones, ChenzwBot, Jklukas, Theozzfancometh, Skippy le Grand Gourou, Luckas-bot, Yobot, Grebaldar, AnomieBOT, Icalanise, Citation bot, Xqbot, Blennow, Franco3450, ⁇⁇, Paine Ellsworth, Jonesey95, Three887, Plasticspork, 3ph, Miracle Pen, RjwilmsiBot, TjBot, Ripchip Bot, EmausBot, Dcirovic, Suslindisambiguator, Quondum, Rezabot, Helpful Pixie Bot, Bibcode Bot, BG19bot, Sudsguest, YFdyh-bot, Redcliffe maven, TwoTwoHello, Akro7, Nøkkenbuer, KasparBot, JPPepper, QzPhysics and Anonymous: 48

- **Photon** *Source:* https://en.wikipedia.org/wiki/Photon?oldid=684257917 *Contributors:* AxelBoldt, WojPob, Mav, Bryan Derksen, The Anome, Tarquin, Koyaanis Qatsi, Ap, Josh Grosse, Ben-Zin~enwiki, Heron, Youandme, Spiff~enwiki, Bdesham, Michael Hardy, Ixfd64, TakuyaMurata, NuclearWinner, Looxix~enwiki, Snarfies, Ahoerstemeier, Stevenj, Julesd, Glenn, AugPi, Mxn, Smack, Pizza Puzzle, Wikiborg, Reddi, Lfh, Jitse Niesen, Kbk, Laussy, Bevo, Shizhao, Raul654, Jusjih, Donarreiskoffer, Robbot, Hankwang, Fredrik, Eman, Sanders muc, Altenmann, Bkalafut, Merovingian, Gnomon Kelemen, Hadal, Wereon, Anthony, Wjbeaty, Giftlite, Art Carlson, Herbee, Xerxes314, Everyking, Dratman, Michael Devore, Bensaccount, Foobar, Jaan513, DÅ,ugosz, Zeimusu, LucasVB, Beland, Setokaiba, Kaldari, Vina, RetiredUser2, Icairns, Lumidek, Zondor, Randwicked, Eep², Chris Howard, Zowie, Naryathegreat, Discospinster, Rich Farmbrough, Yuval madar, Pjacobi, Vsmith, Ivan Bajlo, Dbachmann, Mani1, SpookyMulder, Kbh3rd, RJHall, Ben Webber, El C, Edwinstearns, Laurascudder, RoyBoy, Spoon!, Dalf, Drhex, Bobo192, Foobaz, I9Q79oL78KiL0QTFHgyc, La goutte de pluie, Zr40, Apostrophe, Minghong, Rport, Alansohn, Gary, Sade, Corwin8, PAR, UnHoly, Hu, Caesura, Wtmitchell, Bucephalus, Max rspct, BanyanTree, Cal 1234, Count Iblis, Egg, Dominic, Gene Nygaard, Ghirlandajo, Kazvorpal, UTSRelativity, Falcorian, Drag09, Boothy443, Richard Arthur Norton (1958-), Woohookitty, Linas, Gerd Breitenbach, StradivariusTV, Oliphaunt, Cleonis, Pol098, Ruud Koot, Mpatel, Nakos2208~enwiki, Dbl2010, Ch'marr, SDC, CharlesC, Alan Canon, Reddwarf2956, Mandarax, BD2412, Kbdank71, Zalasur, Sjakkalle, Rjwilmsi, Саша Стефановић, Strait, MarSch, Dennis Estenson II, Trlovejoy, Mike Peel, HappyCamper, Bubba73, Brighterorange, Cantorman, Egopaint, Noon, Godzatswing, FlaBot, RobertG, Arnero, Mathbot, Nihiltres, Fresheneesz, TeaDrinker, Srleffler, BradBeattie, Chobot, Jaraalbe, DVdm, Elfguy, EamonnPKeane, YurikBot, Bambaiah, Splintercellguy, Jimp, RussBot, Supasheep, JabberWok, Wavesmikey, KevinCuddeback, Stephenb, Gaius Cornelius, Salsb, Trovatore,

Długosz, Tailpig, Joelr31, SCZenz, Randolf Richardson, Ravedave, Tony1, Roy Brumback, Gadget850, Dna-webmaster, Enormousdude, Lt-wiki-bot, Oysteinp, JoanneB, Ligart, John Broughton, GrinBot~enwiki, Sbyrnes321, Itub, SmackBot, Moeron, Incnis Mrsi, KnowledgeOfSelf, CelticJobber, Melchoir, Rokfaith, WilyD, Jagged 85, Jab843, Cessator, AnOddName, Skizzik, Dauto, JSpudeman, Robin Whittle, Ati3414, Persian Poet Gal, MK8, Jprg1966, Complexica, Sbharris, Colonies Chris, Ebertek, WordLife565, Vladislav, RWincek, Aces lead, Stangbat, Cybercobra, Valenciano, EVula, A.R., Mini-Geek, AEM, DMacks, N Shar, Sadi Carnot, FlyHigh, The Fwanksta, Drunken Pirate, Yevgeny Kats, Lambiam, Harryboyles, IronGargoyle, Ben Moore, A. Parrot, Mr Stephen, Fbartolom, Dicklyon, SandyGeorgia, Mets501, Ceeded, Am-buj.Saxena, Ryulong, Vincecate, Astrobayes, Newone, J Di, Lifeverywhere, Tawkerbot2, JRSpriggs, Chetvorno, Luis A. Veguilla-Berdecia, CalebNoble, Xod, Gregory9, CmdrObot, Wafulz, Van helsing, John Riemann Soong, Rwflammang, Banedon, Wquester, Outriggr (2006-2009), Logical2u, Myasuda, Howardsr, Cydebot, Krauss, Kanags, A876, WillowW, Bvcrist, Hyperdeath, Hkyriazi, Rracecarr, Difluoroethene, Edgerck, Michael C Price, Tawkerbot4, Christian75, Ldussan, RelHistBuff, Waxigloo, Kozuch, Thijs!bot, Epbr123, Opabinia regalis, Markus Pössel, Mglg, 24fan24, Headbomb, Newton2, John254, J.christianson, Escarbot, Stannered, AntiVandalBot, Luna Santin, Jtrain4469, Nor-manmargolus, Tyco.skinner, TimVickers, NSH001, Dodecahedron~enwiki, Tim Shuba, Gdo01, Sluzzelin, Abyssoft, CosineKitty, AndyBloch, Bryanv, ScottStearns, Hroðulf, Bongwarrior, VoABot II, B&W Anime Fan, SHCarter, Lgoger, I JethroBT, Dirac66, Hveziris, Maliz, Lord GaleVII, TRWBW, Shijualex, Glen, DerHexer, Patstuart, Gwern, Taborgate, MartinBot, MNAdam, Jay Litman, HEL, Ralf 58, J.delanoy, DrKiernan, Trusilver, C. Trifle, AstroHurricane001, Numbo3, Pursey, CMDadabo, Kevin aylward, UchihaFury, Pirate452, H4xx0r, Iamthe-walrus35, Iamthewalrus36, Gee Eff, Chimpy07, Dirkdiggler69, Lk69, Hallamfm, Annoying editter, Yehoodig, Acalamari, Foreigner1, McSly, Samtheboy, Tarotcards, Rominandreu, ARTE, Tanaats, Potatoswatter, Y2H, Divad89, Scott Illini, Stack27, THEblindwarrior, VolkovBot, AlnoktaBOT, Hyperlinker, DoorsAjar, TXiKiBoT, Oshwah, Cosmic Latte, The Original Wildbear, Davehi1, Chiefwaterfall, Vipinhari, Hqb, Anonymous Dissident, HansMair, Predator24, BotKung, Luuva, Calvin4986, Improve~enwiki, Kmhkmh, Richwil, Antixt, Gorank4, Fal-con8765, GlassFET, Cryptophile, MattiasAndersson, AlleborgoBot, Carlodn6, NHRHS2010, Relilles~enwiki, Tpb, SieBot, Timb66, Graham Beards, WereSpielChequers, ToePeu.bot, JerrySteal, Android Mouse, Likebox, RadicalOne, Paolo.dL, Lightmouse, PbBot, Spartan-James, Duae Quartunciae, Hamiltondaniel, StewartMH, Dstebbins, ClueBot, Bobathon71, The Thing That Should Not Be, Mwengler, EoGuy, Jagun, RODERICKMOLASAR, Wwheaton, Dmlcyal8er, Razimantv, Mild Bill Hiccup, Feebas factor, J8079s, Rotational, MaxwellsLight, Awick-ert, Excirial, PixelBot, Sun Creator, NuclearWarfare, PhySusie, El bot de la dieta, DerBorg, Shamanchill, PoofyPeter99, J1.grammar natz, Laserheinz, TimothyRias, XLinkBot, Jovianeye, Petedskier, Hess88, Addbot, Mathieu Perrin, DOI bot, DougsTech, Download, James thir-teen, AndersBot, LinkFA-Bot, Barak Sh, AgadaUrbanit, Тиверополник, Dayewalker, Quantumobserver, Kein Einstein, Legobot, Luckas-bot, Yobot, Kilom691, Allowgolf~enwiki, AnomieBOT, Ratul2000, Kingpin13, Materialscientist, Citation bot, Xqbot, Ambujarind69, Mananay, Emezei, Sharhalakis, Shirik, RibotBOT, Rickproser, SongRenKai, Max derner, Merrrr, A. di M., ⁇⁇, CES1596, Paine Ellsworth, Gsthae with tempo!, Nageh, TimonyCrickets, WurzelT, Steve Quinn, Spacekid99, Radeksonic, Citation bot 1, Pinethicket, I dream of horses, HRoest-Bot, Tanweer Morshed, Eno crux, Tom.Reding, Jschnur, RedBot, IVAN3MAN, Gamewizard71, FoxBot, TobeBot, Earthandmoon, Please-Stand, Marie Poise, RjwilmsiBot, Антон Гліністы, Ripchip Bot, Ofercomay, Chemyanda, EmausBot, Bookalign, WikitanvirBot, Roxbreak, Word2need, Gcastellanos, Tommy2010, Dcirovic, K6ka, Hhhippo, Cogiati, 1howadsr1, StringTheory11, Waperkins, Jojojlj, Access De-nied, Quondum, AManWithNoPlan, Raynor42, L Kensington, Maschen, HCPotter, Haiti333, RockMagnetist, Rocketrod1960, ClueBot NG, JASMEET SINGH HAFIST, Schicagos, Snotbot, Vinícius Machado Vogt, Helpful Pixie Bot, SzMithrandir, Bibcode Bot, BG19bot, Roberti-cus, Paolo Lipparini, Wzrd1, Rifath119, Davidiad, Mark Arsten, Peter.sujak, Wikarchitect, Hamish59, Caypartisbot, Penguinstorm300, KSI ROX, Bhargavuk1997, Chromastone1998, TheJJJunk, Nimmo1859, EagerToddler39, Dexbot, EZas3pt14, Webclient101, Chrisanion, Van-quisher.UA, Tony Mach, PREMDASKANNAN, Meghas, Reatlas, Profb39, Zerberos, Thesuperseo, The User 111, Eyesnore, Ybidzian, Tenti-nator, Illusterati, JustBerry, Celso ad, Quenhitran, Manul, DrMattV, Anrnusna, Wyn.junior, K0RTD, Monkbot, Vieque, BethNaught, Mark-mizzi, Garfield Garfield, Smokey2022, Zargol Rejerfree, RAL2014, Shahriar Kabir Pavel, Sdjncskdjnfskje, Anshul1908, Professor Flornoy, Thatguytestw, Tetra quark, Harshit100, KasparBot, Chinta 01, Geek3, TheKingOfPhysics and Anonymous: 499

- **Force carrier** *Source:* https://en.wikipedia.org/wiki/Force_carrier?oldid=651865824 *Contributors:* Camembert, Looxix~enwiki, Bevo, Al-ison, Urvabara, Mal~enwiki, Euyyn, Matt McIrvin, Keenan Pepper, TenOfAllTrades, Bubba73, Complexica, QFT, Sadi Carnot, Alan.ca, Nethac DIU, Myasuda, Headbomb, Trevyn, Alphachimpbot, Macboots, Maurice Carbonaro, VolkovBot, Antoni Barau, FDominec, SieBot, Jim E. Black, ClueBot, Anon126, WikiHead, Truthnlove, Addbot, Mac Dreamstate, Barak Sh, Luckas-bot, Yobot, Flewis, Bci2, ArthurBot, Carlog3, Armando-Martin, JSquish, NatNapoletano, ClueBot NG, Bibcode Bot, The Illusive Man, Monkbot, Tetra quark and Anonymous: 21

- **Spontaneous symmetry breaking** *Source:* https://en.wikipedia.org/wiki/Spontaneous_symmetry_breaking?oldid=678321175 *Contributors:* AxelBoldt, Bryan Derksen, XJaM, Edward, Michael Hardy, Lexor, Charles Matthews, Timwi, Reddi, Phys, Bevo, Dusik, Nagelfar, Giftlite, Lethe, Alison, JeffBobFrank, Jcobb, Gotanda, Gadfium, DragonflySixtyseven, Lumidek, FT2, Hidaspal, Ascánder, Mal~enwiki, Bender235, Clement Cherlin, PhilHibbs, MPS, Shenme, Physicistjedi, Kocio, StuTheSheep, Linas, Jmhodges, Dennis Estenson II, Salix alba, Jehochman, BjKa, Chobot, YurikBot, Ugha, Bambaiah, Archelon, Zzuuzz, Reyk, Roques, RupertMillard, SmackBot, Maksim-e~enwiki, Complexica, Colonies Chris, Jmnbatista, Lagrangian, Akriasas, P199, JarahE, Hetar, Dan Gluck, JMK, Harej bot, Ezrakilty, Thijs!bot, Barticus88, Head-bomb, Arcresu, Hillarryous, Dougher, Gökhan, JAnDbot, Yuksing, Attarparn, Jpod2, R'n'B, Natsirtguy, Lseixas, BernardZ, Cuzkatzimhut, Holme053, TXiKiBoT, Red Act, Michael H 34, Pamputt, Moose-32, SieBot, Wing gundam, Renatops, Denisarona, Mastertek, BlueDevil, MelonBot, Truthnlove, Addbot, Yakiv Gluck, Zahd, LaaknorBot, SpBot, OlEnglish, Luckas-bot, Yobot, Yotcmdr, Christopher Pritchard, Zim-boz Montizawooba, Obersachsebot, False vacuum, Waleswatcher, Gsard, A. di M., ⁇⁇, CES1596, Freddy78, Pmokeefe, RobinK, Mary at CERN, Marie Poise, Slightsmile, Quondum, Shovkovy, Maschen, Boris Breuer, Vatsal19, Helpful Pixie Bot, Bibcode Bot, Ahhaha, Kalmiop-siskid, Fraulein451, Dexbot, Lugia2453, CMTdrew, Mparisi90, LudicrousTripe and Anonymous: 85

- **1964 PRL symmetry breaking papers** *Source:* https://en.wikipedia.org/wiki/1964_PRL_symmetry_breaking_papers?oldid=680679590 *Con-tributors:* Nealmcb, AnonMoos, Auric, Giftlite, Kaldari, Rich Farmbrough, FT2, David Schaich, Count Iblis, GregorB, Rjwilmsi, Tim!, Teply, Jmnbatista, CmdrObot, Cydebot, Headbomb, Nick Number, SamIAmNot, Vanished user ty12kl89jq10, Guillaume2303, Duncan.Hull, Lamro, Falcon8765, Moose-32, Ptrslv72, Randy Kryn, NuclearWarfare, XLinkBot, Good Olfactory, OlEnglish, Yobot, Anypodetos, AnomieBOT, Citation bot, Omnipaedista, Fortdj33, Meier99, Mary at CERN, Magmalex, RjwilmsiBot, DASHBot, GoingBatty, KHamsun, LHC Tommy, Brendanpbehan, Helpful Pixie Bot, Bibcode Bot, EuroAgurbash, Dexbot, Hmainsbot1, SFK2, Stamptrader, Karthikprabhu22 and Anonymous: 25

- **Inflation (cosmology)** *Source:* https://en.wikipedia.org/wiki/Inflation_(cosmology)?oldid=684461201 *Contributors:* Bryan Derksen, The Anome ,Diatarn iv~enwiki, Roadrunner, David spector, Hephaestos, Stevertigo, Edward, Nealmcb, Boud, Michael Hardy, Tim Starling, Dcljr, Cyde,

Ellywa, William M. Connolley, Theresa knott, Jeff Relf, Mxn, Timwi, Rednblu, Bartosz, Pierre Boreal, Raul654, Chuunen Baka, Robbot, Gandalf61, Rursus, Ancheta Wis, Giftlite, Barbara Shack, Mikez, Lethe, Dratman, Curps, Jcobb, Just Another Dan, Andycjp, HorsePunchKid, Beland, Elroch, JDoolin, Burschik, Shadypalm88, Eep², Mike Rosoft, DanielCD, Noisy, Rich Farmbrough, FT2, Pjacobi, Luxdormiens, Dbachmann, Bender235, AdamSolomon, Pt, Worldtraveller, Art LaPella, Orlady, Drhex, Guettarda, I9Q79oL78KiL0QTFHgyc, Jeodesic, Rsholmes, Anthony Appleyard, Plumbago, JHG, Schaefer, EmmetCaulfield, Cgmusselman, Dirac1933, Oleg Alexandrov, Matevzk, Yeastbeast, StradivariusTV, BillC, Bluemoose, Wdanwatts, Joke137, Rnt20, Malangthon, Ketiltrout, Drbogdan, Rjwilmsi, Zbxgscqf, Mattmartin, Strait, Eyu100, Jehochman, Ems57fcva, Bubba73, FlaBot, Nihiltres, Itinerant1, Phoenix2~enwiki, Chobot, Hermitage, Bgwhite, YurikBot, Wavelength, Supasheep, Ytrottier, Gaius Cornelius, Anomalocaris, NawlinWiki, LiamE, Davemck, JonathanD, Enormousdude, 2over0, Arthur Rubin, Argo Navis, Physicsdavid, Profero, Luk, SmackBot, Haza-w, KnowledgeOfSelf, Lawrencekhoo, Onsly, Jdthood, Salmar, Jefffire, Hve, QFT, Vanished User 0001, Stevenmitchell, BIL, Lostart, Ligulembot, Yevgeny Kats, Byelf2007, Lambiam, Rcapone, JorisvS, Heliogabulus, Dan Gluck, Spebudmak, JoeBot, UncleDouggie, Fsotrain09, Oshah, JRSpriggs, Chetvorno, Friendly Neighbour, Drinibot, Vanished user 2345, Brownlee, SuperMidget, Cydebot, BobQQ, Mortus Est, Cyhawk, Ttiotsw, Julian Mendez, Dr.enh, Michael C Price, Kozuch, LilDice, Thijs!bot, Headbomb, Z10x, Jklumker, Alfredr, Dawnseeker2000, Rico402, Lfstevens, Gmarsden, JAnDbot, Olaf, LinkinPark, GurchBot, Magioladitis, Jpod2, Vanished user ty12kl89jq10, Rickard Vogelberg, Dr. Morbius, Bhenderson, TomS TDotO, Tarotcards, Wesino, Student7, Potatoswatter, Ollie 9045, Ja 62, Useight, Idioma-bot, Sheliak, Tokenhost, VolkovBot, ABF, ColdCase, Philip Trueman, TXiKiBoT, Calwiki, Thrawn562, Gobofro, SwordSmurf, Northfox, PaddyLeahy, SieBot, Wing gundam, OpenLoop, Likebox, Flyer22, Mimihitam, Hockeyboi34, Lightmouse, Sunrise, Southtown, Hamiltondaniel, Epistemion, ClueBot, Niceguyedc, ChandlerMapBot, Jusdafax, ResidueOfDesign, Ploft, Scog, Schreiber-Bike, TimothyRias, Katsushi, MidwestGeek, Addbot, Roentgenium111, DOI bot, Blethering Scot, Ronhjones, Glane23, Deamon138, TStein, Barak Sh, Tassedethe, Zorrobot, Ben Ben, Legobot, Yinweichen, Luckas-bot, Amirobot, Aldebaran66, Amble, Isotelesis, Magog the Ogre, AnomieBOT, Pyrrhon8, Rubinbot, Piano non troppo, Collieuk, Ulric1313, Citation bot, Xqbot, Plastadity, Capricorn42, P14nic997, False vacuum, Waleswatcher, Ignoranteconomist, Bigger digger, Chatul, 김정, CES1596, FrescoBot, Mesterhd, Paine Ellsworth, Schnufflus, Charles Edwin Shipp, Bbhustles, Ahnoneemoos, Pinethicket, Tom.Reding, Σ, Aknochel, Mercy11, Trappist the monk, Jordgette, Wdanbae, Aabaakawad, Michael9422, CobraBot, Deathflyer, Mathewsyriac, EmausBot, Thucyd, GoingBatty, Wikipelli, Kiatdd, Italia2006, Werieth, ZéroBot, Chasrob, Wackywace, Bamyers99, Suslindisambiguator, AManWithNoPlan, RaptureBot, Maschen, HCPotter, Crux007, RockMagnetist, Whoop whoop pull up, ClueBot NG, J kay831, Law of Entropy, Supermint, Helpful Pixie Bot, Bibcode Bot, Lowercase sigmabot, BG19bot, Negativecharge, MSgtpotter, Badon, BML0309, Zedshort, Hamish59, Minsbot, BattyBot, SupernovaExplosion, ChrisGualtieri, JYBot, Rfassbind, Ikjyotsingh, Astroali, Lepton01, Pkanella, Chwon, Rolf h nelson, Comp.arch, Kogge, Hilmer B, Anrnusna, Stamptrader, Dodi 8238, Epaminondas of Thebes, Man of Steel 85, Abitslow, Monkbot, Accnln, BradNorton1979, YeOldeGentleman, Tetra quark, Sleepy Geek, Anand2202, Quasiopinionated, Phseek and Anonymous: 212

• **Coupling (physics)** *Source:* https://en.wikipedia.org/wiki/Coupling_(physics)?oldid=676102617 *Contributors:* Charles Matthews, Karol Langner, Qutezuce, Ntmatter, I9Q79oL78KiL0QTFHgyc, Passw0rd, YurikBot, Grafen, Leptictidium, Esprit15d, KasugaHuang, SmackBot, Srnec, DabMachine, Yaris678, WISo, Addbot, ArthurBot, Erik9bot, ZéroBot, Kleopatra, Gingerspice.14 and Anonymous: 4

• **Cosmological constant** *Source:* https://en.wikipedia.org/wiki/Cosmological_constant?oldid=684416542 *Contributors:* AxelBoldt, Magnus Manske, Vicki Rosenzweig, Bryan Derksen, The Anome, Ed Poor, Enchanter, William Avery, Roadrunner, Schewek, Hephaestos, Boud, Bcrowell, Lquilter, TakuyaMurata, Minesweeper, Stevenj, Kimiko, Samw, Timwi, Reddi, Asar~enwiki, Dogface, Bevo, Anupamsr, Johnleemk, BenRG, Phil Boswell, Robbot, Goethean, Wereon, Giftlite, Bobblewik, Jonel, Rjpetti, Icairns, Rgrg, Burschik, JimJast, 4pq1injbok, Pjacobi, Vsmith, StephanKetz, Pavel Vozenilek, Dmr2, Bender235, RJHall, Pt, El C, Frankenschulz, RoyBoy, Rbj, I9Q79oL78KiL0QTFHgyc, Knucmo2, Jumbuck, Falcorian, Angr, OwenX, Linas, StradivariusTV, Kzollman, Mpatel, Joke137, Wisq, Christopher Thomas, Rnt20, Ashmoo, Coneslayer, Rjwilmsi, Coemgenus, Nightscream, RE, Itinerant1, Chobot, PointedEars, YurikBot, Hillman, RussBot, Ytrottier, SpuriousQ, Gaius Cornelius, Salsb, Sir48, Muu-karhu, DeadEyeArrow, Helge Rosé, Petri Krohn, KasugaHuang, SmackBot, Incnis Mrsi, WilyD, Nickst, Cush, Colonies Chris, Avb, Cybercobra, Ligulembot, Yevgeny Kats, Lambiam, Matt489, Paladinwannabe2, Ckatz, Onionmon, Basicdesign, Newone, Sirwhiteout, Chetvorno, CmdrObot, Orannis, Hardrada, Mlsmith10, MaxEnt, Phatom87, Forthommel, Frostlion, Dr.enh, Michael C Price, Tawkerbot4, Clovis Sangrail, Christian75, Thijs!bot, Mathmoclaire, Peter Gulutzan, Gnarlyocelot, Escarbot, AntiVandalBot, Tim Shuba, JAnDbot, LinkinPark, .anacondabot, WolfmanSF, SHCarter, Ling.Nut, Jlerner, DAGwyn, Nikopopl, MartinBot, Mschel, Morris729, Lantonov, BobEnyart, Jorfer, Blckavnger, Fylwind, Atheuz, TXiKiBoT, Rei-bot, Mathwhiz 29, Thrawn562, Venny85, SieBot, El Wray, Puzhok, Gerakibot, BartekChom, OKBot, ClueBot, The Thing That Should Not Be, Frdayeen, Excirial, Bender2k14, Brews ohare, Kentgen1, Scog, Panos84, Louis925, Alphatronic, XLinkBot, DCCougar, Sesquihypercerebral, Torchflame, Addbot, DOI bot, Zahd, Delaszk, Legobot, Luckas-bot, Yobot, Aldebaran66, Amble, Perusnarpk, AnomieBOT, Materialscientist, Citation bot, Louelle, Srich32977, Waleswatcher, A. di M., 김정, Paine Ellsworth, Citation bot 1, Newt Scamander, Gil987, Tom.Reding, BlackHades, Jordgette, Michael9422, Earthandmoon, Vekov, RjwilmsiBot, Racerx11, Solomonfromfinland, Italia2006, Hhhippo, ZéroBot, Liquidmetalrob, Arbnos, Quondum, Ewa5050, Iiar, Zueignung, Khestwol, ClueBot NG, Astrocog, Frietjes, Jhmmok, Rezabot, Const.S, Helpful Pixie Bot, Bibcode Bot, Rascal Sage, Jeffloiselle, Hippokrateszholdacskai, RiseUpAgain, Makecat-bot, Kryomaxim, Wjs64, Andyhowlett, Jp4gs, Blackbombchu, Prokaryotes, Inanygivenhole, Kogge, Paspaspas, Christophe1946, RandomAgentNation, Monkbot, Tetra quark, KasparBot, Maha Abdelmoneim and Anonymous: 139

33.9.2 Images

• **File:080998_Universe_Content_240_after_Planck.jpg** *Source:* https://upload.wikimedia.org/wikipedia/commons/b/b6/080998_Universe_Content_240_after_Planck.jpg *License:* Public domain *Contributors:* http://map.gsfc.nasa.gov/media/080998/index.html updated data from http://www.nasa.gov/mission_pages/planck/news/planck20130321.html *Original artist:* NASA, Modified by User:김정

• **File:2-photon_Higgs_decay.svg** *Source:* https://upload.wikimedia.org/wikipedia/commons/3/32/2-photon_Higgs_decay.svg *License:* CC BY-SA 3.0 *Contributors:* Own work *Original artist:* Parcly Taxel

• **File:3gluon.png** *Source:* https://upload.wikimedia.org/wikipedia/commons/a/a8/3gluon.png *License:* Public domain *Contributors:* Transferred from en.wikipedia by SreeBot *Original artist:* Bambaiah at en.wikipedia

• **File:4-lepton_Higgs_decay.svg** *Source:* https://upload.wikimedia.org/wikipedia/commons/b/b2/4-lepton_Higgs_decay.svg *License:* CC BY-SA 3.0 *Contributors:* Own work *Original artist:* Parcly Taxel

- **File:PiPlus_muon_decay.svg** *Source:* https://upload.wikimedia.org/wikipedia/commons/6/69/PiPlus_muon_decay.svg *License:* CC0 *Contributors:* Own work *Original artist:* Krishnavedala

- **File:Portal-puzzle.svg** *Source:* https://upload.wikimedia.org/wikipedia/en/f/fd/Portal-puzzle.svg *License:* Public domain *Contributors:* ? *Original artist:* ?

- **File:Qcd_fields_field_(physics).svg** *Source:* https://upload.wikimedia.org/wikipedia/commons/4/41/Qcd_fields_field_%28physics%29.svg *License:* CC0 *Contributors:* Own work *Original artist:* Maschen

- **File:Queryensdf.jpg** *Source:* https://upload.wikimedia.org/wikipedia/commons/5/5e/Queryensdf.jpg *License:* Public domain *Contributors:* Own work *Original artist:* Minivip

- **File:Question_book-new.svg** *Source:* https://upload.wikimedia.org/wikipedia/en/9/99/Question_book-new.svg *License:* Cc-by-sa-3.0 *Contributors:*
Created from scratch in Adobe Illustrator. Based on Image:Question book.png created by User:Equazcion *Original artist:* Tkgd2007

- **File:Right_left_helicity.svg** *Source:* https://upload.wikimedia.org/wikipedia/commons/a/a9/Right_left_helicity.svg *License:* Public domain *Contributors:* en:Image:Right left helicity.jpg *Original artist:* en:User;HEL, User:Stannered

- **File:SatyenBose1925.jpg** *Source:* https://upload.wikimedia.org/wikipedia/commons/f/fe/SatyenBose1925.jpg *License:* Public domain *Contributors:* Picture in Siliconeer *Original artist:* Unknown

- **File:Science.jpg** *Source:* https://upload.wikimedia.org/wikipedia/commons/5/54/Science.jpg *License:* Public domain *Contributors:* ? *Original artist:* ?

- **File:Spin_One-Half_(Slow).gif** *Source:* https://upload.wikimedia.org/wikipedia/commons/6/6e/Spin_One-Half_%28Slow%29.gif *License:* CC0 *Contributors:* Own work *Original artist:* JasonHise

- **File:Spontaneous_symmetry_breaking_(explanatory_diagram).png** *Source:* https://upload.wikimedia.org/wikipedia/commons/a/a5/Spon_symmetry_breaking_%28explanatory_diagram%29.png *License:* CC BY-SA 3.0 *Contributors:* Own work *Original artist:* FT2

- **File:Standard_Model_Feynman_Diagram_Vertices.png** *Source:* https://upload.wikimedia.org/wikipedia/commons/7/75/Standard_Model_Feynman_Diagram_Vertices.png *License:* CC BY-SA 3.0 *Contributors:* I made it in Adobe Illustrator *Original artist:* Garyzx

- **File:Standard_Model_of_Elementary_Particles.svg** *Source:* https://upload.wikimedia.org/wikipedia/commons/0/00/Standard_Model_of_Elementary_Particles.svg *License:* CC BY 3.0 *Contributors:* Own work by uploader, PBS NOVA [1], Fermilab, Office of Science, United States Department of Energy, Particle Data Group *Original artist:* MissMJ

- **File:Stimulatedemission.png** *Source:* https://upload.wikimedia.org/wikipedia/commons/8/8a/Stimulatedemission.png *License:* CC-BY-SA-3.0 *Contributors:* en:Image:Stimulatedemission.png *Original artist:* User:(Automated conversion),User:DrBob

- **File:Strong_force_charges.svg** *Source:* https://upload.wikimedia.org/wikipedia/commons/b/b6/Strong_force_charges.svg *License:* CC BY-SA 3.0 *Contributors:* Own work, Created from Garret Lisi's Elementary Particle Explorer *Original artist:* Cjean42

- **File:Stylised_Lithium_Atom.svg** *Source:* https://upload.wikimedia.org/wikipedia/commons/e/e1/Stylised_Lithium_Atom.svg *License:* CC-BY-SA-3.0 *Contributors:* based off of Image:Stylised Lithium Atom.png by Halfdan. *Original artist:* SVG by Indolences. Recoloring and ironing out some glitches done by Rainer Klute.

- **File:Symmetricwave2.png** *Source:* https://upload.wikimedia.org/wikipedia/commons/1/1d/Symmetricwave2.png *License:* CC BY 3.0 *Contributors:* Own work *Original artist:* TimothyRias

- **File:Text_document_with_page_number_icon.svg** *Source:* https://upload.wikimedia.org/wikipedia/commons/3/3b/Text_document_with_page_number_icon.svg *License:* Public domain *Contributors:* Created by bdesham with Inkscape; based upon Text-x-generic.svg from the Tango project. *Original artist:* Benjamin D. Esham (bdesham)

- **File:Text_document_with_red_question_mark.svg** *Source:* https://upload.wikimedia.org/wikipedia/commons/a/a4/Text_document_with_red_question_mark.svg *License:* Public domain *Contributors:* Created by bdesham with Inkscape; based upon Text-x-generic.svg from the Tango project. *Original artist:* Benjamin D. Esham (bdesham)

- **File:The_incomplete_circle_of_everything.svg** *Source:* https://upload.wikimedia.org/wikipedia/commons/0/0d/The_incomplete_circle_of_everything.svg *License:* CC BY 3.0 *Contributors:* Own work *Original artist:* Zhitelew

- **File:VFPt_charges_plus_minus_thumb.svg** *Source:* https://upload.wikimedia.org/wikipedia/commons/e/ed/VFPt_charges_plus_min.svg *License:* CC BY-SA 3.0 *Contributors:* This plot was created with VectorFieldPlot *Original artist:* Geek3

- **File:VFPt_minus_thumb.svg** *Source:* https://upload.wikimedia.org/wikipedia/commons/d/d7/VFPt_minus_thumb.svg *License:* CC BY-SA 3.0 *Contributors:* This plot was created with VectorFieldPlot *Original artist:* Geek3

- **File:VFPt_plus_thumb.svg** *Source:* https://upload.wikimedia.org/wikipedia/commons/9/95/VFPt_plus_thumb.svg *License:* CC BY-SA 3.0 *Contributors:* This plot was created with VectorFieldPlot *Original artist:* Geek3

- **File:Vertex.png** *Source:* https://upload.wikimedia.org/wikipedia/en/d/db/Vertex.png *License:* Cc-by-sa-3.0 *Contributors:* ? *Original artist:* ?

- **File:Vertex_correction.svg** *Source:* https://upload.wikimedia.org/wikipedia/commons/8/87/Vertex_correction.svg *License:* Public domain *Contributors:* ? *Original artist:* User:Harmaa

- **File:VisibleEmrWavelengths.svg** *Source:* https://upload.wikimedia.org/wikipedia/commons/e/e2/VisibleEmrWavelengths.svg *License:* Public domain *Contributors:* created by me *Original artist:* maxhurtz

- **File:Weak_Decay_(flipped).svg** *Source:* https://upload.wikimedia.org/wikipedia/commons/4/4b/Weak_Decay_%28flipped%29.svg *License:* CC BY-SA 3.0 *Contributors:* Created in Inkscape based on :File:Weak decay diagram.svg *Original artist:* Niamh O'C

33.9.3 Content license

28181407R00152

Made in the USA
San Bernardino, CA
22 December 2015